A variety of models can be used to study nuclear structure. This book gives a comprehensive overview of these various models, concentrating in particular on a description of deformed and rotating nuclei. Emphasis is given throughout to the important physical features, rather than to esoteric theoretical topics.

Beginning with a treatment of the semi-empirical mass formula and nuclear stability, the liquid-drop model is then described and its use in the study of nuclear deformation and fission is discussed. The spherical nuclear one-particle potential is introduced and developed to cover the case of deformed nuclei. The main features of the shell correction method are described, with applications to nuclear deformation, fission, superheavy elements and rotation. A detailed discussion of terminating rotational bands and superdeformation is included. Finally, the nucleon–nucleon interaction is briefly described and the main features of the nuclear pairing interaction are discussed.

As well as treating important experimental and theoretical aspects of this fundamental subject, many problems and solutions are included, which help to illustrate key concepts. The book will be invaluable to graduate students of nuclear physics and to anyone engaged in research in this field.

T0192699

SHAPES AND SHELLS IN NUCLEAR STRUCTURE

SHAPES AND SHELLS IN NUCLEAR STRUCTURE

SVEN GÖSTA NILSSON and INGEMAR RAGNARSSON

Department of Mathematical Physics, Lund Institute of Technology

CAMBRIDGE
UNIVERSITY PRESS

CAMBRIDGE UNIVERSITY PRESS
Cambridge, New York, Melbourne, Madrid, Cape Town, Singapore, São Paulo

Cambridge University Press
The Edinburgh Building, Cambridge CB2 2RU, UK

Published in the United States of America by Cambridge University Press, New York

www.cambridge.org
Information on this title: www.cambridge.org/9780521373777

First published 1995
This digitally printed first paperback version 2005

A catalogue record for this publication is available from the British Library

Library of Congress Cataloguing in Publication data
Nilsson, Sven Gösta.
Shapes and shells in nuclear structure / Sven Gösta Nilsson
and Ingemar Ragnarsson.
p.   cm.
Includes index.
ISBN 0 521 37377 8
1. Nuclear shell theory.   2. Nuclear shapes.   3. Nuclear spin.
4. Nuclear structure.   I. Ragnarsson, Ingemar.   II. Title.
QC793.3.S8R34   1995
539.7′4–dc20   94-25634   CIP

ISBN-13 978-0-521-37377-7 hardback
ISBN-10 0-521-37377-8 hardback

ISBN-13 978-0-521-01966-8 paperback
ISBN-10 0-521-01966-4 paperback

Contents

Contents

Preface

It is our intention in this volume to describe in an elementary way some of the achievements of the nuclear shell model, especially in its form used for deformed nuclei and rotating nuclei. In recent applications on nuclear deformation, fission and rotation, the microscopic shell model is merged with macroscopic models, which are thus also briefly discussed here. We try to concentrate on physical features rather than theoretical methods, not introducing more sophisticated models than are needed to understand the basic principles.

We have tried to put the presentation on such a level that the book should be suitable as a first course in theoretical nuclear physics. The reader is supposed to be familiar with the very basic concepts of experimental nuclear physics and to have some knowledge of quantum mechanics including central forces and angular momentum, the spin formalism and some perturbation theory. We also believe that this volume should be useful for people doing research in nuclear physics, not least for experimentalists.

The introductory chapter defines in a very elementary way the building blocks of nuclei, how these building blocks are held together and how nuclei might decay. It then also becomes natural to discuss the boundaries of nuclear stability and the abundancy of different nuclei.

The subjects of the three following chapters relate to the macroscopic properties of nuclei. The size and average matter distribution of nuclei are discussed in chapter 2. The semi-empirical mass formula is introduced in chapter 3 and the liquid-drop model of nuclear deformation and fission is treated in chapter 4.

The single-particle concept is then introduced in chapter 5, making comparisons between the electron system and the nucleon system. The mean field concept leading to the introduction of a single-particle potential is discussed. The single-particle potential of a spherical nucleus is treated in some detail in chapter 6. As the reader is not assumed to be familiar with the quantum-mechanical formalism for coupling of angular momentum vectors (the Clebsch–Gordan formalism), this formalism is presented as an appendix. Some elementary applications of the spherical single-particle potential are taken up in chapter 7 where the simplest static moments of nuclei with few particles outside closed shells are calculated.

The measured quadrupole moments indicate that it is not enough to study a spherical potential. Thus, in chapter 8, the orbitals of a deformed single-particle potential are discussed, i.e. this chapter deals with the so-called Nilsson model and related subjects, which probably more than anything else have made Sven Gösta Nilsson's name well-known among nuclear physicists.

In chapter 9, it is shown how realistic calculations on nuclear ground state properties can be carried out if the single-particle model and the macroscopic model are combined. The methods introduced here make it possible to calculate the nuclear energy as a function of the most important shape degrees of freedom. The energy surfaces can be used as input for barrier penetration calculations of nuclear fission. This is demonstrated in chapter 10 where alpha-decay, which can be treated by similar methods, is also discussed. The success of these methods is demonstrated on heavy and superheavy nuclei. The hunt for an 'island' of superheavy nuclei has stimulated the imagination of nuclear structure physicists for many years and the possibility of such an island is also discussed in other parts of the present volume.

The most recent success of the macroscopic–microscopic method has been the application to fast nuclear rotation and this subject is treated in some detail in chapters 11 and 12. It is described how a unifying picture of single-particle excitations and collective rotation emerges from straightforward generalisations of the methods introduced in earlier chapters. In particular, recent applications on band terminations and superdeformation are discussed within a cranking formalism, which is first illustrated on the conceptionally much simpler sd-shell nuclei. This chapter leads up to the present research front with a discussion on identical rotational bands and how they might be described.

We felt that even though it was not our primary goal, we should still present some basic concepts about the nucleon–nucleon interaction, which is thus the subject of chapter 13. This chapter is only intended for the inexperienced reader to make the present volume reasonably self-contained.

The final chapter deals with the pairing interaction in a way that we hope should be understandable even for readers who have no previous knowledge of the so-called second quantisation formalism. This chapter should, we hope, clarify some of the discussion in earlier parts of the book, where the importance of pairing has been mentioned and some consequences of pairing have been anticipated.

The text is accompanied by a number of problems, some trivial and some that might be quite tough for the ordinary reader. In some cases we have felt that the text becomes more transparent without too many derivations. Thus, some simple derivations are put as problems, which we also hope will encourage the reader to try to carry them through by him-/herself. In other problems the more general formalism of the text might be applied to more concrete cases. If not very trivial, solutions are given to the problems.

In the applications of the shell model, we have used those models that we think demonstrate best the surprising success of the concept of individual particle motion in an average field. We have thus concentrated on the so-called modified oscillator potential in order not to hide the simple physical arguments with too much mathematical complexity. We are convinced that, having understood the physical arguments, the reader will be well prepared to find his way through more 'realistic' potentials of e.g. Woods–Saxon or Hartree–Fock type.

The reference list is far from complete. We have only put in the papers whose results and ideas we specifically refer to and some more general papers so that the reader should easily find his way through the current literature. Except for some older and generally recognised papers, we have thus not tried to trace the origin of many of the arguments put forth here. Naturally, the reference list is weighted toward our own papers and those of our closest collaborators.

This volume has grown from a course given at the Institute of Technology in Lund. Many people in our Department have contributed at different stages of this course and thus also studied and given constructive critisism to the manuscript. In particular, I want to thank Göran Andersson, Per Arve, Tord Bengtsson, Ikuko Hamamoto, Stig Erik Larsson, Georg Leander, Peter Möller and Sven Åberg. I also want to express my gratitude to Edith Halbert, Witold Nazarewicz and Zdzisław Szymański who have read major parts of the manuscript and made many valuable suggestions. Several people have assisted in typing the manuscript and preparing the figures; I want to thank Ulla Jacobsson, Sigurd Madison and Ewa Westberg for their important contributions and especially Pia Bruhn who has done a lot of careful work on the munuscript over the years and finally converted it all to LaTeX files.

Sven Gösta Nilsson died in 1979. As a student, colleague and friend of his, it is with a deep sense of loss that I remember his warm and enthusiastic personality and all the inspiration I have had from numerous discussions with him. To bring the present volume to an end has been a much harder job than I realised when I started. The long delay means that several developments from the 1980s are covered. However, these subjects are all natural continuations of the original manuscript. I hope that the present book is in line with Sven Gösta's vision and that it will be useful and inspiring for its readers.

Ingemar Ragnarsson

1

Naturally occurring and artificially produced nuclei

Nature is very rich. It uses a great number of different building blocks, atoms, to compose the different chemical compounds of the organic and inorganic world. For chemistry the electron shells, surrounding the atoms, are the most important. The stability of the nucleus is, however, decisive for the existence of the atom. The problem of which atoms may exist is therefore reduced to the problem of what nuclei may exist. In this introductory chapter, we will consider in a qualitative manner the following problem. What is the range in neutron and proton number for the existing nuclei? This problem will be treated more quantitatively within the liquid drop model in chapter 3.

To define a nucleus, one must specify its proton number Z (i.e. the chemical element) and in addition its neutron number N or alternatively its mass number $A = N + Z$. The most common notation is for example ^{208}Pb for the nucleus with $A = 208$ and $Z = 82$. Sometimes, the neutron and proton numbers are explicitly given, i.e. $^{208}_{82}$Pb$_{126}$.

The question of which combinations of N and Z are stable and which are not is governed by the principle of minimisation of the total relativistic energy, where a mass M corresponds to an energy Mc^2, c being the velocity of light. If an N, Z combination can find a lower energy state by decaying in one way or another, this decay will generally take place with some probability. There are, however, certain constraints, the most important one being that the total number of nucleons, i.e. neutrons and protons, stays constant (preservation of baryon number). A possible decay occurs according to the radioactive decay law, the number of decays being proportional to the number v of nuclei of a specific kind in a sample

$$\frac{\mathrm{d}v}{\mathrm{d}t} = -\lambda v$$

1

which leads to the exponential decay law

$$v = v_0 e^{-\lambda t}$$

The decay constant λ is typical for the decay in question. From this constant, one defines the (mean) life-time, $\tau = 1/\lambda$ and the half-life, $t_{1/2}$, corresponding to the time it takes (on the average) for half of the nuclei to decay, $t_{1/2} = (\ln 2/\lambda) = 0.693/\lambda$.

The stability of nuclei is closely related to the forces holding them together. From our macroscopic world we know two kinds, namely gravitational and electromagnetic forces. It is the Coulomb force (of electromagnetic origin) that ties the electrons to the nucleus, forming the atoms. The nucleus, on the other hand, is built of protons of like charges (in addition to neutrons). This means that the Coulomb forces are repulsive and the attractive gravitational forces, being orders of magnitude weaker, can by no means compensate for this.

Indeed, it turns out that the nuclei are held together by an interaction of a different origin, the *strong* (nuclear) *interaction*. This interaction is strongly attractive at the internucleon distance in a nucleus ($\simeq 1$ fm $= 10^{-15}$ m) but it has a very short range and it becomes more or less unimportant when the nucleons are only a few femtometres apart (cf. chapter 13). One could compare this with the $1/r$ dependence for the potential energy of the gravitational as well as the electromagnetic interaction. This means that these interactions are important at all distances and could never be said to go to zero.

Also a fourth type of interaction is known, namely the *weak interaction*†. It is of short range but it is orders of magnitude weaker than the strong interaction. Thus, its contribution to the nuclear binding energy is negligible. However, for example electrons do not feel the strong interaction and the weak interaction could then become important.

A final key factor for an elementary understanding of the stability of nuclei is the Pauli principle, an effect of quantum mechanics. The Pauli principle forbids a proton to be in the same quantal state as any other proton. This means that if we start from the ground state of one nucleus and try to add a proton, this proton will be placed in a state having a higher energy than those already present. The same holds true for the neutron states and the neutrons while the proton and neutron states are essentially independent (this independence between protons and neutrons is often expressed by an additional quantum number, the isospin (chapter 13)). For a nucleus having

---

† In recent years, a unifying description of the electromagnetic and weak interactions has been achieved, a result for which S.L. Glashow, A. Salam and S. Weinberg won the 1979 Nobel Prize for Physics.

proton excess, the 'last proton' will be in a state having a considerably higher energy than the state of the 'last neutron'. Consequently, the total energy will be lower if one proton is transformed to a neutron. The net result is thus that the most stable nuclei are those having about an equal number of neutrons and protons. The repelling Coulomb forces acting between the protons will modify this tendency toward making the number of protons somewhat smaller than the number of neutrons, especially for heavy nuclei.

The discussion above leads to the conclusion that the stability of nuclei is determined by competition between the attractive nuclear forces (the strong interaction) and the disruptive Coulomb forces, with the Pauli principle taken into account. As the Coulomb interaction has an infinite range, its destabilising effects will dominate for large enough mass numbers, which means that the number of nucleons in a nucleus will be limited. It turns out that this limit is reached around mass number $A = 240$, i.e. for heavier nuclei the half-lives are so small that no such nuclei have been found in terrestrial matter. The most important decay processes for heavy elements are fission, i.e. division into two more or less equal fragments, and alpha-decay (α-decay), i.e. emission of an alpha-particle consisting of two protons and two neutrons. The alpha-particle, which is identical to a $^4_2\text{He}_2$ nucleus, has an unusually high binding energy (i.e. it has a low mass or equivalently a low total energy) which explains the importance of the alpha-decay mode.

The stability of lighter elements is determined by β-decay (beta-decay). A free neutron is not a stable particle but decays into a proton, an electron (a β-particle) and a particle of vanishing or almost vanishing mass called a neutrino, ν_e (or in this case rather an antineutrino, $\bar{\nu}_e$)

$$\text{n} \rightarrow \text{p}^+ + \text{e}^- + \bar{\nu}_e + \text{energy}$$

The half-life of this process is 12 min. The energy released comes out in the form of kinetic energy of the emitted particles. It equals 0.78 MeV corresponding to the difference in mass between the particles on the left and right hand sides. If some energy is added, the process

$$\text{energy} + \text{p}^+ \rightarrow \text{n} + \text{e}^+ + \nu_e$$

also takes place, where the positron, e^+, is the so called antiparticle of the electron. A process of similar kind is electron capture

$$\text{energy} + \text{p}^+ + \text{e}^- \rightarrow \text{n} + \nu_e$$

All these processes are governed by the weak interaction and are the most clear-cut examples where this interaction becomes important.

One now realises that nuclei with constant A (often referred to as isobars) but different numbers of protons and neutrons may decay into one another and that the different binding energies may provide the energy necessary for this. Indeed, the difference in mass between a proton and a neutron, 1.294 MeV, is small compared with typical differences in binding energies. The total mass of the electron is even smaller, 0.511 MeV. Thus, as a first crude approximation, only that (N, Z) combination which for constant A gives the strongest binding is stable. For example, for mass number $A = 91$, the strongest binding is observed for $^{91}_{40}Zr_{51}$ with weaker bindings for $^{91}_{39}Y_{52}$ and $^{91}_{41}Nb_{50}$ etc. (cf. fig. 3.4 below). This means that the decays

$$^{91}_{39}Y_{52} \rightarrow {}^{91}_{40}Zr_{51} + e^- + \bar{\nu}_e + \text{energy}$$

and

$$^{91}_{41}Nb_{50} + e^- \rightarrow {}^{91}_{40}Zr_{51} + \nu_e + \text{energy}$$

are energetically possible, leaving ^{91}Zr as the only stable isobar. In many cases, however, the binding energy is far from a smooth function for a chain of isobars and therefore two or more isobars are often stable. The most important factor leading to a staggering binding energy is the odd–even effect, see chapters 3 and 14.

One might have expected that the nuclear stability was lost first at the so-called neutron (or proton) drip lines (see chapter 3) when for one nucleus, the next neutron (or proton) added becomes unbound. Because of the processes discussed above, this is, however, not the case even though the drip lines define some kind of ultimate limit beyond which one cannot really talk about a nucleus as one entity. On the other hand, β-unstable nuclei might survive for quite some time and it is only if the energy gain of the decay process becomes large that they are really short-lived. Similarly, alpha- and fission-unstable nuclei might survive for long or very long times. Indeed, many of the nuclei we consider as stable are fission- or alpha-unstable but with such long half-lives (much longer than the age of the universe) that we cannot possibly observe their decay.

Let us now consider the nuclear periodic table. For elements up to Bi with $Z = 83$, there is at least one stable isotope (nuclei with the same number of protons are generally referred to as isotopes and those with the same number of neutrons as isotones) with the exception of $_{43}Tc$ and $_{61}Pm$. Of the latter elements, the most long-lived isotopes have half-lives of 2.6×10^6 years (for ^{97}Tc) and 18 years (for ^{145}Pm). Naturally radioactive isotopes are the elements between $_{92}U$ and $_{84}Po$. These elements belong to the famous radioactive series, decaying by α- and β-particle emission, explored by Marie and Pierre Curie and their followers. Many of these have very short half-lives

Table 1.1. *Distribution of elements in the solar system.*

Element	Fraction according to mass	Fraction according to number of atoms
$_1$H	0.71	0.94
$_2$He	0.27	0.06
$_3$Li, $_4$Be, $_5$B	10^{-8}	10^{-9}
$_6$C, $_7$N, $_8$O, $_{10}$Ne	1.8×10^{-2}	10^{-3}
Si-group: $_{11}$Na, ...$_{22}$Ti	2×10^{-3}	10^{-4}
Fe-group: $50 \lesssim A \lesssim 62$	2×10^{-4}	4×10^{-6}
Intermediate nuclei: $63 \lesssim A \lesssim 100$	10^{-6}	10^{-8}
Heavy nuclei: $A \gtrsim 100$	10^{-7}	10^{-9}

compared with the age of the earth ($\simeq 4.6 \times 10^9$ years) and they are present in terrestrial matter only because of the existence of the relatively long-lived radioactive parent nuclei ^{235}U ($t_{1/2} = 7 \times 10^8$ years), ^{238}U ($t_{1/2} = 4.5 \times 10^9$ years) and ^{232}Th ($t_{1/2} = 1.4 \times 10^{10}$ years).

In 1971 it was announced that a research team at the Los Alamos laboratories in New Mexico, USA, had identified in terrestrial ores nuclei of $^{244}_{94}$Pu, with a half-life of 80×10^6 years. This was a remarkable discovery. If these nuclei were formed when or before the earth was formed, it means that since their formation there have passed $4600/80 \approx 58$ half-lives. This means that of the original atoms only $1 : 2^{58}$ or one in 10^{17} are left. The minute amounts left, even in an originally pure sample, just escape *normal* detection. The frequency of occurrence of ^{244}Pu in bastnasite ore might even seem to necessitate that the ^{244}Pu atoms arrived on the earth considerably later than the time when the bulk of terrestrial matter was assembled.

To the elements mentioned we have then to add the artificially produced elements $_{93}$Np, $_{95}$Am, $_{96}$Cm, $_{97}$Bk, $_{98}$Cf, $_{99}$Es, $_{100}$Fm, $_{101}$Md, $_{102}$No, $_{103}$Lw and those with proton numbers $Z = 104$–109. If we add the neutron we end up with 110 presently known elements. To each Z correspond usually several N-values, or isotopes. We have thus 280 stable nuclei, 68 naturally radioactive nuclei, and presently almost 3000 artificially produced ones.

Many of the elements are rare in terrestrial matter and even more so in the solar system as a whole. Table 1.1 exhibits the mass distribution of atoms of various groups of atomic species in the solar system. Note that this table does not apply to the distribution of elements in terrestrial, lunar matter or meteoritic matter. The difference in abundances between the earth and the solar system, for example, is due to the fact that the earth has too weak a gravitational field to prevent light volatile elements such as H, He etc. escaping.

It is obvious from this table that an overwhelming amount of the material is associated with H and He with a few weight percent distributed over the C, N, O, Ne group, which latter has shown itself very important for the development of organic material.

2

Charge and matter distribution in nuclei

The simplest property of a nucleus is the volume or alternatively the radius. The average density of nuclear matter is found to be roughly the same in all nuclei, i.e. the volume is found to be proportional to A or the radius to $A^{1/3}$ over the entire table of nuclei. The average central density is found to have the value

$$\rho(0) \approx 0.17 \text{ nucleons fm}^{-3}$$

or

$$\rho(0) \approx 1.7 \times 10^{44} \text{ nucleons m}^{-3}$$

The average matter distribution is, as shown in fig. 2.1, such that there is an inner region of roughly constant density surrounded by a diffuse region where the density gradually falls to zero. The diffuseness thickness d is measured as the distance from 90% of the central density to 10% of this. One has empirically $d \approx 2.5$ fm. It appears that d is roughly the same for all intermediate or heavy nuclei but with a variation of approximately 10% due to shell structure. When we speak about density, we may distinguish between total nucleon density, proton density or neutron density. The one most accessible to measurement is the proton density (or charge density). However, due to the strong coupling between neutrons and protons, the neutron and proton density distributions are grossly the same.

To describe quantitatively what has been said qualitatively above one often expresses the average nucleon density distribution in terms of a Fermi function

$$\rho(r) = \rho_0 \left[1 + \exp \left(\frac{r - R}{a} \right) \right]^{-1}$$

with

$$R \approx 1.1 \cdot A^{1/3} \text{ fm}$$

7

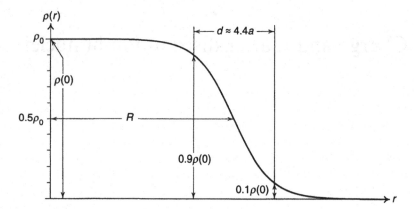

Fig. 2.1. The radial dependence, $\rho(r)$, generally used for the average matter density (or the proton density or the neutron density) in a nucleus. The quantity d marks the diffuseness depth or the distance between the 90% and 10% density radii.

and $\rho(0) \approx \rho_0$ because $\exp(-R/a) \ll 1$. The quantity d of fig. 2.1 is related to a as $d = 4a \ln(3) \approx 4.4a$ (cf. problem 2.1), i.e. to $d = 2.5$ fm corresponds the value $a \approx 0.57$ fm.

Besides the radius parameter, R, the so-called root-mean-square radius is also frequently used in the literature. It is defined as

$$R_{\text{rms}} = \left(\frac{5}{3} \langle r^2 \rangle \right)^{1/2}$$

where

$$\langle r^2 \rangle = \frac{\int \rho(r) r^2 \, d^3 r}{\int \rho(r) \, d^3 r}$$

As shown in problem 2.2, for a Fermi function one can find the following expansion in (a/R):

$$\langle r^2 \rangle = \frac{3}{5} R^2 \left[1 + \frac{7\pi^2}{3} \left(\frac{a}{R} \right)^2 + \ldots \right]$$

and therefore

$$R_{\text{rms}} = R \left[1 + \frac{7\pi^2}{6} \left(\frac{a}{R} \right)^2 + \ldots \right]$$

As (a/R) approaches zero, obviously the two radius parameters, R and R_{rms}, agree exactly. For intermediately heavy nuclei ($R \approx 6$ fm) the empirical value of a ($a = 0.55$–0.60 fm) gives the second order term an order of magnitude of 10%. Thus, $R = 1.1 \cdot A^{1/3}$ fm leads to $R_{\text{rms}} \approx 1.2 \cdot A^{1/3}$ fm.

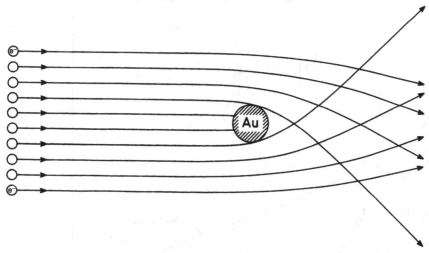

Fig. 2.2. Schematic illustration showing results of bombardment of a Au nucleus with electrons. From the alteration of the electron trajectories due to the Coulomb attraction, the size and character of the Au charge distribution can be inferred.

Information about the proton distribution comes mainly from measurements on particles scattered against nuclei (e.g. Hofstadter, 1963). The particles are thus deflected differently depending on how close they pass relative to the nucleus. A requirement associated with the wave nature of matter is that the wavelength of the probes has to be small compared with the size of the object probed, i.e. the nucleus. Ordinary light with a wavelength of about 10^{-7} m obviously will not suffice. One instead uses for example protons of 20 MeV or so, or electrons with energies in excess of 100 MeV. If electrons are scattered as indicated in fig. 2.2, only the Coulomb effects due to the protons are measured. Because of our complete knowledge of the Coulomb field, it is then possible to deduce the position in space occupied by the protons from the scattering data. In order to measure the total matter distribution or the neutron distribution, it becomes necessary to use some strongly interacting particles like protons, α-particles or pions. We are then faced with the problem that the interaction is only partly known and therefore, much less is known about the matter distribution than about the proton distribution.

Fig. 2.3 shows the so-called differential cross section $d\sigma/d\Omega$ (proportional to the number of particles scattered into a certain space angle $d\Omega$) as a function of the angle between the incoming probes and the direction from the target to detector. The data in this figure are obviously sufficient to

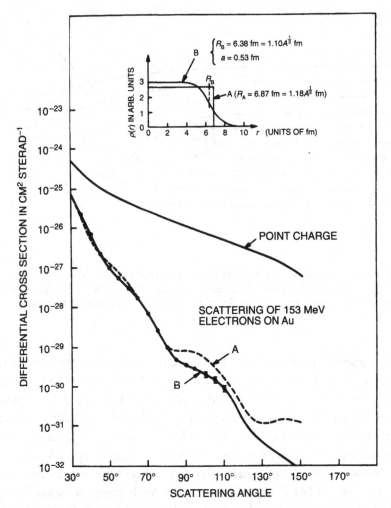

Fig. 2.3. The so-called differential cross section for scattering of 153 MeV electrons against a stable $_{79}$Au target, i.e. the relative number of scattered electrons counted over a certain space angle ($d\Omega = \sin\theta\,d\theta\,d\varphi$). The theoretical angular distribution corresponding to a hypothetical central point charge is given by the solid upper curve. The dashed curve A corresponds to a homogeneous, constant charge distribution in Au out to a radius R_A. Better agreement with the data points is obtained by the curve B based on the assumption of a finite charge diffuseness (from A. Bohr and B.R. Mottelson, 1969, *Nuclear Structure*, © 1969 by W.A. Benjamin, Inc. Reprinted with permission of Addison-Wesley Publishing Company, Inc.)

conclude that the charge distribution is rather the one marked by the solid line, i.e. the one with the diffuse surface, than the one marked by a dashed curve, i.e. a sharp surface.

Information on the charge distribution in nuclei is also obtained from the bound electrons. As the electron wave function penetrates the nucleus, its energy will depend on the extension of the nuclear charge. This disturbance is relatively small but can still be measured with a high accuracy. One property which has received much interest recently is the so-called isotope shift (e.g. Otten, 1989; Schuessler, 1981), which involves variation of the charge distribution when neutrons are added. One elementary conclusion from such studies is for example that, when neutrons are added, the charge radius increases only by about 50% of the value suggested by the formula $R = r_0 A^{1/3}$. Similarly, if only protons are added the charge radius increases on the average by a factor of 1.5 relative to the simple $A^{1/3}$ estimate. It thus also seems clear that the equalness between the proton and neutron distributions can only be approximately valid, and indeed, much work has gone into the problem of determining the differences between the two distributions (e.g. Barrett and Jackson, 1977).

In muonic atoms, one electron is replaced by a muon having a mass about 200 times that of an electron. Thus, the Bohr radius of the muon becomes 200 times smaller than that of the electron and in a heavy nucleus, the muon may spend half of its time inside the nucleus. The energy spectrum will thus be very strongly influenced also by small variations in the nuclear charge radius and muonic atoms have frequently served as tools to measure the charge distribution in nuclei (e.g. Wu and Wilets, 1969; Devons and Duerdoth, 1969).

In recent years it has been possible to perform electron scattering at considerably higher energies (e.g. Heisenberg and Blok, 1983) than used for example when constructing fig. 2.3. The technique has also been refined in other aspects and, combined with muonic atom data, it is now possible to determine the nuclear charge distribution and its radial variation with a very high accuracy. A sample of recent results is collected in fig. 2.4 where the 'line width' represents the experimental uncertainty. It is evident that the average properties are well described by the Fermi function of fig. 2.1 but also that there are important local variations. Theoretically, these local effects are understood as shell effects but one notes that even with the most recent theoretical tools we are not able to describe them fully.

As will become apparent in the coming chapters, the comparison between theory and experiment in fig. 2.4 illustrates a rather typical situation in nuclear physics. Thus, the observed features are qualitatively understood

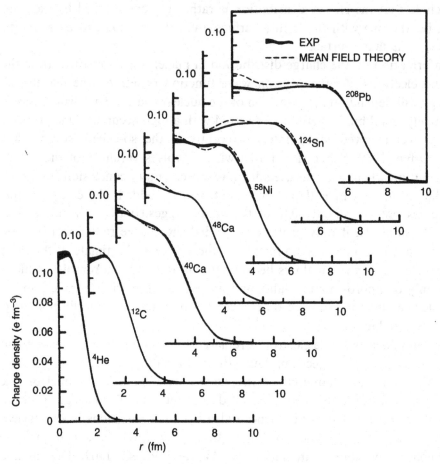

Fig. 2.4. Nuclear ground state charge distribution as measured for a sample of nuclei throughout the periodic table. The experimental uncertainty is indicated by the widths of the solid lines while the dashed lines show the result of a state-of-the-art calculation (from B. Frois, *Proc. Int. Conf. Nucl. Phys., Florence*, 1983, eds. P. Blasi and R.A. Ricci (Tipografia Compositori, Bologna) vol. 2, p. 221).

but, on the detailed level, discrepancies between theory and experiment might still be quite large. This indicates the difficulty of the many-body problem in general and especially the nuclear many-body problem where the forces are not only very complicated but, in addition, they are only partly known.

Exercises

2.1   Prove for the Fermi distribution function that the entering diffuseness constant a is related to the skin thickness d as

$$d \approx (4 \ln 3) \cdot a$$

where d is the radial difference between the 90% and 10% density surfaces.

2.2   Show that $\langle r^2 \rangle$ for a spherical Fermi distribution can be expanded in (a/R) according to the expression†

$$\langle r^2 \rangle = \frac{3}{5} R^2 \left[1 + \frac{7\pi^2}{3} \left(\frac{a}{R} \right)^2 + \cdots \right]$$

and that for a general power of n the formula generalises to

$$\langle r^n \rangle = \frac{3}{n+3} R^n \left[1 + \frac{n(n+5)}{6} \pi^2 \left(\frac{a}{R} \right)^2 + \cdots \right]$$

2.3   Show that the Coulomb energy for a spherical Fermi distribution can be expanded in (a/R) as

$$E_C = \frac{3}{5} \cdot \frac{Z^2 e^2}{4\pi\varepsilon_0 R} \left[1 - \frac{7\pi^2}{6} \left(\frac{a}{R} \right)^2 + \cdots \right]$$

† Hint for problems 2 and 3: To treat integrals over the Fermi distribution function it is convenient to use a mathematical trick according to e.g. p. 394 of R. Reif *Fundamentals of Statistical and Thermal Physics* (McGraw Hill, 1965).

3

The semi-empirical mass formula and nuclear stability

3.1 The mass formula

The nuclear binding energy $B(N,Z)$ is defined in the following way

$$m(N,Z) = \frac{1}{c^2}E(N,Z) = NM_n + ZM_H - \frac{1}{c^2}B(N,Z)$$

where $m(N,Z)$ is the atomic mass corresponding to neutron number N and proton number Z, and M_n and M_H the free neutron and hydrogen atom masses. The binding energy is thus the energy gained (or the energy saved) by amalgamating the neutrons and protons instead of keeping them apart. From an analysis of available nuclear masses already in the 1930s von Weizsäcker (1935) and Bethe and Bacher (1936) were able to identify four leading terms that accounted relatively well for the variation of B with N and Z:

$$B = a_{vol}A - a_{surf}A^{2/3} - \frac{1}{2}a_{sym}\frac{(N-Z)^2}{A} - \frac{3}{5}\cdot\frac{Z^2e^2}{4\pi\varepsilon_0 R_c}$$

The first two terms, the volume and surface (binding) energies, are formally the same as those employed to describe a liquid drop. Provided the liquid is homogeneously charged, there is also a term as the last one in the liquid-drop case, the Coulomb repulsion energy. A straightforward generalisation of this formula to describe shapes of the nucleus other than the spherical one is usually called the liquid-drop model of the nucleus. (We shall come back to this case in the next chapter.)

Let us discuss briefly each one of these four terms. The first and dominant term, the volume energy, reflects the nearly linear A-dependence of the nuclear volume or the A-independence of the nuclear density. Every nucleon appears to interact basically with the nearest of its neighbours. Most authors give $a_{vol} = a_v \simeq 16$ MeV. This is the binding energy per particle of nuclear matter, the latter so defined that surface effects are negligible and Coulomb interaction non-existent. Furthermore, by definition, there are equal numbers of neutrons and protons in nuclear matter.

The second term, the surface energy, is proportional to the nuclear surface area and represents the loss of binding suffered by the particles in the surface layer due to the lower density (fewer neighbours) there. Actually, the kinetic energy is also increased due to the special surface conditions, such as a rapid fall-off of the nuclear density. A fit to mass data gives $a_{surf} = a_s \simeq 17\text{--}20$ MeV (i.e. $a_s \approx a_v$, see problem 3.4).

The third term, the symmetry energy, reflects the fact that nuclear forces favour equal numbers of neutrons and protons, or $N = Z$. In the case $N = Z$, the limitations brought about by the Pauli principle are reduced to a minimum. This term determines the width of the mass peninsula (see below) and together with the Coulomb term the path of the stability line through the N, Z plane. One may write the volume and the symmetry terms together as

$$B_v = a_v A \left[1 - \kappa_v \left(\frac{N - Z}{A} \right)^2 \right]$$

expressing the fact that this is really a volume symmetry term. A reasonable numerical value is $a_{sym} = 50\text{--}60$ MeV corresponding to $\kappa_v = 1.5\text{--}2.0$.

The last term in the binding energy formula, the electric repulsion term, corresponds to the electrostatic energy of a homogeneously charged sphere of radius R_C. One may regard the radius R_C as an available mass formula parameter. Its value may hide a certain granularity in the charge distribution and also a neglected surface diffuseness. One should, however, expect R_C to approximate the value normally assumed for the nuclear radius.

In chapter 2 we discussed the complications in the concept of a radius from the fact that the nuclear surface is diffuse. For a Fermi distribution of charge one, the Coulomb energy is

$$E_C = -B_C = \frac{1}{2} \cdot \frac{e^2}{4\pi\varepsilon_0} \int \frac{\rho(r)}{|\mathbf{r}_1 - \mathbf{r}_2|} d^3 r_1 \, d^3 r_2 = \frac{3}{5} \cdot \frac{e^2}{4\pi\varepsilon_0 R_C} \left[1 - \frac{\pi^2}{2} \frac{5}{3} \left(\frac{a}{R_C} \right)^2 \cdots \right]$$

where R_C is the 'equivalent radius' of a homogeneous sphere of the central density (cf. problem 2.3 where the formula looks somewhat different because of the differing definitions of the radii, R_C and R).

In fact a second correction enters, which also tends to reduce the Coulomb energy. This is the so-called exchange correction. As the protons have to obey the Pauli principle, the immediate vicinity of one proton will be forbidden territory to the other protons. We will not go further into the derivation

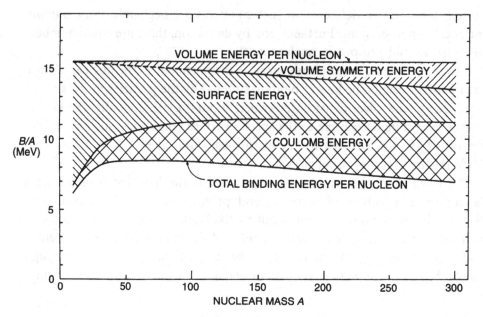

Fig. 3.1. Numerical value of specific binding energy B/A, according to the semi-empirical mass formula. The constant volume energy enters with opposite sign to all the other contributions, which together reduce the binding down to the lower curve, fitted to empirical mass data.

of this term (see Bohr and Mottelson (1969), pp. 149–152). With this term included, the formula for the electrostatic energy of a spherical nucleus becomes

$$E_C = \frac{3}{5} \frac{Z^2 e^2}{4\pi\varepsilon_0 R_C} \left[1 - \frac{\pi^2}{2} \frac{5}{3} \left(\frac{a}{R_C} \right)^2 - \frac{0.76}{Z^{2/3}} \right]$$

In the often cited mass fit by Myers and Swiatecki (1967), the exchange correction term does not occur. The authors then cite values of the equivalent radius and the diffuseness, respectively as $R_C = 1.2249 \times A^{1/3}$ fm and $a = 0.544$ fm. The large value of R_C hints that the Coulomb energy term in their semi-empirical mass formula simulates some additional effect otherwise not explicitly accounted for.

A survey of the effect of the different terms in the complete mass formula as functions of A is displayed in fig. 3.1. The top of the vertical scale is represented by the (negative) volume energy. From this is subtracted the surface, Coulomb and symmetry energy for a sequence of masses selected along the stability line. The energy is counted per particle.

The binding energy per nucleon for different nuclei along β-stability is exhibited in fig. 3.2. Note that for $A \simeq 40$, a roughly constant value of about

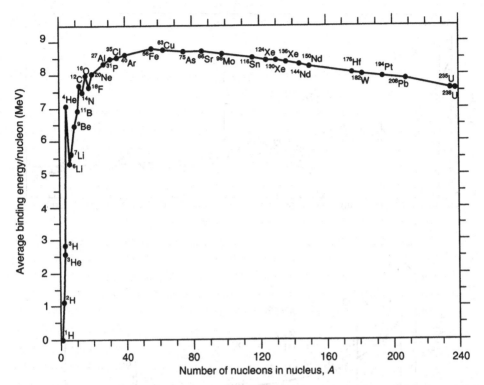

Fig. 3.2. 'Specific binding energy' or binding energy per nucleon B/A for nuclei along the stability line. The largest value of about 8.7 MeV is exhibited by $^{56}_{26}$Fe.

8 MeV is reached. It becomes clear from fig. 3.1 that below this A-value, the surface energy dominates. The negative slope of the total binding energy for $A \simeq 60$ reflects the growing importance of the Coulomb energy term.

It has been argued that not only the volume energy, but also the surface energy should be isospin-dependent. Thus one has added a symmetry-dependent term to the surface energy, which is now written

$$E_s = -B_s = a_s' A^{2/3} \left[1 - \kappa_s \left(\frac{N-Z}{A} \right)^2 \right]$$

where Myers and Swiatecki (1967) give $a_s' = 17.944$ MeV and $\kappa_s = 1.7826$. (In this reference furthermore $a_v = 15.494$ MeV, and $\kappa_s = \kappa_v$ leading to $a_{\text{sym}} = 55.24$ MeV.)

In the systematic study of masses one also notices obvious odd–even effects. In comparing odd–odd, odd-Z and odd-N nuclei with their even–even neighbours one observes (on the average) the following relations:

$$E(\text{odd-}N) - E(\text{even–even}) = \Delta_{\text{n}}$$

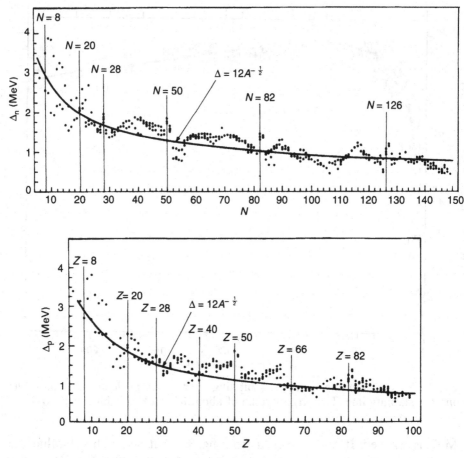

Fig. 3.3. Odd–even mass differences for odd-N and odd-Z nuclei. The Δ_n and Δ_p values are calculated from the experimental masses by the formulae given in problem 3.5 (from Bohr and Mottelson, 1969).

$$E(\text{odd-}Z) - E(\text{even–even}) = \Delta_p$$
$$E(\text{odd–odd}) - E(\text{even–even}) = \Delta_p + \Delta_n - E_{np}$$

where empirically†

$$\Delta_n \simeq \Delta_p \simeq 12 \times A^{-1/2} \text{ MeV}$$

while

$$E_{np} \simeq 20 \times A^{-1} \text{ MeV}$$

The systematic odd–even differences Δ_n and Δ_p all over the periodic table

† Formulae to calculate Δ_n, Δ_p and E_{np} in practical cases are given in problem 3.5.

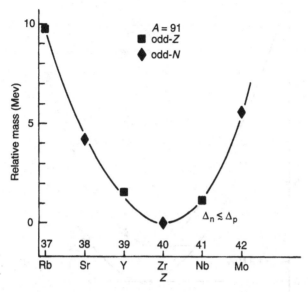

Fig. 3.4. Relative masses of $A = 91$ isobars from ${}^{91}_{37}$Rb to ${}^{91}_{42}$Mo. Note that all nuclei fall roughly on one parabola, implying $\Delta_n \approx \Delta_p$.

are plotted in fig. 3.3. One may observe that the data may be taken to indicate a slightly larger Δ_p than Δ_n. This also is consistent with the fact that there are altogether 68 stable odd-N nuclei compared with 53 stable odd-Z ones. In fig. 3.4 the relative masses of a series of $A = 91$ isobars are plotted. Every other one, as ${}^{91}_{38}$Sr, ${}^{91}_{40}$Zr and ${}^{91}_{42}$Mo, is an odd-N element while ${}^{91}_{37}$Rb, ${}^{91}_{39}$Y and ${}^{91}_{41}$Nb are odd-Z elements. They apparently all fall on the same parabola, implying $\Delta_p \simeq \Delta_n$.

In fig. 3.5 a similar plot is made for $A = 92$ isobars. Of these ${}^{92}_{38}$Sr, ${}^{92}_{40}$Zr and ${}^{92}_{42}$Mo are even–even elements, while ${}^{92}_{37}$Rb, ${}^{92}_{39}$Y, ${}^{92}_{41}$Nb and ${}^{92}_{43}$Tc are odd–odd ones. The displacement of the two parabolae now apparent is a measure of the quantity $\Delta_n + \Delta_p - E_{np}$.

To the other terms of the mass formula there is usually added a pairing energy term $P(N, Z)$. The energy E_{np} is often neglected and $P(N, Z)$ taken equal to $(-\Delta, 0, \Delta)$ for even–even, odd-A, and odd–odd nuclei, respectively, where $\Delta \approx (12/\sqrt{A})$ MeV. This term is to account grossly for the odd–even effects.

Recent experiments on nuclei far from stability have made it meaningful to consider Δ not only as a function of mass number A but also as a function of neutron excess, $N - Z$ (Jensen *et al.*, 1984; Madland and Nix, 1988). For example, in the version of Madland and Nix, the following expressions for

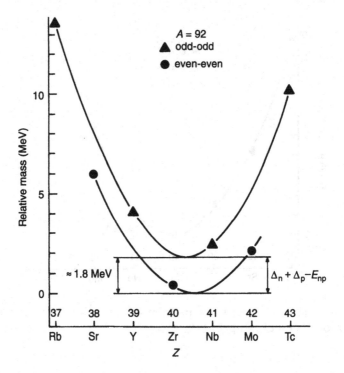

Fig. 3.5. Relative masses of $A = 92$ isobars. The energy distance between the two very distinct parabolas should equal $\Delta_n + \Delta_p - E_{np}$, where E_{np} represents the extra coupling energy between the unpaired neutron and the unpaired proton.

the average pairing gaps are used:

$$\Delta_n = \frac{r}{N^{1/3}} \exp\left[-s\left(\frac{N-Z}{A}\right)^2 - t\left(\frac{N-Z}{A}\right)^4\right]$$

$$\Delta_p = \frac{r}{Z^{1/3}} \exp\left[s\left(\frac{N-Z}{A}\right)^2 - t\left(\frac{N-Z}{A}\right)^4\right]$$

A fit to odd–even mass differences extracted from measured masses gives $r = 5.72$ MeV, $s = 0.119$ and $t = 7.89$.

The liquid-drop formula can be seen as a first-order description in terms of two small expansion parameters: the ratio of the surface diffuseness to the size of the system ($\propto A^{-1/3}$) and the square of the relative neutron excess $[(N - Z)/A]^2$. In the 'droplet model' (see e.g. Myers, 1977) this expansion has been carried to one higher order than in the liquid-drop model. It seems questionable whether the droplet model gives any improvement in the description of the nuclear masses. The model has, however, found

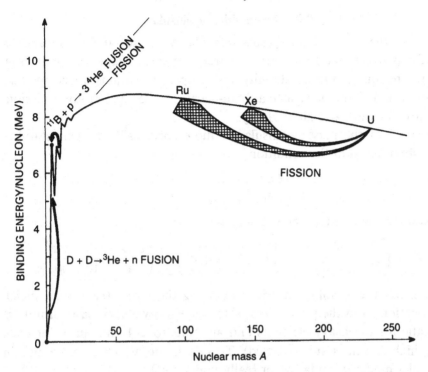

Fig. 3.6. Binding energy gains through fission or fusion. Local variations in the binding-energy curve are exploited in the fission and fusion processes marked in the figure.

applications to a larger variety of nuclear problems than the liquid-drop model. Furthermore, in the droplet model, the value of the Coulomb radius R_C is reduced to a more reasonable value.

One interesting application of the mass formula is that it forms a basis for a discussion of nuclear power as illustrated in fig. 3.6. The most stable species of nuclei are centred around Fe and Ni. In fact $^{62}_{28}$Ni is the element with the most favourable B/A ratio. Elements to the right of this peak may gain total binding by subdivision or fission, elements to the left of this peak by fusion. In fact all elements heavier than $A = 110$–120 are in principle fission-unstable. The fission barriers are, however, so large that the resulting decay rates are far below the observation limit. In fact, for elements along the stability line, the fission process is unimportant for nuclei with $A < 230$.

3.2 The stability peninsula

In terms of the semi-empirical mass formula, where so far no allowance is made for possible deviation from a purely spherical shape, one may now study the region of nuclear stability with respect to various break-ups and transformation phenomena such as beta-decay, neutron and proton emission, spontaneous fission and alpha-decay.

The trend of the mass peninsula, or more specifically the beta-stability line, is obtained from the condition

$$\left(\frac{\partial m}{\partial N}\right)_{A=\text{const}} = 0$$

This leads to the relation below for $\kappa_s = 0$

$$N - Z = \left(\frac{3}{10} \cdot \frac{e^2}{4\pi\varepsilon_0 R_{\text{C}}} A - \frac{1}{2}(M_n - M_{\text{H}})c^2\right) \bigg/ \left(\frac{a_{\text{sym}}}{A} + \frac{3}{10} \cdot \frac{e^2}{4\pi\varepsilon_0 R_{\text{C}}}\right)$$

In fig. 3.7 the stability line is drawn together with the stable nuclei. The deviation from the predictions of the Myers–Swiatecki mass formula is systematic for the heavy elements and amounts to 2–4 units in N for each Z (too high N-values are predicted). This deviation, which is found also in the droplet model formula, is not really understood.

The heaviest attainable neutron isotope of a given element corresponds to zero neutron separation energy, or

$$\left(\frac{\partial B}{\partial N}\right)_{Z=\text{const}} = 0$$

This so-called neutron 'drip line' is of great astrophysical interest. The corresponding proton 'drip line' corresponds to the condition

$$\left(\frac{\partial B}{\partial Z}\right)_{N=\text{const}} = 0$$

The use of these equations is exemplified in problem 3.7 and calculated neutron and proton 'drip lines' are exhibited in fig. 3.7.

Instability with respect to alpha-decay and spontaneous fission set additional limits to the availability of nuclei in terrestrial matter. In particular, the limit of availability for a heavy-A nucleus is largely decided by the spontaneous-fission process. As already mentioned, the fission process first becomes important for normal nuclei for $A > 230$. The spontaneous fission half-life of $^{230}_{90}\text{Th}$ is approximately 10^{17} y, while the corresponding half-life for $^{240}_{94}\text{Pu}$ is $\simeq 10^{11}$ y, for $^{254}_{102}\text{No}$ 10 s, and for $^{260}104$ 0.3 s. The fall-off of half-lives with A is thus very rapid. The liquid-drop model (see chapter 4) makes Z^2/A the one relevant parameter for the fission process, and it

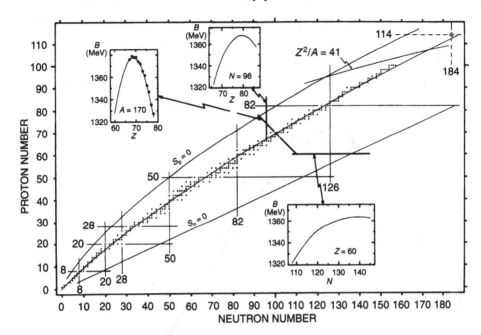

Fig. 3.7. Stable elements versus Z and N together with some quantities calculated from the Myers and Swiatecki (1967) mass formula. For nuclei beyond ^{208}Pb, where no stable elements exist, the nucleus having the lowest mass for fixed A is plotted. The line of beta-stability, and the proton ($S_p = 0$) and neutron ($S_n = 0$) drip lines according to the mass formula, are indicated. Furthermore, the binding energies are drawn for three cuts having constant proton number ($Z = 60$), constant mass number ($A = 170$) and constant neutron number ($N = 96$), respectively. For constant Z and N the slope of these curves gives the one-particle separation energy per added neutron and proton, respectively. This quantity is around 8 MeV at β-stability and goes to zero at the drip lines. The beta-stability line is defined as the extremum point of the mass when plotted for constant A. Owing to the difference of the neutron and hydrogen atom masses, this point is slightly different from the extremum of the binding energy. For some nuclei with $A = 170$, the experimental binding energies (A.H. Wapstra and G. Audi, *Nucl. Phys.* **A432**, 1 (1985)) are shown by solid points. The small differences relative to the semi-empirical mass formula indicate that, for these nuclei, the shell effects are small. Stronger shell effects are expected around the closed shells, which are drawn in the figure. For nuclei beyond the line $Z^2/A = 41$, very short fission half-lives ($\lesssim 1$ s) are expected. However, shell effects like those at the predicted superheavy islands around ^{298}114 could locally change the half-lives by many orders of magnitude.

has been inferred that the quantity Z^2/A being ≈ 41 (see fig. 3.7) should correspond to half-lives in the region of seconds (which in the $Z = 100$ region is a rather appropriate estimate).

Use of the semi-empirical mass formula defines only rough boundaries to the regions of long-lived or stable nuclei. The boundaries of the stability

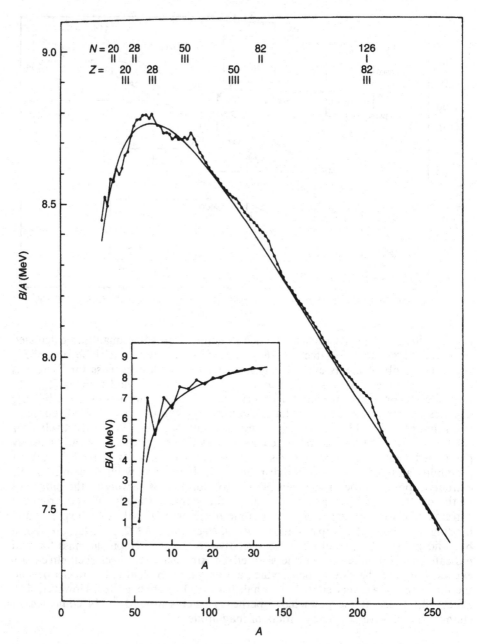

Fig. 3.8. Empirical specific binding energies (B/A) compared with the average curve representing the semi-empirical mass formula. Deviations mark shell structure effects. The nuclei are chosen along the line of β-stability while the inset shows $N = Z$ nuclei (from Bohr and Mottelson, 1969).

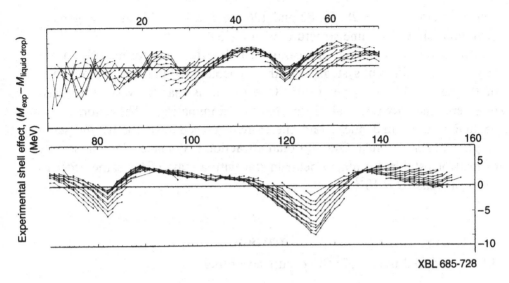

Fig. 3.9. A more quantitative diagram of the detailed differences between the measured nuclear masses and the semi-empirical mass formula in the 1966 Myers–Swiatecki version. Isotopic masses are connected by lines. Effects due to shell closures at $N = 28, 50, 82$ and 126 are clearly apparent.

peninsula are in their details connected with nuclear shell structure that gives modifications in the binding energy relative to the predictions of the semi-empirical mass formula of the order of 10 MeV. For the stable nuclei exhibited in fig. 3.7, one can discern small irregularities around the magic proton and neutron numbers (closed shells), $Z,N = 8, 20, 28, 50, \ldots$. In a similar way, the drip lines should be irregular due to shell effects and also due to odd–even effects. The problem of a possible island of relative stability beyond the short-lived heavy nuclei of elements with $Z = 107$–109, which so far are the heaviest produced, is directly connected with the existence of nuclear magic numbers, in this case $Z = 114$ and $N = 184$ (see chapter 10).

A comparison of empirical nuclear masses with a semi-empirical mass formula is exhibited in fig. 3.8 for nuclei along β-stability. Except for the very light nuclei, the largest deviations are connected with neutron numbers $N = 28, 50, 82$ and 126 and proton numbers $Z = 28, 50$ and 82. The deviation is indicative of shell structure and we shall return to this important point later.

The deviations obtained between the measured nuclear masses and the liquid drop fit of Myers and Swiatecki (1966) are shown in fig. 3.9. The plot is constructed in terms of neutron number with the isotopes connected. The neutron numbers $N = 28, 50, 82$ and 126 are clearly visible and the general deviations show the same structure as in fig. 3.8.

Although it is tempting, one cannot employ the mass formula for much larger or very different systems than those encountered in the neighbourhood of the beta-stable mass peninsula. One reason is that for very large mass numbers, the gravitational forces become dominating. These forces are negligible in normal nuclei and therefore not accounted for by the semi-empirical mass formula. Gravitational attraction may make a large assembly of neutrons stable, forming a neutron star with a mass of the same order as the solar mass, 2×10^{32} g.

Exercises

3.1   Induced fission of ^{235}U occurs according to

$$n + {}^{235}U \rightarrow {}^{236}U \rightarrow {}^{91}Kr + {}^{142}Ba + 3n + \text{energy}$$

From fig. 3.2 it is apparent that for ^{235}U the binding energy per nucleon, $B/A \simeq 7.6$ MeV, while for Kr and Ba, $B/A \simeq 8.6$ MeV and 8.3 MeV, respectively. The energy gain is thus approximately $(142 \times 0.7 + 91 \times 1.0 - 2 \times 7.6)$ MeV per U atom. Compare this value with the value calculated from the measured masses if (in units related to the mass of ^{12}C):

$$m\left({}^{235}_{92}U\right) = 235.0439$$
$$m\left({}^{91}_{36}Kr\right) = 90.9232$$
$$m\left({}^{142}_{56}Ba\right) = 141.9165$$
$$m(n) = 1.0087$$

3.2   Calculate how many kilogrammes of ^{235}U are consumed per year in a reactor producing 2000 MW thermal energy (about 600 MW electric). 1 MeV $= 1.60 \times 10^{-13}$ J.

3.3   A fusion–fission reaction has been proposed to occur according to

$$p + {}^{11}_{5}B \rightarrow 3\left({}^{4}He\right) + E$$

Calculate the energy E. How many kilogrammes of ^{11}B are needed per year for a reactor with 2000 MW thermal effect. Use the following values for the nuclear masses: $m(^{4}He) = 4.0026$, $m(^{11}B) = 11.0093$, $m(p) = 1.0078$.

3.4 With the assumption that a nucleon in a nucleus interacts only with its nearest neighbours, one can show that for the coefficients of the semi-empirical mass formula, $a_{\text{vol}} \approx a_{\text{surf}}$. Do this! (No detailed calculations.)

3.5 Calculate the quantities Δ_n, Δ_p and E_{np} for nuclei around $^{170}_{70}\text{Yb}$. Use the following formulae in the respective cases (try to justify these formulae):

(a) $\Delta_n =$
$\pm\frac{1}{4}[B(N-2,Z)-3B(N-1,Z)+3B(N,Z)-B(N+1,Z)]$
($Z = $ even, $+$ for $N = $ even, $-$ for $N = $ odd).

(b) $\Delta_p =$
$\pm\frac{1}{4}[B(N,Z-2)-3B(N,Z-1)+3B(N,Z)-B(N,Z+1)]$
($N = $ even, $+$ for $Z = $ even, $-$ for $Z = $ odd).

(c) $\Delta_p - E_{np} =$
$\pm\frac{1}{4}[B(N,Z-2)-3B(N,Z-1)+3B(N,Z)-B(N,Z+1)]$
($N = $ odd, $+$ for $Z = $ even, $-$ for $Z = $ odd).

3.6 The nucleus $^{242}_{94}\text{Pu}$ ($t_{1/2} = 3.8 \times 10^5$ years) decays through emission of alpha particles, $E(\alpha_0) = 4.903$ MeV, to $^{238}_{92}\text{U}$. Show that this corresponds to an energy difference $Q_\alpha = 4.985$ MeV. Furthermore show that Q_α is in agreement with the Q-value obtained from $m(^{242}_{94}\text{Pu})$ $= 242.058739$, $m(^{238}_{92}\text{U}) = 238.050786$ and $m(^4_2\text{He}) = 4.002603$.

3.7 Assume the semi-empirical formula in the form

$$m(N,Z) = NM_n + ZM_H - \frac{1}{c^2}B(N,Z)$$

with

$$B(N,Z) = a_v A - a_s A^{2/3} - a_C\frac{Z^2}{A^{1/3}} - a_{\text{sym}}\frac{(N-Z)^2}{2A}$$

Furthermore the constants (A.E.S. Green and N.A. Engler, *Phys. Rev.* **91**, (1953) 40)

$a_v = 15.56$ MeV $a_s = 17.23$ MeV $a_C = 0.697$ MeV $a_{\text{sym}} = 46.57$ MeV

give a good fit to known masses.

(a) Calculate from the above formula an expression for the β-stability line. The masses of the hydrogen atom and the neutron, respectively are $M_H c^2 = 938.77$ MeV and $M_n c^2 = 939.55$ MeV. Derive the equation for the line in the limit of $A \to \infty$.

(b) The neutron and proton 'drip' lines are characterised by

$$\left[\frac{\partial B}{\partial N}\right]_{Z=\text{const}} = 0 \qquad \left[\frac{\partial B}{\partial Z}\right]_{N=\text{const}} = 0$$

respectively. Derive expressions of the form $N = N(A)$ and $Z = Z(A)$ for these curves.

(c) Calculate numerical values according to the formulae derived for $A = 100$, 200 and 300.

4

Nuclear fission and the liquid-drop model

Fission was the third of the modes of nuclear decay to be discovered. Alpha and beta decay had been studied since their discovery in 1895. Nuclear fission was not discovered until 1938, when O. Hahn and F. Strassmann (1938, 1939) bombarded uranium with neutrons and definitely identified barium atoms in the products in the reaction resulting from neutron capture in uranium. This reaction had been studied extensively (see Amaldi, 1984, for a review of these years of physics) since shortly after the discovery of the neutron in 1932. However, as the reaction was expected to yield *transuranium* elements it was not until the work by Hahn and Strassmann that it was realised that a new and entirely unexpected nuclear phenomenon was being observed.

Hahn's coworker Lise Meitner and her nephew O. Frisch (Meitner and Frisch, 1939) coined the word fission for the new phenomenon discovered by Hahn and Strassmann. The word is a loan from biology. In close cooperation with Niels Bohr the former authors also correctly interpreted the reaction as a break-up of the nucleus into two smaller fragments and offered the first qualitative explanation of the phenomenon in terms of competition between the disrupting trend of Coulomb repulsion and the shape-stabilising effect of surface tension. The effect should have been anticipated from the liquid-drop model. Thus, for heavy nuclei the large number of positively charged protons repel each other so strongly that the small amount of energy added to the nucleus by the impinging neutron (adding its binding energy) is sufficient to make it break apart. The liquid-drop theory was now systematically extended to non-spherical shapes to explain the process of fission in a, by now, classic paper by Bohr and Wheeler (1939). The ideas put forth in this paper still form a basis for the present theories of nuclear fission.

4.1 Some basic fission phenomena

The fission phenomenon studied by Hahn and Strassmann was so-called
neutron-induced fission, the process of division occurring first after absorption
of an impinging neutron. A very important conclusion was reached first
by Niels Bohr (1939). On the basis of seemingly contradictory findings, he
realised that only one isotope of natural uranium, ^{235}U (with 3% abundance)
was responsible (almost exclusively) for the observed fission, while the other
naturally occurring isotope ^{238}U (with 97% abundance) was inactive.

All this was neutron-induced fission. In 1940, Petrzhak and Flerov discovered that uranium also *undergoes fission spontaneously* without added energy
from any external agent. Later the heavier artificially produced actinides
were also found to undergo fission spontaneously. The spontaneous fission
half-life is perhaps the most directly measurable quantity associated with
fission. The longest half-life determined is 10^{17} years for one isotope of U.
One of the shortest measured is 10^{-10} years or 1 ms for ^{258}Fm (see fig. 10.2
below). There is thus a drastic time variation with charge number Z.

Another striking phenomenon associated with fission is the mass distribution of the fission fragments. Fig. 4.1 shows this distribution for spontaneous
fission of $^{252}_{98}$Cf, $^{256}_{100}$Fm, $^{257}_{100}$Fm and for neutron-induced fission of $^{257}_{100}$Fm. The
y-axis shows the percentage yield. For the lighter mass numbers in fig. 4.1,
an outstanding feature is the mass asymmetry, i.e. the fact that the two
fragment nuclei have unequal masses. The mass asymmetry becomes even
more pronounced for lighter actinides and for $_{92}$U and $_{94}$Pu, the fragment
masses centre around $M_1 \simeq 100$ and $M_2 \simeq 140$. The peak to valley ratio
also increases with decreasing mass, i.e. the number of asymmetric divisions increases relative to the number of symmetric divisions. For ^{257}Fm
the asymmetry has almost disappeared, as is seen in fig. 4.1. Furthermore,
with increasing excitation energy the asymmetry also decreases. This explains the difference between the two curves for ^{257}Fm in fig. 4.1 because,
for neutron-induced fission, the excitation energy of the nucleus undergoing
fission becomes approximately equal to the neutron binding energy. For
higher excitation energies symmetric fission dominates in all cases.

Another quantity that is measured experimentally is the fission barrier
height. In fig. 4.2 a typical fission barrier is plotted. When the nuclear shape
deviates from sphericity, the surface energy (which is assumed proportional
to the nuclear surface) increases and the Coulomb energy decreases. The
total potential energy in the liquid-drop model which is illustrated in fig. 10.6
below, is determined by the sum of these terms. To get agreement between
measured and calculated fission barriers it is necessary to take shell effects

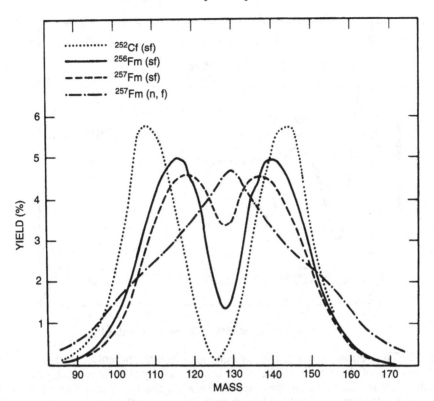

Fig. 4.1. Mass distributions in terms of the fission fragment masses for spontaneous fission of $^{252}_{98}$Cf, $^{256}_{100}$Fm and $^{257}_{100}$Fm and for neutron-induced fission of ^{257}Fm. Note the trend toward symmetric fission with increasing mass and in addition the larger number of symmetric events for neutron-induced than for spontaneous fission (from R. Vandenbosch and J.R. Huizenga, *Nuclear Fission* (Academic Press, New York and London, 1973)).

into account. These shell effects are responsible for the complicated structure of the barrier shown in fig. 4.2. The ground state minimum for deformed shape, the secondary minimum and the very shallow third minimum arise mainly because, at various distortions, specific nucleon numbers give rise to particularly high binding energy (cf. chapter 9).

In spontaneous fission the nucleus is said to 'penetrate' the barrier. Experimentally a typical value of the fission barrier height is 6 MeV in the actinide region. Nuclei can also exist in a fission isomeric state, trapped in the second minimum (cf. fig. 4.2). They have then a much thinner barrier to penetrate and their fission half-lives are consequently many orders of magnitude shorter. Known fission isomeric states have half-lives in the microsecond to picosecond range (see fig. 10.3 below).

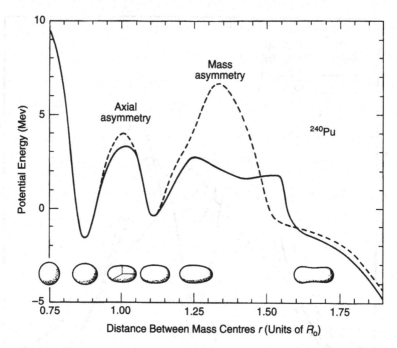

Fig. 4.2. Theoretical fission barrier or equivalently total 'potential' energy for ^{240}Pu as a function of deformation. The latter is indicated by the shapes along the x-axis which is graded in terms of the distance between the two centres of mass of the two nascent fragments. The liquid-drop barrier is a smooth function of deformation as illustrated for ^{244}Pu in fig. 10.6 below. If the shell effects are added, the dashed curve is obtained if only rotation-symmetric and reflection-symmetric nuclear shapes are considered (the reflection plane is perpendicular to the symmetry axis). Finally, the solid curve results if more general nuclear shapes are considered. The first and second barriers are lowered due to the inclusion of axial asymmetry and reflection asymmetry, respectively. Note that the reflection asymmetry naturally leads to an asymmetric fission mass distribution (from P. Möller and J.R. Nix, *Proc. Third IAEA Symp. on Physics and Chemistry of Fission*, Rochester, New York (IAEA, Vienna, 1974) vol. 1, p. 103).

After penetrating the barrier the nucleus arrives at the point where the fragments separate, the so-called scission point, which lies some 30–40 MeV below the ground state as indicated in fig. 4.3. After scission the fragments are accelerated by the Coulomb repulsion between the two separated charged fragments. Most of the energy released is associated with this 'Coulomb push'. The fragments acquire an energy from fission of almost 200 MeV for ^{240}Pu (cf. fig. 4.3), most of it in the form of kinetic energy. In fact the Coulomb push after scission is almost entirely transferred to kinetic energy. For thermal fission of ^{235}U one obtains a kinetic energy for the fragments

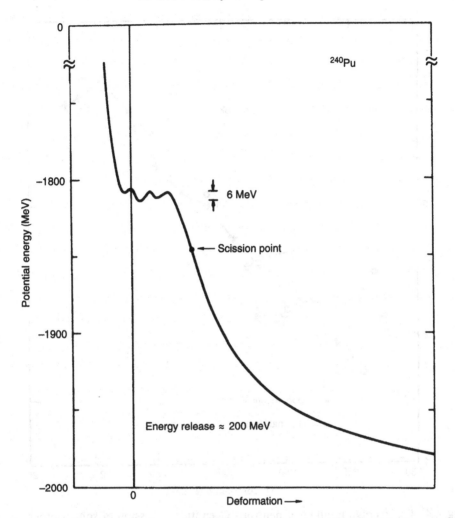

Fig. 4.3. A total view of the extended $^{240}_{94}$Pu fission 'barrier'. The total binding energy is plotted against deformation. Note that scission (fragmentation) is reached about 35 MeV below the equilibrium ground state energy. The additional fission net energy in excess of 160 MeV is recovered from the Coulomb repulsion between the two charged fragments (from M. Bolsterli, E.O. Fiset, J.R. Nix and J.L. Norton, *Phys. Rev.* **C5** (1972) 1050).

of about 170 MeV and for ^{239}Pu a kinetic energy of about 178 MeV (see problem 3.1).

In the fission process, the remaining energy is dissipated in the form of separation energy and kinetic energy of the emitted neutrons and of emitted gamma rays from the excited fragment nuclei. The number of emitted neutrons is found to grow strongly with charge and mass of the actinides

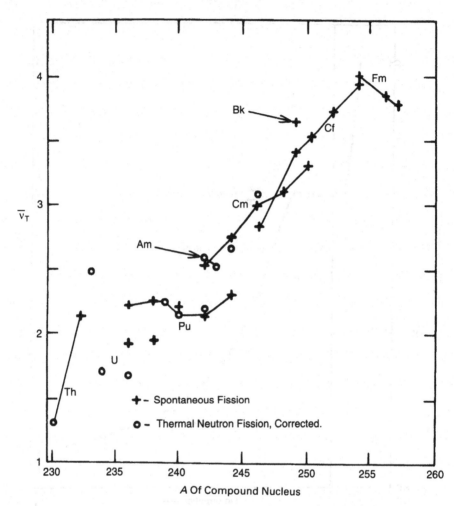

Fig. 4.4. The average number of neutrons $\bar{\nu}_T$ emitted on fission of some isotopes of elements between $_{90}$Th and $_{100}$Fm as a function of mass. Note the fast growth in neutron number with Z. In fact for $Z = 114$ the number of emitted neutrons in the fission process is extrapolated to be 10–12 (from D.C. Hoffman and M.M. Hoffman, Reproduced with permission from the *Annual Review of Nuclear Science*, Vol. **24**, ©1974 by Annual Reviews Inc.)

undergoing fission, as is shown in fig. 4.4. The increase in ν, the number of neutrons emitted, is from about 2 for $_{92}$U and $_{94}$Pu to about 4 for the $_{100}$Fm isotopes. Extrapolating to the predicted super-heavy-element region around $Z = 114$ and $N = 184$, one predicts the emission of about 12 neutrons in the fission of ^{298}114. The number of neutrons emitted in the fission process is very important for the criticality of the associated chain reaction. A chain reaction is started if, on average, at least one neutron from a fission event induces a second fission event etc. A high v-value generally implies a small critical mass. (The 'critical mass' is the smallest mass for which the chain reaction is self-sustaining.)

4.2 The liquid-drop model and the fission barrier

In the liquid-drop model expression that was given in chapter 3 for the nuclear mass or energy, the relevant terms for the description of the fission barrier as a function of deformation, d, can be written

$$E(d) = E_C(d) + E_s(d) = a_C \frac{Z^2}{A^{1/3}} \left(\frac{E_C(d)}{\overset{0}{E}_C} \right) + a_s \left(1 - \kappa_s I^2 \right) A^{2/3} \left(\frac{E_s(d)}{\overset{0}{E}_s} \right)$$

where $I = (N - Z)/(N + Z)$ and zero over a character indicates that the quantity is calculated for spherical shape. The formula has been generalised to deformed shapes by the inclusion of the quantities $E_C(d)/\overset{0}{E}_C$ and $E_s(d)/\overset{0}{E}_s$. They are the ratios of the Coulomb and surface energies of the deformed nucleus to the Coulomb and surface energies of the spherical nucleus. For a study of the fission barrier, these quantities are calculated for some sequence of nuclear shapes. One shape parametrisation corresponds to an expansion of the nuclear radius in spherical harmonics with the shape parameters given by $\alpha_{\lambda\mu}$:

$$R(\theta, \varphi) = R_\alpha \left\{ 1 + \sum_{\lambda=1}^{\infty} \sum_{\mu=-\lambda}^{\lambda} \alpha_{\lambda\mu} Y_{\lambda\mu}(\theta, \varphi) \right\}$$

The requirement that the radius should be a real number, $R(\theta, \varphi) = R^*(\theta, \varphi)$, and the property of the spherical harmonics,

$$Y_{\lambda\mu}^* = (-1)^\mu Y_{\lambda-\mu}$$

leads to

$$\alpha_{\lambda\mu}^* = (-1)^\mu \alpha_{\lambda-\mu}$$

The deformation-dependent radius R_α is related to its spherical counterpart R_0 by the condition of volume conservation

$$\frac{4\pi}{3} R_0^3 = \int dV = \int d\Omega \int^{R(\theta,\varphi)} r^2 \, dr = \frac{4\pi}{3} R_\alpha^3 \left(1 + \frac{3}{4\pi} \sum |\alpha_{\lambda\mu}|^2 + O\left(\alpha^3\right) \right)$$

where the orthogonality of the spherical harmonics has been used

$$\int Y_{\lambda\mu}^* Y_{\lambda'\mu'} \, d\Omega = \delta_{\lambda\lambda'} \delta_{\mu\mu'}$$

Using this general expansion one can to second and lowest order in $\alpha_{\lambda\mu}$ derive the following expressions for the surface and Coulomb energies. For the surface energy, one has

$$E_s = \overset{0}{E_s}\left(1 + \frac{1}{8\pi}\sum_{\lambda,\mu}(\lambda-1)(\lambda+2)\,|\alpha_{\lambda\mu}|^2\right)$$

with

$$\overset{0}{E_s} = a_s\left(1 - \kappa_s I^2\right)A^{2/3}$$

The Coulomb energy is determined from

$$E_C = \frac{1}{2}\left(\frac{1}{4\pi\varepsilon_0}\int\rho\,(\mathbf{r}_1)\;d\tau_1\int\rho\,(\mathbf{r}_2)\;d\tau_2\frac{1}{|\mathbf{r}_1-\mathbf{r}_2|}\right)$$

where a homogeneous charge distribution corresponds to

$$\rho(\mathbf{r}) = \begin{cases} \rho_0, & r \leq R(\theta,\varphi) \\ 0, & r > R(\theta,\varphi) \end{cases}$$

For a spherical nucleus one finds by straightforward integration (cf. problem 2.3)

$$\overset{0}{E_C} = \frac{3}{5}\frac{Z^2e^2}{4\pi\varepsilon_0}\frac{1}{R_C} = a_C\frac{Z^2}{A^{1/3}}$$

For a deformed nucleus one obtains after somewhat more elaborate calculations, also to second order,

$$E_C = \overset{0}{E_C}\left(1 - \frac{5}{4\pi}\sum_{\lambda,\mu}\frac{\lambda-1}{2\lambda+1}\,|\alpha_{\lambda\mu}|^2\right)$$

For an approximate description of the whole fission barrier, it is necessary to treat the deformation parameters $\alpha_{\lambda\mu}$ at least to third order. It is furthermore reasonable to assume that the nucleus undergoing fission remains symmetric with respect to rotation around the deformation axis, chosen as the z-axis. We thus do not need the $Y_{\lambda\mu}(\theta,\varphi)$ angular base but can confine ourselves to an expansion in $P_\lambda(\cos\theta)$. We have

$$Y_{\lambda 0}(\theta,\varphi) = \left(\frac{2\lambda+1}{4\pi}\right)^{1/2}P_\lambda(\cos\theta)$$

One then writes the radius vector in the form

$$R(\theta) = R_\beta\left(1 + \sum_\lambda\beta_\lambda Y_{\lambda 0}(\theta,\varphi)\right) = R_\beta\left[1 + \left(\frac{2\lambda+1}{4\pi}\right)^{1/2}\sum_\lambda\beta_\lambda P_\lambda(\cos\theta)\right]$$

where thus

$$\beta_\lambda = \alpha_{\lambda 0}$$

The lowest-λ Legendre polynomials $P_\lambda(x)$ are

$$P_0(x) = 1 \qquad\qquad P_4(x) = \tfrac{1}{8}\left(35x^4 - 30x^2 + 3\right)$$
$$P_1(x) = x \qquad\qquad P_5(x) = \tfrac{1}{8}\left(63x^5 - 70x^3 + 15x\right)$$
$$P_2(x) = \tfrac{1}{2}\left(3x^2 - 1\right) \qquad P_6(x) = \tfrac{1}{16}\left(231x^6 - 315x^4 + 105x^2 - 5\right)$$
$$P_3(x) = \tfrac{1}{2}\left(5x^3 - 3x\right)$$

In fact the β_2 coordinate is the most important for small deformations. Next in importance comes β_4. For the illustrative calculation to be carried out in more detail in problem 4.2 we shall confine ourselves to $\beta_2 \neq 0$ only. Furthermore, to simplify the notation, we introduce $a_2 = (5/4\pi)^{1/2}\beta_2$. We have then

$$R(\theta, a_2) = R_a(a_2)\left(1 + a_2 P_2(\theta)\right)$$

To calculate the surface element one needs the product $\mathbf{n} \cdot \mathbf{e}_r$, where $\mathbf{n}$ is the unit normal vector of the surface and $\mathbf{e}_r$ the unit radius vector. The surface element can then be written as

$$dS = \frac{R^2 \, d\Omega}{\mathbf{n} \cdot \mathbf{e}_r} = \frac{R^3 \, d\Omega}{\mathbf{n} \cdot \mathbf{R}}$$

The unit normal vector $\mathbf{n}$ is obtained as follows. Consider the radius vector definition in terms of a_2 above. It can be considered as a surface in (r, θ, φ) space. Each new R_a value defines a new surface

$$R_a = \frac{r}{1 + a_2 P_2(\cos\theta)}$$

The normal is then obtained as

$$\mathbf{n} = \frac{\nabla R_a(r, \theta)}{|\nabla R_a|}$$

where

$$\nabla = \mathbf{e}_r \frac{\partial}{\partial r} + \mathbf{e}_\theta \frac{1}{r}\frac{\partial}{\partial \theta} + \mathbf{e}_\varphi \frac{1}{r\sin\theta}\frac{\partial}{\partial\varphi}$$

One then obtains

$$\nabla R_a = \frac{1}{1 + a_2 P_2}\mathbf{e}_r - \frac{a_2}{(1 + a_2 P_2)^2}\frac{\partial P_2}{\partial\theta}\mathbf{e}_\theta$$

It is then easy to derive the integral

$$S = \int \int dS = R_a^2 \int \int (1 + a_2 P_2)^2 \left[1 + \frac{a_2^2}{(1 + a_2 P_2)^2} \left(\frac{\partial P_2}{\partial \theta} \right)^2 \right]^{1/2} \sin \theta \, d\theta \, d\varphi$$

Expanding in a_2 and retaining terms to third order one obtains

$$S = 4\pi R_a^2 \left(1 + \frac{4a_2^2}{5} + \ldots \right)$$

Using the normalisation of R_a implied from volume conservation or

$$R_a = R_0 \left(1 - \frac{1}{5}a_2^2 - \frac{2}{105}a_2^3 + \ldots \right)$$

one obtains finally the following expression for the surface area

$$S = 4\pi R_0^2 \left(1 + \frac{2}{5}a_2^2 - \frac{4}{105}a_2^3 + \ldots \right)$$

The calculation of the Coulomb energy term to third order in a_2 is a more formidable calculational problem, for which we here give only the final result

$$E_C = \overset{0}{E}_C \left(1 - \frac{1}{5}a_2^2 - \frac{4}{105}a_2^3 \right)$$

Introducing the conventional fissility parameter x as

$$x = \frac{\overset{0}{E}_C}{2 \overset{0}{E}_s}$$

on whose meaning we shall comment below, we obtain for the change in the sum of the surface and Coulomb energies, i.e. the total deformation energy

$$E_s + E_C - \overset{0}{E}_s - \overset{0}{E}_C = \Delta E = \overset{0}{E}_s \left(\frac{2}{5}(1 - x)a_2^2 - \frac{4}{105}(1 + 2x)a_2^3 + \ldots \right)$$

The deformation energy contains a second- and a third-order term. The first dominates for small a_2, i.e. near spherical shape. For $x < 1$ the deformation energy has a positive curvature at $a_2 = 0$ and is thus stable against small deformations. For $x > 1$ the nucleus is already unstable against small deformations. Also for $x < 1$ and sufficiently large a_2 the deformation energy again becomes negative. The third-order expression in a_2 therefore generates a fission barrier. One may now ask whether the expansion is convergent. In fact it is not so beyond $a_2 \simeq 0.6$. Still, this third-order expansion can be used for an estimate with amazing accuracy for $0.7 \lesssim x$. It provides an example of a semiconvergent series. It is tempting to use the

Table 4.1. *Height of fission barriers and β_2 (a_2) deformations at the top of the barrier calculated in the liquid-drop model using a third-order expansion for different values of the fissility parameter x.*

	Z^2/A	x	E_{barr} (MeV)	a_2^{barr}	β_2^{barr}
$^{209}_{83}$Bi	32.96	0.700	17.9	0.88	1.40
$^{232}_{90}$Th	34.91	0.753	9.7	0.69	1.09
$^{238}_{92}$U	35.56	0.769	7.8	0.64	1.01
$^{242}_{94}$Pu	36.51	0.787	6.0	0.58	0.92
$^{254}_{100}$Fm	39.37	0.841	2.4	0.41	0.65
$^{294}_{110}$X	41.16	0.912	0.4	0.22	0.35

expression derived for an estimate of the fission barrier. We then calculate the maximum point of the third-order polynomial by considering

$$\frac{\partial \Delta E}{\partial a_2} = 0 = \overset{0}{E}_{\text{s}} \left(\frac{4}{5}(1-x)a_2 - \frac{4}{35}(1+2x)a_2^2 \right)$$

This equation has two roots: $a_2 = 0$ and $a_2 = 7(1-x)/(1+2x)$. The first corresponds to the spherical minimum, the second to the barrier maximum. The barrier maximum is obtained as

$$E_{\text{barr}} = \frac{98}{15} \frac{(1-x)^3}{(1+2x)^2} \cdot \overset{0}{E}_{\text{s}}$$

It is useful to remind oneself of the fact that all these expressions are valid only for small values of $|1-x|$.

The fissility parameter x is thus a measure of the 'readiness' of the nucleus with respect to division. We see that for $x > 1$ there is no barrier towards fission. For the liquid-drop model parameters of Myers and Swiatecki (1967), $a_C = 0.7053$ ($R_C = 1.2249 \cdot A^{1/3}$ fm), $a_s = 17.944$ and $\kappa_s = 1.7826$, we have

$$x = 0.01965 \frac{Z^2}{A} \frac{1}{(1 - 1.7826I^2)}$$

As illustrated in table 4.1, one may evaluate this expression for a few nuclei

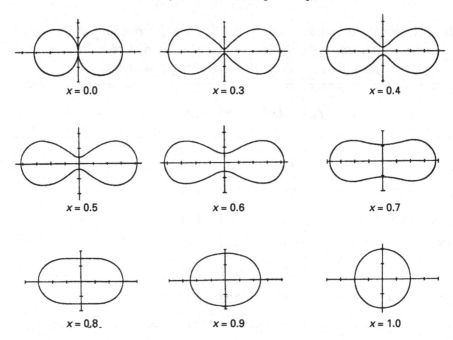

Fig. 4.5. Shapes at the liquid-drop saddle point, i.e. at the top of the fission barrier, for different values of the fissility parameter x (from S. Cohen and W.J. Swiatecki, *Ann. Phys.* **22** (1963) 406).

along the β-stability line. We see from table 4.1 that actinide nuclei have $x < 1$ although empirically they are observed to be unstable to spontaneous fission. However, x is close to 1, which means that the barriers are rather low. The barrier energies according to the simple third-order deformation energy expression are given in the fourth column. The nuclear shape corresponding to the barrier maximum is indicated in the last columns. The lighter elements, which have longer fission half-lives, are thus characterised by a higher and thicker fission barrier. In the liquid-drop model approximation, it is also possible to find the exact nuclear shape at the barrier maximum as shown in fig. 4.5 for different values of x. An important feature of the figure is the strong 'necking' for small x-values. Such shapes cannot be described by the simple expansion of the radius using β_2 deformations only.

In the region of nuclei where x is close to 1, shell effects give rise to important modifications of the barriers. As was anticipated in fig. 4.2 and as we shall see in chapter 9, in the more refined calculations one considers the barrier energy to be the sum of two parts, a 'macroscopic' part taken from the liquid-drop model and a microscopic part, or shell correction, related to the motion of the individual nucleons.

4.3 Alternative models for the macroscopic energy

The surface energy term in the liquid-drop model is a correction to the volume energy term and corrects for the fact that the nucleons at the surface have less interaction energy as they have fewer neighbours than the nucleons at the centre (see problem 3.4). They thus bind to nucleons only on one side of the surface. An analysis shows that the surface energy term in our above-discussed liquid-drop model gives an adequate description only for nuclear forces of zero range (or for systems of very large dimensions relative to the range of the forces). Thus, for systems with strong variations in the nuclear surface the 'standard' surface energy term may be inadequate as exemplified by the dividing nucleus at the point of scission. In the unmodified liquid-drop model the theoretical surface energy will get an unphysically large contribution from the area in the neck region. This is so because the surface nucleons in the neck region obtain some binding due to their interaction with the nucleons on the other side of the neck (provided the distance across is less than the range of the nuclear force, or 1–2 fm). Thus, the unmodified liquid-drop surface energy is consequently overly sensitive to 'wrinkles' in the surface. Equivalently, distortions of high multipole order give large positive contributions to the surface energy. This corresponds to the fact that the coefficient in front of $\alpha_{\lambda\mu}^2$ in the expansion of the surface energy is proportional to λ^2.

The above considerations show the need to account for corrections due to the finite range of the nuclear interaction as well as the diffuseness of the nuclear surface. This led Krappe and Nix (1973) to propose the following modification of the liquid-drop energy. One starts with consideration of an effective interaction term

$$E = -\frac{V_0}{4\pi a^3} \int d^3r \, d^3r' \frac{\exp\left(-|\mathbf{r}-\mathbf{r}'|/a\right)}{|\mathbf{r}-\mathbf{r}'|/a}$$

This six-fold integral represents the total interaction energy over the entire nucleus due to an attractive two-body interaction of a so-called Yukawa-type (the Yukawa function, $e^{-\mu r}/r$, is often used to describe the radial dependence of the nuclear force, see appendix 13A). The range of the two-body interaction is defined by the quantity a in the exponential. The normalisation of the integral expression will be discussed below. The quantity V_0 measures the interaction strength. The volume of integration is specified by the nuclear-radius parameter R_0. A straightforward evaluation of the above integral in spherical coordinates yields for the case of a pure sphere

$$E = V_0 \left[-\frac{4\pi}{3}R_0^3 + 2\pi a R_0^2 - 2\pi a^3 + 2\pi a (R_0 + a)^2 \exp\left(-\frac{2R_0}{a}\right) \right]$$

The first term $-(4\pi/3)V_0R_0^3$, being proportional to A, should be considered a volume contribution. It is not *a priori* obvious that this term can be used to determine the strength parameter V_0 in terms of previously introduced parameters such as a_v. The reason is that no account is taken of exchange contributions, which come from the anti-symmetrisation of the nuclear wave function, and are considered decisive for description of nuclear saturation. The latter are thus needed to account for the nuclear equilibrium density. Also the repulsive core of the nuclear interaction is not included in this first variant of the Krappe–Nix calculations. Still, with $R_0 = r_0A^{1/3}$, we may attempt to write the condition

$$\frac{4\pi}{3}V_0r_0^3A = a_vA\left(1 - \kappa_vI^2\right)$$

The remaining terms can be written in the form:

$$2\pi aV_0R_0^2\left[1 - \left(\frac{a}{R_0}\right)^2 + \left(1 + \frac{a}{R_0}\right)^2\exp\left(-\frac{2R_0}{a}\right)\right]$$

Requiring that the usual (i.e. zero-range) liquid-drop model is obtained in the limit $a/R_0 \to 0$, we arrive at the following relation between V_0, a and the usual surface energy expression

$$\overset{0}{E}_s = 2\pi aV_0R_0^2 = a_sA^{2/3}\left(1 - \kappa_sI^2\right)$$

For $\kappa_v = \kappa_s$ one obtains the following relation between a_v and a_s by elimination of V_0:

$$\frac{3}{2}\frac{a}{r_0}\frac{a_v}{a_s} = 1 \qquad a \simeq r_0 \Rightarrow a_v \simeq a_s$$

For the non-spherical case the integral can still be evaluated provided that the radius can be expanded in spherical harmonics. With expansion coefficients $\alpha_{\lambda\mu}$, lengthy analytical calculations lead to the following elegant expression:

$$E_s = \overset{0}{E}_s\left(1 + \frac{1}{4\pi}\sum_{\lambda\mu}C_\lambda|\alpha_{\lambda\mu}|^2\right)$$

where

$$C_\lambda = \left(\frac{R_0}{a} + 1\right)\left[\frac{R_0}{a} - 1 + \left(\frac{R_0}{a} + 1\right)\exp\left(-2R_0/a\right)\right]$$
$$- 2\left(\frac{R_0}{a}\right)^3 I_{\lambda+1/2}\left(\frac{R_0}{a}\right)K_{\lambda+1/2}\left(\frac{R_0}{a}\right)$$

Here $I_{\lambda+1/2}$ and $K_{\lambda+1/2}$ are the modified Bessel and Hankel functions. For higher multipoles λ, the expression for the stiffness constant C_λ becomes independent of multipole order because, for large λ, the second term of C_λ can be neglected as seen from the limiting relation

$$I_{\lambda+1/2}(Z)K_{\lambda+1/2}(Z) \to \frac{1}{2\lambda+1}$$

This is to be contrasted with the quadratic increase with multipole order for the stiffness constant calculated with the usual liquid-drop model. It illustrates the insensitivity of the modified liquid-drop formula to fine wrinkles of the surface.

The incorporation of the modified surface term into the liquid-drop model makes re-determination of the parameters necessary. The two new parameters a and r_0 are determined from interaction barrier heights in nucleus–nucleus collisions and from electron scattering data. The parameters a_s and κ_s have been determined from a fit to fission barriers and a_v and κ_v from a fit to nuclear masses in the region $A > 165$ with shell effects included. Preferably all these effects should be considered simultaneously and the mass fit should include the whole region of known nuclei. In the less ambitious determination described above (Krappe and Nix, 1973; Möller *et. al.*, 1974), the results were

$$r_0 = 1.16\,\text{fm}$$
$$a = 1.4\,\text{fm}$$
$$a_s = 24.7\,\text{MeV}$$
$$\kappa_s = 4.0$$
$$a_v = 16.485\,\text{MeV}$$
$$\kappa_v = 2.324$$

In this determination the relation between the ratios a/r_0 and a_s/a_v derived above is not enforced. It is found to be fulfilled rather accurately, however. We obtain $(3/2) \cdot (a/r_0) \cdot (a_v/a_s) \simeq 1.21$.

In fig. 4.6 the fission barriers predicted by the liquid-drop model and by the generalised model with the effects of the finite range of nuclear forces included are compared. For lighter nuclei ($A \simeq 100$), there is a difference of the order of 10 MeV between the predictions of the models as to the fission barrier heights. From recent experimental data, it seems clear that the fission barriers predicted by the liquid-drop model are too high and that the generalised model gives an improved description of the fission barrier heights.

Subsequently, the alternative liquid-drop model has been further developed

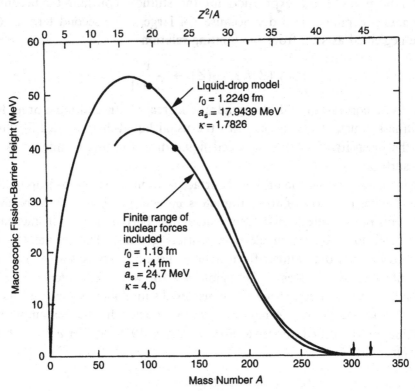

Fig. 4.6. Macroscopic barrier height (in MeV) as a function of Z^2/A for the Myers–Swiatecki liquid-drop model and the modified surface energy model of Krappe and Nix, respectively. Note that the barrier height culminates for $Z^2/A \simeq 15$ or $A \simeq 100$ in both models (from Krappe and Nix, 1973).

(Krappe, Nix and Sierk, 1979) and the parameters have been fitted to masses and fission barriers of heavy as well as light nuclei (Möller *et al.* 1992).

Exercises

4.1   Define the fissility parameter

$$x = \overset{0}{E}_C \,/2\, \overset{0}{E}_s$$

where $\overset{0}{E}_C$ and $\overset{0}{E}_s$ are the Coulomb energy and surface energy, respectively, for a spherical distribution.

   (a) Show that division into infinitely separated spheres, i.e. binary fission, is energetically possible if

$$x > 0.35$$

(b) division into three, i.e. ternary fission, is possible if

$$x > 0.43$$

(c) and division into n equal parts is possible if

$$2 \times \left(1 - n^{-2/3}\right) > \left(n^{1/3} - 1\right)$$

Comments?

4.2   Assume a nuclear shape given by

$$R(\theta) = R_a \left[1 + a_2 P_2(\cos \theta)\right]$$

where P_2 is a Legendre polynomial of second order and where R_a is determined through volume conservation.

(a) Sketch the nuclear shape for $a_2 = 0.5$ ($\beta = 0.79$).
(b) Show the surface energy to be given by

$$E_s = \overset{0}{E_s} \left(1 + \frac{2}{5}a_2^2 - \frac{4}{105}a_2^3 + \ldots\right)$$

(c) For the Coulomb energy one can show

$$E_C = \overset{0}{E_C} \left(1 - \frac{1}{5}a_2^2 - \frac{4}{105}a_2^3 + \ldots\right)$$

Use this to derive an expression for the deformation energy as

$$\Delta E = E_s + E_C - \overset{0}{E_s} - \overset{0}{E_C}$$
$$= \overset{0}{E_s} \left[(1 - x) \cdot \frac{2}{5} \cdot a_2^2 - \frac{4}{105}(1 + 2x)a_2^3 + \ldots\right]$$

(d) Determine the fission barrier height as

$$E_{\text{barr}} = \frac{98}{15} \overset{0}{E_s} \frac{(1 - x)^3}{(1 + 2x)^2} \qquad x < 1$$

Evaluate this expression for ^{238}U.

(e) For $x > 1$ one finds a minimum for $a_2 < 0$ (so-called oblate shape). Can there exist stable oblate nuclei for $x > 1$?

5

Shell structure and magic numbers

5.1 Closed shells in atoms and nuclei

In the 1940s one spoke about 'magic' numbers among the nuclei. Such numbers are for the protons and neutrons, respectively

$$Z = 2, 8, 20, 28, 50, 82$$
$$N = 2, 8, 20, 28, 50, 82 \text{ and } 126$$

These series of numbers one now wants to extend to $Z = 114$, 164 for protons and $N = 184$ and maybe 228 for neutrons. Nuclei containing the above-listed numbers of neutrons or/and protons are the ones found in figs. 3.8 and 3.9 to be associated with large extra binding energies relative to the predictions of the semi-empirical mass formula. It is natural to associate this extra stability with the filling of nucleon shells. Nuclei with N or Z magic are usually called single-closed-shell nuclei, while those with both N and Z 'magic' are called doubly closed shell nuclei or 'double-magic' nuclei. Examples of the latter are $^4_2\text{He}_2$, $^{16}_8\text{O}_8$, $^{40}_{20}\text{Ca}_{20}$, $^{48}_{20}\text{Ca}_{28}$, $^{56}_{28}\text{Ni}_{28}$, $^{132}_{50}\text{Sn}_{82}$ and $^{208}_{82}\text{Pb}_{126}$. A nucleus that from theoretical extrapolations is predicted to have similar character is $^{298}114$, which has $Z = 114$, $N = 184$, see chapter 10.

The whole notion of closed shells comes from atomic theory. There the noble gases represent particularly stable and 'inactive' electron configurations. The corresponding 'magic' numbers in the electron case are

$$Z = 2, 10, 28, 36, 54, 86$$

to which correspond the atoms

$$\text{He, Ne, Ar, Kr, Xe, Rn}$$

It should be well known that the Hamiltonian describing the motion of one particle in a spherical symmetric potential can be separated into the radial

46

Table 5.1. *Closed shells calculated from the pure Coulomb potential.*

Shell	ℓ	States $(n+1)\ell$	Number of electrons	Total	Exp.
1	0	1s	2	2	2
2	0, 1	2s, 1p	2+6	10	10
3	0, 1, 2	3s, 2p, 1d	2+6+10	28	18
4	0, 1, 2, 3	4s, 3p, 2d, 1f	2+6+10+14	60	36
5	0, 1, 2, 3, 4	5s, 4p, 3d, 2f, 1g	2+6+10+14+18	110	54

and angular degrees of freedom. The corresponding quantum numbers are n, the number of nodes in the radial wave function, the angular momentum ℓ and its projection m_ℓ. As the energy eigenvalues are independent of m_ℓ, one finds subshells of degeneracy $2(2\ell+1)$ where the two possible spin directions give a factor 2. This degeneracy is partly broken if spin-dependent forces are present (see chapter 6). The subshells of different ℓ quantum numbers are generally denoted by letters, s, p, d, f, g, ... for $\ell = 0, 1, 2, 3, 4, \ldots$.

For a Coulomb potential there is an even higher degeneracy as the energy eigenvalues depend only on the $N = n + \ell + 1$ quantum number.† Thus for 'hydrogen-like' atoms with the potential

$$V_C = -\frac{Ze^2}{4\pi\varepsilon_0 r}$$

the energy eigenvalues are given as

$$E_N = -\frac{1}{2}mc^2\alpha^2\frac{Z^2}{N^2}$$

where

$$\alpha = \frac{\left(e^2/4\pi\varepsilon_0\right)}{\hbar c} \simeq \frac{1}{137}$$

is the fine structure constant. With the assumption that, in the many-electron case, the interaction between the different electrons is negligible, it is now straightforward to construct table 5.1. In table 5.1, the first two closed-shell numbers in the next to the last column agree with the empirically known magic numbers of the last column; the other three theoretical numbers are too large. This is mainly due to the fact that the Coulomb potential is smaller than assumed as the inner electrons 'shield' the outer electrons from

---

† We use a notion customary in nuclear physics, with n being the radial quantum number and N the principal quantum number.

Table 5.2. *Approximate closed shells resulting from the Coulomb potential with shielding included.*

Shell	ℓ	States $(n+1)\ell$	Number of electrons	Total	Exp.
1	0	1s	2	2	2
2	0, 1	2s, 1p	2+6	10	10
3	0, 1	3s, 2p	2+6	18	18
4	0, 1, 2	4s, 3p, 1d	2+6+10	36	36
5	0, 1, 2	5s, 4p, 2d	2+6+10	54	54
6	0, 1, 2, 3	6s, 5p, 3d, 1f	2+6+10+14	86	86
7	0, 1, 2, 3	7s, 6p, 4d, 2f	2+6+10+14	118	?
8	0, 1, 2, 3, 4	8s, 7p, 5d, 3f, 1g	2+6+10+14+18	168	?

the nuclear charge. Far away from the nucleus, the Coulomb potential, owing to electrons in deep orbitals, is much weaker than the formula implies. With the shielding included we obtain table 5.2 (where the degeneracy of the shells is only approximate). In this case one has (with a theory based on a Hartree–Fock calculation, see below) succeeded in reproducing the 'magic' numbers associated with the electron shells. The noble gases are predicted at their proper places in the periodic system. The effect of shielding is the largest for orbitals of high ℓ-values. Thus the 1d orbital is displaced from the third shell down to the fourth shell, 2d and 3d appear near the s and p levels of the fifth and sixth shells, respectively. Similarly the 1f orbital occurs first with the s and p levels of the sixth shell etc.

Of considerable interest is the prediction of a new closed shell at $Z = 118$, which latter should be another noble gas (it is barely a gas at room temperature as it has a point of condensation just below room temperature). The electron closed shell $Z = 118$ happens to be close to the island of stability near $Z = 114$ predicted for the atomic nucleus. For this reason there is some remote possibility that $Z = 118$ might be produced. Soon after $Z = 118$, a new chemistry should begin with the filling of the first atomic g-shell.

We have so far appealed to qualitative ideas of the inertness of a noble gas. To give a more quantitative measure of the shell structure let us study fig. 5.1 of the atomic ionisation energy of the chemical elements of the periodic table. The ionisation energy is the cost in energy of removing the last (and least bound) electron. According to a theorem from Hartree–Fock theory (Koopman's theorem), the ionisation energy is very near to the

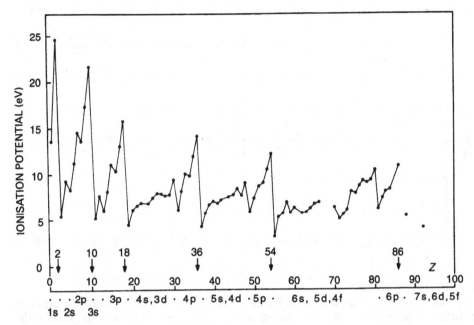

Fig. 5.1. Experimental value of the atomic ionisation potential as a function of Z, exhibiting the strong effects of closed shells at $Z = 2$, 10, 18, 36, 54 and 86 (from Bohr and Mottelson, 1969).

single-electron energy of the valence electron that is removed when the atom becomes singly ionised.

The corresponding quantity in the nuclear case is the neutron separation energy or the proton separation energy, respectively. Similarly to the ionisation energy in the electron case, the neutron separation energy is the minimum energy it costs to remove one neutron from a nucleus. This latter quantity is plotted in fig. 5.2 as a function of neutron number N. In particular the $N = 82$ and $N = 126$ shells are clearly visible. The effects of shells are, however, somewhat less pronounced in the nucleon case than in the electron case.

To sum up the philosophy implicit in the foregoing discussion, the following note should be made. In the atomic case the sudden discontinuities in certain atomic properties over the periodic table can be understood as effects of quantal states in an atomic potential varying smoothly with Z. In the nuclear case the analogous discontinuities in masses and separation energies can again be understood from quantal shells in a nuclear potential. We just have to find the potential.

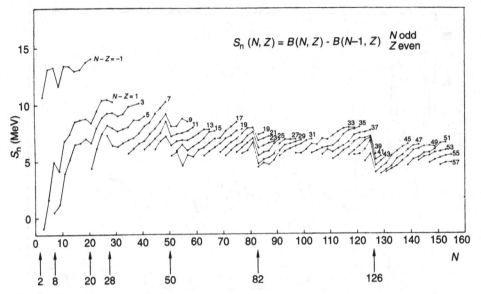

Fig. 5.2. Neutron separation energies for even-Z, odd-N elements as a function of neutron number N. Note the sudden drop in separation energies in passing the closed neutron shells at $N = 50$, 82 and 126 (from Bohr and Mottelson, 1969).

5.2 The atomic and nuclear one-body potentials

In the atomic case it is very apparent that the concept of a one-body potential in which the electrons move is a very successful concept. This is so in spite of the fact that many electrons are involved in addition to the central nuclear charge, assumed to rule alone according to the most elementary considerations. The more refined concept of a screened central nuclear charge is a further improvement, enough to account for the closed shells.

A few words may be needed to clarify further the concept of a one-body potential. This implies that one particle at a time can be considered present in a static potential whose only coordinates are those of the one particle in question, as position **r**, momentum **p** and spin **s**. Formally, such a one-body potential is obtained from the self-consistent field method proposed by Hartree (1928). For the atom the Hartree or Hartree–Fock (Fock, 1930) method is well established. For nucleons the procedure is highly similar though complicated by the fact of the immense complication of the basic two-body nuclear interaction.

Around the atom there are really only electrons. In the nucleus the question can in principle be raised of whether it is really a good approximation to talk about neutrons and protons. We shall briefly come back to these problems in chapter 13. Let us, however, assume that neutrons and protons are enough and that we can assume that the two-body interaction derived from neutron–proton and proton–proton scattering can be used unmodified in the nucleus.

Let us first consider the electron case. The total wave function can be written as a product wave function

$$\psi = \prod_i \phi_i$$

where the one-electron wave functions are given by ϕ_i. We now consider the potential seen by electron i under the influence of the nucleus and the remaining $Z - 1$ electrons placed statically in the nucleus:

$$V_i = -\frac{Ze^2}{4\pi\varepsilon_0 r_i} + \sum_{j \neq i} \frac{1}{4\pi\varepsilon_o} \int \phi_j^* \frac{e^2}{|\mathbf{r}_i - \mathbf{r}_j|} \phi_j \, d^3 r_j$$

leading to the one-particle Schrödinger equation

$$\left(-\frac{\hbar^2}{2m}\nabla_i^2 + V_i \right) \phi_i = e_i \phi_i$$

with the energy eigenvalues e_i.

Many wave functions and energy levels are obtained and in each level we may place one particle in accordance with the Pauli principle. These ϕ_j are then used to determine new ϕ_j functions or new densities $\rho_j = \phi_j^* \phi_j$. One problem then is how to obtain the initial ϕ_j or ρ_j terms. The answer is that one may 'guess' initial ϕ_j or alternatively initial potentials. The solutions can then be iterated.

In the nuclear case, there is no central potential and furthermore, the two-particle interaction is not fully determined. Still, it turns out to be feasible to use a similar approach with a mean-field potential

$$V_i = \sum_j \langle \phi_j | V_{ij} | \phi_j \rangle$$

where the two-particle interactions V_{ij} in the general case not only depend on distance but also on e.g. momentum and spin. As in the electron case, if the wave function is properly symmetrised, the product of ϕ_i is replaced by a determinant and the expressions above are somewhat generalised (Hartree–Fock method).

Another problem is that the empirical nucleon–nucleon two-body interaction corresponds to an infinite repulsion at very small distances. This leads to convergence problems. To overcome the latter, elaborate technical schemes

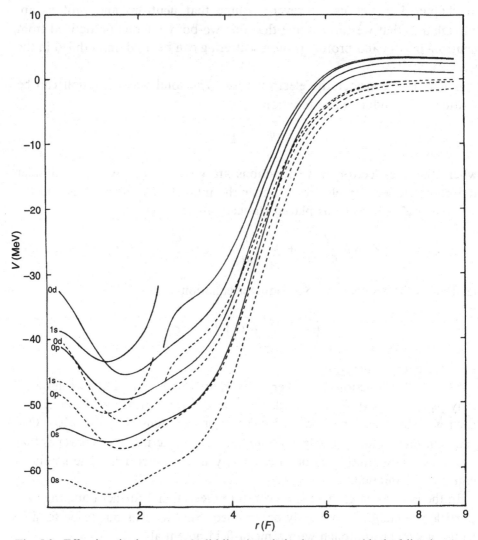

Fig. 5.3. Effective single-proton (solid lines) and single-neutron (dashed lines) potentials for ^{40}Ca obtained in Hartree–Fock calculations based on the so-called Skyrme interaction. Note that to each orbital corresponds a slightly different potential shape. All these potentials show a radial dependence similar to that of the Woods–Saxon potential (from J.W. Negele, *Phys. Rev.* **C1** (1970) 1260).

have been devised by K. Brueckner and his coworkers (1958). Thanks to this development one is therefore presently capable of rather 'realistic' Hartree and Hartree–Fock calculations in nuclear physics.

Based on a more or less 'realistic' nucleon–nucleon interaction, it has been possible to obtain Hartree–Fock one-body potentials for large series of nuclei as exemplified in fig. 5.3 for ^{40}Ca. Note that there is one potential for each state. The fact that the Coulomb interaction modifies the orbitals of the protons relative to those of the neutrons leads to a nuclear potential for the protons different from that of the neutrons. The one-body potential can thus at least roughly be derived from the two-body interaction. As a general observation one may note that the potential depth is about 50 MeV in the case illustrated and the radius approximately the same as the matter radius.

We will follow the path taken historically, however. Thus we shall assume a parameter form of a purely empirical potential. The parameters are subsequently determined to fit data as well as possible. This procedure is outlined in the following chapter.

Exercises

5.1   Consider a three-dimensional anisotropic harmonic oscillator potential

$$V = \frac{M\omega_x^2}{2}x^2 + \frac{M\omega_y^2}{2}y^2 + \frac{M\omega_z^2}{2}z^2$$

The corresponding Hamiltonian is separable into the x-, y- and z-directions and the energy eigenvalues are easily obtained as

$$E = \hbar\omega_x \left(n_x + \frac{1}{2}\right) + \hbar\omega_y \left(n_y + \frac{1}{2}\right) + \hbar\omega_z \left(n_z + \frac{1}{2}\right)$$

(a) In the case of spherical symmetry ($\omega_x = \omega_y = \omega_z$) a large degeneracy is obtained. Calculate the lowest 'magic' numbers. Are there any ℓ-shells that are degenerate?

(b) Another case of large degeneracy is $\omega_x = \omega_y = 2\omega_z$. Calculate the 'magic' numbers also in this case (the fission isomeric states as well as the superdeformed high-spin states (chapter 12) have a deformation approximately corresponding to this frequency ratio).

6

The nuclear one-particle potential in the spherical case

In accordance with the discussion in the preceding chapter, our problem is to find the wave functions of the Hamiltonian,

$$H = -\frac{\hbar^2}{2M}\Delta + V(r)$$

where $V(r)$ is to represent the nuclear one-body potential (sometimes called 'one-body field') and M is the average nucleon mass. In spherical coordinates one may write

$$H = -\frac{\hbar^2}{2M}\frac{1}{r}\frac{\partial^2}{\partial r^2}r + \frac{\ell^2(\theta,\varphi)}{2Mr^2} + V(r)$$

Exploiting the fact that we have spherical symmetry we can postulate a wave function $\psi = R(r)Y_{\ell m}(\theta,\varphi)$, where $Y_{\ell m}$ is an eigenfunction of the angular momentum operator ℓ^2:

$$\ell^2 Y_{\ell m}(\theta,\varphi) = \hbar^2 \ell(\ell+1)Y_{\ell m}(\theta,\varphi)$$

We can now write the Schrödinger equation for the radial wave function $R(r)$:

$$\left(-\frac{\hbar^2}{2M}\frac{1}{r}\frac{d^2}{dr^2}r + \frac{\hbar^2\ell(\ell+1)}{2Mr^2} + V(r) - E\right)R(r) = 0$$

A commonly considered type of radial potential is the Woods–Saxon potential

$$V(r) = -\frac{V_0}{1 + \exp\left[(r-R)/a\right]}$$

which in many ways reminds one of fig. 5.3. The Schrödinger equation for this potential can only be solved numerically and therefore it is useful to consider some simpler potentials for which analytic solutions exist. It then turns out that the radial shape of the Woods–Saxon potential falls somewhere between two such potentials, namely the harmonic oscillator

$$V(r) = \frac{M\omega_0^2}{2}r^2$$

and the infinite square well

$$V(r) = \begin{cases} -V_0 & \text{for } r \le R \\ +\infty & \text{for } r > R \end{cases}$$

As these two potentials, contrary to a more realistic nuclear potential, go to infinity for large values of r, it might sometimes be useful to consider also the finite square well

$$V(r) = \begin{cases} -V_0 & \text{for } r \le R \\ 0 & \text{for } r > R \end{cases}$$

The solutions of the harmonic oscillator and infinite square well potentials, respectively, are given in appendix 6A and the eigenvectors are plotted in fig. 6.1. In the oscillator, there is a degeneracy in addition to that caused by the spherical symmetry. Thus, by a remarkable 'coincidence' the second root, 2s, of the $\ell = 0$ potential exactly 'coincides' with the first root, 1d, of the $\ell = 2$ potential. This 'coincidence' is understood first in terms of the SU$_3$ group. This SU$_3$ invariance implies that the Hamiltonian operator is invariant under the eight Elliot (1958) SU$_3$ operators.

In the infinite square-well potential the ℓ-degeneracy of the oscillator shells is split such that 1d falls below 2s, 1g below 2d, the latter in turn below 3s. This splitting is easily understood if one considers the effective radial potential where the centrifugal potential has been added to $V(r)$:

$$V_{\text{eff}} = V(r) + \frac{\hbar^2}{2Mr^2}\ell(\ell+1)$$

This is the potential entering the separated radial equation. We may plot V_{eff} for different ℓ-values as done for the harmonic oscillator in fig. 6.2. In the high-ℓ case the wave functions are pushed towards the region of large r-values. Thus, as the square well is 'deeper than the oscillator' close to the 'nuclear surface', it leads to a relative depression of high-ℓ relative to low-ℓ states.

As one might expect, the level ordering in the Woods–Saxon case falls somewhere between the two extremes, the soft-surface harmonic oscillator, and the hard-surface square well. The same level ordering is obtained by the addition of a term

$$V_{\text{corr}} = -\mu'\hbar\omega_0\ell(\ell+1)$$

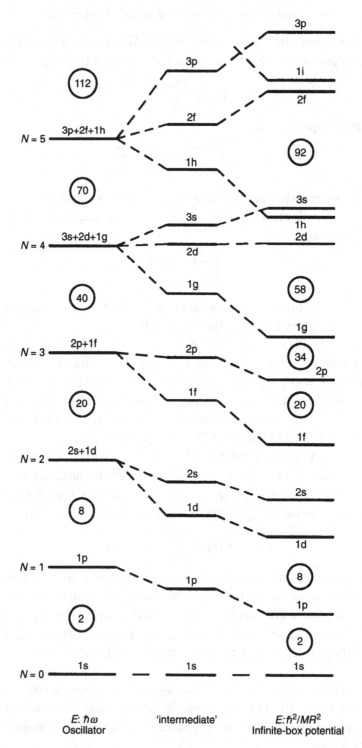

Fig. 6.1. Single-particle subshells for the spherical nuclear potential (without spin–orbit coupling). To the left in the figure the levels are shown for the pure harmonic-oscillator case, to the right the subshells of the infinite square well. In the middle the interpolated case is exhibited. This level order may be obtained if an ℓ^2-term is added to the harmonic oscillator potential.

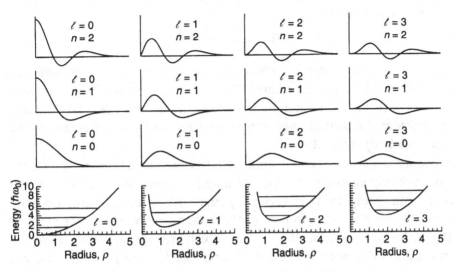

Fig. 6.2. Effective radial potential for a spherical pure oscillator, i.e. with the centrifugal term $\hbar^2\ell(\ell+1)/(2Mr^2)$ added, for $\ell = 0, 1, 2$ and 3. The radial wave functions, $R(\rho)$ are shown above the radial potential in each ℓ case and for $n = 0, 1, 2$, see appendix 6A. The corresponding eigenenergies are shown inside the potentials.

to the harmonic oscillator potential, as you can easily verify for yourself (cf. fig. 6.1). However, before discussing how such a potential can be used in applications, we will introduce the spin–orbit term, which is necessary to get the empirical shell gaps at the proper N- and Z-values.

6.1 The spin–orbit term

In the middle spectrum of fig. 6.1, we want to reproduce e.g. the observed gap for 50 protons or neutrons. This can be obtained if the lg-shell splits into $g_{9/2}$ and $g_{7/2}$. In a similar way, the splitting of the lh-shell into $h_{11/2}$ and $h_{9/2}$ leads to a gap for particle number 82, etc. The energy should thus depend on whether ℓ and s are 'parallel' or 'antiparallel' (we will define these concepts better below). Such a term, $V_{LS} = W(r)\ell \cdot$ s, which obviously has to involve the spin coordinate, was introduced by Haxel, Jensen and Süss (1949) and by M.G. Mayer (1949), for which she and Jensen were awarded the Nobel Prize in 1963.

A spin–orbit term of this type was known in the electron shell. One source of this term is the coupling energy of the electron in the intrinsic magnetic field seen by the electron. Although to the outside observer only a static Coulomb field **E** originating from the nuclear central charge is present, the electron, moving with velocity **v**, experiences a magnetic field **B**, proportional to $\mathbf{v} \times \mathbf{E}$ (see problems),

$$\mathbf{B} = \varepsilon_0 \mu_0 \mathbf{v} \times \mathbf{E}$$

In the nuclear case there is indeed also such a term of electromagnetic origin. The latter is, however, more than an order of magnitude too weak to help us with the magic numbers. Instead the nuclear spin–orbit term is of strong-interaction origin and can be understood from the details of the nuclear two-body interaction.

Phenomenologically one may argue that to construct an invariant quantity out of naturally entering vectors, one might consider, in addition to $\mathbf{s}$, the nucleon spin, being an axial vector†, $\mathbf{p}$, the nucleon momentum, a regular vector, and either $\nabla\rho$, the gradient of the matter density, or ∇V, the gradient of the nuclear central potential, both of which are regular vectors. (Because of this, e.g. $\mathbf{s} \cdot \nabla V$ is not a scalar but rather a pseudoscalar.) It is an empirical fact that the nuclear forces (the strong interaction) preserve parity, which requires that the nuclear potential is scalar (see chapter 13). The simplest conceivable scalar that contains the vector $\mathbf{s}$ seems to be

$$V_{LS} \propto \mathbf{s} \cdot (\mathbf{p} \times \nabla V)$$

Assuming spherical symmetry we have

$$\nabla V = \frac{\mathbf{r}}{r} \frac{\partial V}{\partial r}$$

or

$$V_{LS} \propto \frac{1}{r} \frac{\partial V}{\partial r} \mathbf{s} \cdot \boldsymbol{\ell}$$

In the pure harmonic oscillator case the spin–orbit term becomes especially simple, because

$$\frac{1}{r} \frac{\partial V}{\partial r} = M\omega_0^2 = \text{const.}$$

To calculate the splitting caused by V_{LS} we need to construct new wave functions that are eigenfunctions of $\boldsymbol{\ell} \cdot \mathbf{s}$. To do this we need the Clebsch–Gordan coefficients, which are briefly discussed in appendix 6B. It is, however, instructive to obtain the result by intuitive reasoning. Let us therefore write

$$\boldsymbol{\ell} + \mathbf{s} = \mathbf{j}$$

---

† An axial vector or a pseudovector remains unchanged under space inversion $x_k \to -x_k$ ($k = 1, 2, 3$), in contrast to a regular vector whose components change sign. The concept of pseudovectors (and pseudoscalars) is naturally introduced in connection with the transformation of four-vectors under Lorentz transformations.

and

$$(\ell + s)^2 = j^2$$

i.e.

$$\ell \cdot s = \frac{1}{2}\left(j^2 - \ell^2 - s^2\right)$$

We shall now assume that it is possible to construct wave functions for which ℓ, j and s are good quantum numbers and $\ell \cdot s$ therefore an eigenoperator. Then (expressing ℓ and s in convenient units of $\hbar$):

$$\ell \cdot s \rightarrow \frac{1}{2}\left[j(j+1) - \ell(\ell+1) - s(s+1)\right]$$

We have thus

$$2\langle \ell \cdot s \rangle = \begin{cases} \ell & \text{for} \quad j = \ell + \frac{1}{2} \\ -\ell - 1 & \text{for} \quad j = \ell - \frac{1}{2} \end{cases}$$

(expressing as before ℓ and s in units of $\hbar$). As we shall see later, out of the $2 \cdot (2\ell + 1)$ states with a given ℓ we form one group of states ($2j + 1 = 2\ell + 2$ of them) with $j = \ell + \frac{1}{2}$ and one with $j = \ell - \frac{1}{2}$ ($2j + 1 = 2\ell$ of them). These two groups separate energy-wise. E.g., the g-shell with a degeneracy $2 \cdot (2\ell + 1) = 18$, splits up into $g_{9/2}$ (degeneracy 10) and $g_{7/2}$ (degeneracy 8), as is shown in fig. 6.3 below.

6.2 'Realistic' nuclear one-body potentials

With the addition of the spin–orbit term, the Woods–Saxon (WS) potential takes the form

$$V_{WS} = V(r) + V_{LS} + V_C$$

with

$$V(r) = -\frac{V_0}{1 + \exp\left[(r - R)/a\right]}$$

and

$$V_{LS} = \lambda \frac{1}{r} \frac{\partial V_{SO}(r)}{\partial r} \ell \cdot s$$

where we have used the notation V_{SO} to indicate that one might choose the radial function entering into V_{LS} somewhat different from the central potential, $V(r)$. The Coulomb potential, V_C, enters only for protons and is generated by a charge $(Z - 1)e$, which is uniformly distributed (or possibly with a diffuse surface) inside the nuclear volume.

In practical applications the Woods–Saxon potential has the disadvantage

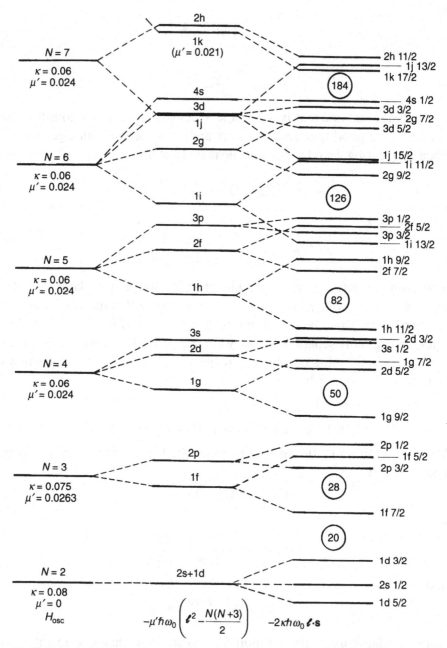

Fig. 6.3. To the left in the figure the pure oscillator shells are exhibited. In the middle graph the effects of an ℓ^2-term proportional to μ' are shown. Finally on the right in the figure the added effect of the $\ell \cdot s$ term is included. The κ- and μ'-values employed, which are different for the different N-shells, are shown on the left.

that it cannot be solved analytically. This is contrary to the so-called modified oscillator (MO) potential (Nilsson 1955, Gustafsson *et al.* 1967)

$$V_{\text{MO}} = \frac{1}{2}\hbar\omega_0\rho^2 - \kappa\hbar\omega_0 \left[2\boldsymbol{\ell}\cdot\mathbf{s} + \mu\left(\ell^2 - \left\langle\ell^2\right\rangle_N\right)\right]; \quad \rho = \left(\frac{M\omega_0}{\hbar}\right)^{1/2} r$$

where, as already suggested in this chapter, the last term in parentheses has the effect of interpolating between the oscillator and the square well and thus reproducing effectively the Woods–Saxon radial shape (in fact, for $\kappa\mu = \mu' = 0.04$, the energy levels order of the infinite square well potential is nearly exactly reproduced, see e.g. fig. 27 of Ragnarsson, Nilsson and Sheline, 1978). The ℓ^2-term alone would result in a general compression of the shells. To avoid this undesired feature, the average value of $\left\langle\ell^2\right\rangle$, taken over each N-shell, is substracted. This average, which is calculated in problem 6.7, takes the value

$$\left\langle\ell^2\right\rangle_N = \frac{N(N+3)}{2}$$

Once the analytic form of the potential is chosen, it remains to fix the parameters. For the Woods–Saxon potential, they have a simple physical meaning. Thus, the nuclear radius R, the diffuseness depth a and the potential depth V_0 should have values in reasonable agreement with Hartree–Fock potentials such as those exhibited in fig. 5.3. The radial function $V_{\text{SO}}(r)$, which enters into V_{LS}, is generally parametrised in a similar way to $V(r)$. Thus, together with the coupling strength λ, another four parameters are added. Furthermore, there is one potential for protons and another for neutrons and in principle also one potential for each combination of Z and N.

In practice, the same potential is used for some limited region of nuclei and furthermore, some of the parameters are more or less arbitrarily put equal. For example, $V_{\text{SO}}(r)$ is often chosen identical to $V(r)$, the same diffuseness depth a and/or spin–orbit coupling strength λ is used for protons and neutrons, etc. The rather small number of remaining free parameters might then be varied within reasonable limits in order to describe different nuclear properties (mainly excitation spectra, see below) as accurately as possible. It then also seems reasonable that the different parameters should vary smoothly, or maybe even stay constant, when considered as functions of mass number (or proton and neutron number). A recent fit of the Woods–Saxon parameters ('universal parameters') is briefly described by Nazarewicz *et al.* (1985). Let us also mention the folded Yukawa potential, which is parametrised in a different way to the Woods–Saxon potential but which appears very similar in practical applications. Some standard parameters for the folded Yukawa potential have been published e.g. by Möller and Nix (1981).

In the modified oscillator potential there are basically three parameters, ω_0, κ and μ for each kind of nucleon. If we for the moment neglect the neutron–proton differences of the three parameters, ω_0 is used to determine the radius of the resulting matter distribution (from the wave functions of the occupied orbitals), μ or rather $\mu' = \kappa\mu$ can be viewed as simulating the surface diffuseness depth, and κ is the spin–orbit coupling strength.

To illustrate further the effect of the $\ell \cdot$ **s**- and ℓ^2-terms, we show on the left in fig. 6.3 the pure harmonic oscillator levels. Next is shown how the term $-\mu'\hbar\omega_0[\ell^2 - N(N+3)/2]$ energetically favours high-ℓ subshells. Finally, in the third column of levels a spin–orbit term $-2\kappa\hbar\omega_0\ell \cdot$ **s** has been added. The values of κ and μ' are roughly those that fit the neutrons in the $^{208}_{82}\mathrm{Pb}_{126}$ region of nuclei as to level order etc. To make the same diagram applicable with somewhat improved accuracy in the lighter-element region, smaller μ'-values and somewhat larger κ-values have been used in the plot for the lower shells, as indicated in the figure. (This is roughly the same fit as made by Nilsson (1955).)

In the applications of the following chapters, we will mainly concentrate on the modified oscillator (MO) potential. This is so because the physical effects we want to illustrate come out in a similar way in all reasonable potentials (or Hartree–Fock calculations) and then we want to add as few calculational difficulties as possible. To conclude this chapter, we will discuss the parameters of the MO potential and the experimental information used for their determination in some detail.

6.3 The nuclear volume parameter

Let us first consider the parameter ω_0 of the MO (modified oscillator) potential. The radial coordinate of the nuclear wave function is $(M\omega_0/\hbar)^{1/2} \cdot r$. The characteristic length is thus $(\hbar/M\omega_0)^{1/2}$. From the wave functions we may calculate a total nuclear density as a sum of all the single-particle densities from which the average radius could be further studied and compared with experiments. A simply accessible quantity for oscillator wave functions (or for any wave function that is given in an oscillator basis) is, however

$$\langle r_i^2 \rangle = \left(N_i + \frac{3}{2}\right)\frac{\hbar}{M\omega_0}$$

(see problem 6.8).

In terms of this we have already in preceding chapters defined an average 'root mean square radius' R_{rms} as

$$R_{rms}^2 = \frac{5}{3}\langle r^2 \rangle = \frac{5}{3}\frac{1}{A}\sum_i \langle r_i^2 \rangle$$

We proceed to evaluate $\langle r \rangle^2$ for the case of filled oscillator shells. The degeneracy of an N'-shell is $(N'+1)(N'+2)$ (see problems), so for equal numbers of neutrons and protons, we have

$$A = 2\sum_{N'=0}^N (N'+1)(N'+2) \simeq 2\sum_{N'=0}^N (N'+3/2)^2$$

$$\simeq 2\int_{-1/2}^{N+1/2}(x+3/2)^2 \, dx \simeq \frac{2}{3}(N+2)^3$$

and

$$A\langle r^2 \rangle = 2\frac{\hbar}{M\omega_0}\sum_{N'=0}^N \left(N'+\frac{3}{2}\right)(N'+1)(N'+2) \simeq \frac{1}{2}\frac{\hbar}{M\omega_0}(N+2)^4$$

From these expressions we immediately obtain the relations

$$N+2 \simeq (3A/2)^{1/3}$$

and

$$R_{rms}^2 = \frac{5}{3}\frac{\hbar}{M\omega_0}\frac{1}{2}\left(\frac{3}{2}\right)^{4/3} \cdot A^{1/3}$$

Using the empirical value $R_{rms} = r_0 A^{1/3}$ with $r_0 = 1.2$ fm we obtain†

$$\hbar\omega_0 = \frac{\hbar^2}{Mr_0^2}\frac{5}{4}\left(\frac{3}{2}\right)^{1/3} \cdot A^{-1/3} \simeq 41 \cdot A^{-1/3} \text{ MeV}$$

The neutron and proton potentials are really different as shown in fig. 6.4. This is so although the neutron–neutron and proton–proton strong

---

† When calculating numerically the following expressions are useful

$$\frac{\hbar^2}{Mr_0^2} = 28.8 \text{ MeV}$$

$$r_0 = 1.2 \text{ fm}$$

$$m_e c^2 = 0.511 \text{ MeV}$$

$$M_p c^2 = 938.3 \text{ MeV}$$

$$M_n c^2 = 939.6 \text{ MeV}$$

$$\frac{e^2}{4\pi\varepsilon_0\hbar c} = \frac{1}{137}$$

$$\frac{e^2}{4\pi\varepsilon_0 r_0} = 1.2 \text{ MeV}$$

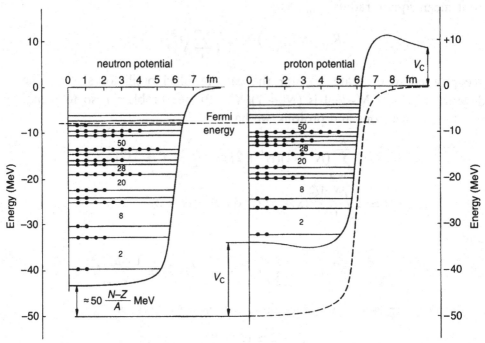

Fig. 6.4. The approximate neutron and proton potentials for ^{114}Sn. As the 50 protons have more neighbours of the attractive opposite kind than the 64 neutrons have, the potential is deeper for the protons by $\simeq 50(N-Z)/A$ MeV. To this one has to add the Coulomb repulsive potential, which raises the total effective proton potential by nearly 20 MeV at the origin. The fact that the Fermi energy becomes the same for protons and neutrons leads to a neutron excess, which also comes out from the semi-empirical mass formula. The symmetry and Coulomb energies of this formula are directly related to the two differences between the proton and neutron potentials illustrated here.

two-body interactions are the same. The difference in the total one-body potential enters in two ways. First only the protons interact via the Coulomb force. In problem 6.11 we calculate the Coulomb potential inside and outside a homogeneous charge of radius $R = r_0 A^{1/3}$. The result is

$$V_C(r) = \begin{cases} \dfrac{Ze^2}{4\pi\varepsilon_0 R}\left(\dfrac{1}{2}\dfrac{r^2}{R^2}-\dfrac{3}{2}\right); & r < R \\[2ex] -\dfrac{Ze^2}{4\pi\varepsilon_0 r} & ; & r > R \end{cases}$$

Inside $r = R$ the Coulomb potential is thus proportional to r^2, apart from a constant. It can thus very well be incorporated into the oscillator potential by a modification of ω_0.

Actually, due to the repulsive Coulomb interaction, nuclei along the stability line have more neutrons than protons. It turns out that this has as a consequence that the protons move in a deeper nuclear potential than the neutrons (this is apart from the electromagnetic Coulomb repulsion). This is so because unlike particles bind each other better than like particles. Owing to the Pauli principle only half of the relative states accessible to unlike particles are also accessible to like particles. The protons furthermore have more unlike neighbours than the neutrons, and consequently a deeper potential results.

Both of these effects, the Coulomb and the Pauli principle effect, can be incorporated into the nuclear potential by choosing

$$\omega_0^N = \omega_0\left(1 + \gamma\frac{N-Z}{A}\right)$$

$$\omega_0^Z = \omega_0\left(1 - \gamma\frac{N-Z}{A}\right)$$

where γ is as yet undetermined.

The simplest way to determine γ is to use the empirically fulfilled requirement that (see problem 6.12)

$$\left\langle r^2\right\rangle_N \simeq \left\langle r^2\right\rangle_Z$$

This leads to $\gamma \simeq 1/3$, and the resulting difference in neutron–proton potential accounts well for the Coulomb potential and the difference in the purely nuclear potential for protons and neutrons when $N \neq Z$.

Actually from the numerical calculations of the nuclear wave functions it is very easy to obtain values of $\left\langle r^2\right\rangle_N$ and $\left\langle r^2\right\rangle_Z$. From these numerical values, $\hbar\omega_0^N$ and $\hbar\omega_0^Z$ can be fitted so that the desired values for the radii of the neutron and proton distributions are reproduced exactly. It turns out that the original estimate of

$$\hbar\omega_0^{N,Z} = 41 \cdot A^{-1/3}\left(1 \pm \frac{1}{3}\frac{N-Z}{A}\right) \text{ MeV}$$

is correct within 1–2% for all nuclei with exception of the very lightest ones.

6.4 Single-particle spectra of closed-shell ± 1 nuclei – the parameters κ and μ′

Nuclei having doubly closed shells apart from one added neutron or one added proton or one missing neutron (a neutron hole) or one missing proton (a proton hole) are the nuclei for which the single-particle picture should be particularly appropriate. This is so first because these nuclei are spherical and we have limited our calculations to the case of good spherical symmetry. Furthermore, with several particles outside a closed shell and occupying an incompletely filled subshell, the 'residual forces' (interactions not taken care of by the average field) are, in spite of their relative smallness, decisive for the order of occurring nuclear spins in the excitation spectrum. We shall therefore take the approach that we first consider only doubly magic ± 1 nuclei. For such nuclei, with the odd particle in the lowest subshell above the gap, the total angular momentum I equals the j-value of this subshell. For example, in $^{17}_{8}O_9$, with the odd neutron in the $d_{5/2}$ shell, $I = 5/2$.

'One-hole' states are also associated with angular momentum j. Thus

$$I(\text{one hole}) = I(\text{one particle}) = j$$

This arises from the following argument: firstly, a closed subshell, having all positions filled, has $I = 0$. Secondly, the last particle of such a closed subshell contributes an angular momentum vector $\mathbf{j}$, which has to balance exactly the vector sum of the angular momentum of all the other particles. This can only be true if these other particles vector couple to $\mathbf{I} = \mathbf{j}$ (cf. fig. 7.3).

For doubly magic $\pm n$ nuclei (where n is a small integer), we first consider the case of an even number of particles of one kind. About these we shall assume that they always couple to total angular momentum zero in the lowest energy state. This is supported by much data.

For an odd number of like particles we shall assume the 'seniority rule' (Racah, 1950) that *two by two particles of the same kind pair off to angular momentum zero*. This rule can be qualitatively understood if one introduces a residual force proportional to $\delta(\mathbf{r}_1 - \mathbf{r}_2)$ operating between the like particles in the partly filled subshell (Mayer, 1950). Then the state favoured energetically (and by definition of 'lowest seniority') is the one with particles pairing off to angular momentum zero. By this rule e.g.

$$I\left[\left(d_{5/2}\right)^3\right] = I\left(d_{5/2}\right) = 5/2$$

The spins resulting for light odd-Z nuclei if the seniority rule were strictly followed are listed in table 6.1. A similar table can also be made for odd-N nuclei. Note that an 'odd-Z' nucleus has $Z = $ odd, $N = $ even, while an 'odd-N' nucleus has $N = $ odd and $Z = $ even.

When comparing predicted and measured spins in table 6.1, one first notes the remarkable agreement. Indeed, the seniority rule is found to hold also for heavier nuclei if $|n|$ is kept reasonably small, say $|n| \lesssim 5$ or 7. For larger

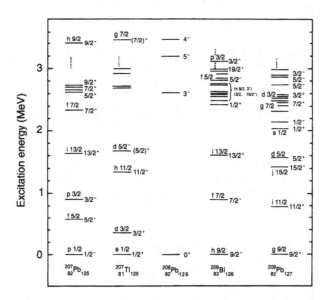

Fig. 6.5. Observed low-energy spectra for ^{208}Pb and neighbouring one-hole and one-particle nuclei. The states are labelled by their spin and parity and, when appropriate, by the corresponding subshell of the particle or hole. The states of ^{209}Bi formed from the $h_{9/2}$ ground state coupled to the collective 3^- state in ^{208}Pb are also indicated (data from *Table of Isotopes*, ed. by C.M. Lederer and V.S. Shirley, John Wiley, 1978)

values of $|n|$, other coupling schemes might take over. For example, the nuclei with $A \simeq 20$–25, where disagreement is found in table 6.1, are described as being deformed (such nuclei are discussed first in chapter 8).

One may go further and also discuss excitation spectra for closed-shell ±1 nuclei. As an example, the observed low-energy spectra for ^{208}Pb and the neighbouring nuclei are given in fig. 6.5. The states with the odd particle or the hole in the j-shells around $Z = 82$ and $N = 126$ are easy to identify. It is first for excitation energies larger than about 2 MeV that other kinds of states are observed, for example where the collective 3^- state of ^{208}Pb couples to the ground state of the odd nucleus. From the spectra of fig. 6.5, the energies of the different subshells can be obtained as shown in fig. 6.6 (the energy gaps for $Z = 82$ and $N = 126$ are most easily extracted from the measured mass of ^{208}Pb relative to ^{209}Bi and ^{209}Pb, respectively). In fig. 6.6 is also shown how well these energies are reproduced in a typical fit based on the Woods–Saxon potential.

We now continue to consider the energy spectra in the region of some other closed nuclei. Then, the position of the neutron subshells around $^{16}_{8}$O$_8$, $^{40}_{20}$Ca$_{20}$, $^{48}_{20}$Ca$_{28}$ and $^{56}_{28}$Ni$_{28}$ are obtained as shown in fig. 6.7. There, we also

Table 6.1. *Shell model subshell occupation for light odd-Z nuclei. Note that the spin resulting from the spherical shell model (Theor, I) is in agreement with observed spin (Exp, I) except for the nuclei ¹⁹F and ²³Na. These latter nuclei can be described as deformed where the valence nucleons fill orbitals that are superpositions of the $d_{5/2}$, $d_{3/2}$ and $s_{1/2}$ orbitals.*

Element	Z	$1s_{1/2}$	$1p_{3/2}$	$1p_{1/2}$	$1d_{5/2}$	$2s_{1/2}$	$1d_{3/2}$	$1f_{7/2}$	Config.	Eq. config.	Theor. I	Exp. I
³H	1	1							$s_{1/2}$		1/2	1/2
⁷Li	3	2	1						$p_{3/2}$		3/2	3/2
¹¹B	5	2	3						$(p_{3/2})^3$	$(p_{3/2})^{-1}$	3/2	3/2
¹⁵N	7	2	4	1					$p_{1/2}$		1/2	1/2
¹⁷F	9	2	4	2	1				$d_{5/2}$		5/2	5/2
¹⁹F	9	2	4	2	1				$d_{5/2}$		5/2	1/2
²³Na	11	2	4	2	3				$(d_{5/2})^3$	$d_{5/2}$	5/2	3/2
²⁷Al	13	2	4	2	5				$(d_{5/2})^5$	$(d_{5/2})^{-1}$	5/2	5/2
³¹P	15	2	4	2	6	1			$s_{1/2}$		1/2	1/2
³⁵Cl	17	2	4	2	6	2	1		$d_{3/2}$		3/2	3/2
³⁷Cl	17	2	4	2	6	2	1		$d_{3/2}$		3/2	3/2
³⁹K	19	2	4	2	6	2	3		$(d_{3/2})^3$	$(d_{3/2})^{-1}$	3/2	3/2
⁴¹K	19	2	4	2	6	2	3		$(d_{3/2})^3$	$(d_{3/2})^{-1}$	3/2	3/2
⁴⁵Sc	21	2	4	2	6	2	4	1	$f_{7/2}$		7/2	7/2
⁵¹V	23	2	4	2	6	2	4	3	$(f_{7/2})^3$	$f_{7/2}$	7/2	7/2
⁵⁵Mn	25	2	4	2	6	2	4	5	$(f_{7/2})^5$	$f_{7/2}$	7/2	7/2
⁵⁹Co	27	2	4	2	6	2	4	7	$(f_{7/2})^7$	$(f_{7/2})^{-1}$	7/2	7/2

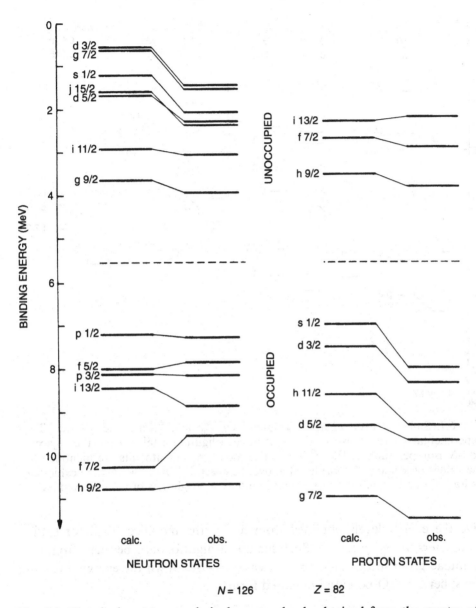

Fig. 6.6. The single-neutron and single-proton levels obtained from the spectra of nuclei near the double-magic nucleus ^{208}Pb (fig. 6.5). The experimental levels are exhibited to the right in each case. They are compared with the single-particle levels obtained from the fit by Blomqvist and Wahlborn (*Arkiv Fysik* **16** (1960) 545) based on a Woods–Saxon potential (from Bohr and Mottelson, 1969).

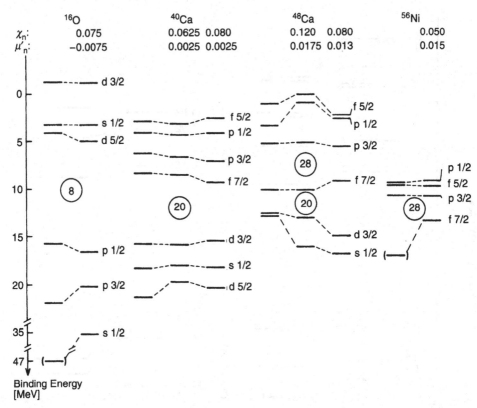

Fig. 6.7. Single-neutron energies around $^{16}_{8}O_8$, $^{40}_{20}Ca_{20}$, $^{48}_{20}Ca_{28}$ and $^{56}_{28}Ni_{28}$. The experimental values shown to the right in each figure are taken from the adjacent odd-N isotopes such as ^{15}O, ^{17}O etc. The modified oscillator fits are identified by the values of κ and μ' ($= \kappa\mu$) listed above the spectra. Where two fits are exhibited, as for ^{40}Ca and ^{48}Ca, they correspond to different weightings of the observed levels.

give the energy levels obtained from fits of the modified oscillator (MO) parameters, κ and μ (or μ'). Such fits are straightforward because, from the formulae given in this chapter, it is easy to write down the energy levels of the spherical MO potential in closed form,

$$E(N\ell j) = \hbar\omega_0 \left[N + \frac{3}{2} - \kappa \left\{ \begin{matrix} \ell \\ -(\ell+1) \end{matrix} \right\} \right.$$
$$\left. - \mu' \left(\ell(\ell+1) - \frac{N(N+3)}{2} \right) \right] \quad \left\{ \begin{matrix} j = \ell + \frac{1}{2} \\ j = \ell - \frac{1}{2} \end{matrix} \right.$$

In fig. 6.7, one could note that, to fit in a satisfactory way the spectra of both ^{40}Ca and ^{48}Ca, the parameter μ' must be chosen substantially larger in the latter nucleus.

Table 6.2. *Values of the modified oscillator parameters κ and μ as suggested by Bengtsson and Ragnarsson (1985). κ is the strength of the* $\ell \cdot$ **s**-term and $\kappa \cdot \mu \ (= \mu')$ *is the strength of the* ℓ^2-term.

N	Protons		Neutrons	
	κ	μ	κ	μ
0	0.120	0.00	0.120	0.00
1	0.120	0.00	0.120	0.00
2	0.105	0.00	0.105	0.00
3	0.090	0.30	0.090	0.25
4	0.065	0.57	0.070	0.39
5	0.060	0.65	0.062	0.43
6	0.054	0.69	0.062	0.34
7	0.054	0.69	0.062	0.26
8...	0.054	0.60	0.062	0.26

By fitting the level spectra of well-established spherical nuclei near the double-magic cases, $(Z = 8, N = 8)$, $(Z = 20, N = 20)$, $(Z = 20, N = 28)$, $(Z = 28, N = 28)$, $(Z = 50, N = 82)$, and finally $(Z = 82, N = 126)$, one obtains sets of κ and μ' for neutrons and protons valid in different parts of the nuclear periodic table. As illustrated in fig. 6.7, however, the agreement between calculated and experimental energies is far from perfect and similar fits can be obtained for rather different values of κ and μ'. Furthermore, the 'observed energies' extracted from closed-shell ±1 nuclei appear to include some correlations and are thus not the bare energies we want to calculate in a single-particle potential. Therefore, and also because of the approximate nature of the MO potential, we can only get a rough determination of κ and μ'.

A better way to obtain κ and μ' is through a fit to known single-particle levels in the well-established deformed regions $20 \leq A \leq 28, 150 \leq A \leq 190$ and $A > 225$ (see e.g. figs. 11.5 and 11.6 below). A criterion for the usefulness of the potential is the fact that the variations of κ and μ' with N and Z are small and fairly continuous (see e.g. Nilsson *et al.*, 1969).

When doing calculations with the MO potential one could use two some-what different strategies. The same values of κ and μ' (or μ) could be used for all shells, leading to a potential only applicable to a limited region of nuclei. The other possibility is to use different values of κ and μ' for different N-shells as exemplified in fig. 6.3. Values from a more recent fit are listed in

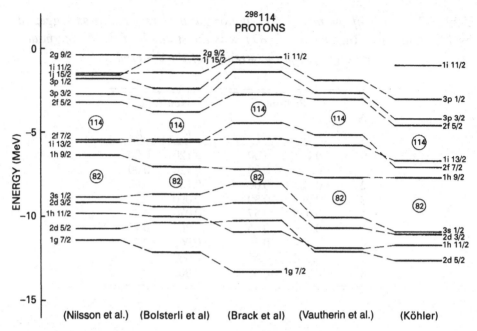

Fig. 6.8. Single-proton levels predicted in the ^{298}114 region. The predictions are based on the modified oscillator potential (S.G. Nilsson *et al., Nucl. Phys.* **A131** (1969) 1), Woods–Saxon type potentials (M. Bolsterli, E.O. Fiset and J.R. Nix, *Physics and Chemistry of Fission*, IAEA, Vienna, 1969, p. 183; M. Brack, J. Damgaard, A. Stenholm-Jensen, H.C. Pauli, V.M. Strutinsky and C.Y. Wong, *Rev. Mod. Phys.*, **44** (1972) 320) and Hartree–Fock calculations (D. Vautherin, M. Véléroni and D.M. Brink, *Phys. Lett.* **33B** (1970) 381; M.S. Köhler, *Nucl. Phys.* **A170** (1971) 88).

table 6.2. With $\kappa = \kappa(N)$ and $\mu = \mu(N)$, it is possible to construct a potential that approximately reproduces the level order around the Fermi surface for all nuclei. As only the orbitals around the Fermi surface influence most measurable properties, the two strategies are essentially equivalent when a limited region of nuclei is studied. The latter strategy, however, has the advantage that the same potential can be used for all nuclei.

6.5 The prediction of nuclear shells at $Z = 114, N = 184$

An interesting problem that presents itself is the possibility of predicting other closed shells beyond neutron and proton numbers corresponding to existing nuclei. The hope is then that the associated shell effects (see chapter 9) are large enough to lead to nuclei with a relative longevity (e.g. $t_{1/2} > 1$ year) with respect to fission and alpha decay.

As seen in fig. 6.8 the next proton shell beyond $Z = 82$ appears to occur

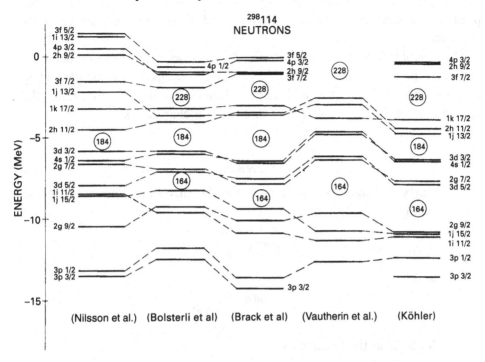

Fig. 6.9. Same as fig. 6.8 but valid for the neutron levels.

for $Z = 114$ and corresponds to a splitting of the 5f shell between $f_{5/2}$ and $f_{7/2}$. This prediction is common to the MO potential, Woods–Saxon type potentials and also the potentials derived from Hartree–Fock calculations. The magnitude of the shell effect is, however, highly dependent on how the spin–orbit strength extrapolates with A. Thus, even if all calculations in fig. 6.8 predict a large 114 gap, they are rather different in other ways, which hints that the uncertainties in the extrapolations are large.

The next neutron shell corresponds to $N = 184$ in most extrapolated nuclear potentials (see fig. 6.9). Here the position of a subshell $h_{11/2}$ above $N = 184$ is critical to the size of the $N = 184$ gap. In the MO potential, the $h_{11/2}$ subshell occurs right in the middle of a gap between $d_{3/2}$ and $k_{17/2}$. The shell $N = 196$, above $h_{11/2}$, is then as important as $N = 184$. In the alternative Woods–Saxon type potentials, as seen in fig. 6.9, the 184 gap dominates clearly over the 196 gap, actually to the point of wiping the latter out.

The fits shown in figs. 6.8 and 6.9 were all obtained around 1970. In the spectrum of Köhler in fig. 6.8, one observes that the $Z = 126$ gap is about as large as the $Z = 114$ gap. Some calculations even give a larger 126 gap than 114 gap. However, the predictions of $Z = 114$ and $N = 184$ as the most probable candidates for closed shells beyond those observed are certainly still valid.

The combination of proton number $Z = 114$ with neutron number $N = 184$ corresponds to the nucleus $^{298}114$. This nucleus or the neighbouring nuclei should then be associated with the longest half-lives. Estimates of the corresponding fission half-lives are discussed in chapter 10.

Exercises

6.1   Carry through the substitution for the three-dimensional harmonic oscillator

$$\rho = \left(\frac{M\omega_0}{\hbar} \right)^{1/2} r$$

and

$$\frac{u(\rho)}{\rho} = R(\rho)$$

Make also the 'ansatz' (why?)

$$u(\rho) = f(\rho)\rho^{\ell+1}e^{-\rho^2/2}$$

to derive the equation

$$\rho f'' + \left(2\ell + 2 - 2\rho^2 \right) f' + \rho \left(\frac{2E}{\hbar\omega_0} - 2\ell - 3 \right) f = 0$$

Make finally the substitution $z = \rho^2$ to show that $f(z)$ satisfies the differential equation of a confluent hypergeometrical function.

6.2   Write out the wave functions for the isotropic three-dimensional harmonic oscillator

$$N = 0 \quad \ell = 0$$
$$N = 1 \quad \ell = 1$$
$$N = 2 \quad \ell = 0, 2$$
$$N = 3 \quad \ell = 1, 3$$

Carry through the normalisations.

6.3   Consider an infinite radial box characterised by $V(r) = -V_0$ for $r \leq R$ and $V(r) = \infty$ for $r > R$ and show that the radial eigenfunctions are given by the spherical Bessel functions $j_\ell(\xi)$ where $\xi = kr$ and $k = [2M (E + V_0) / \hbar^2]^{1/2}$.

6.4   Generate from

$$j_\ell(\xi) = \xi^\ell \left(-\frac{1}{\xi} \frac{d}{d\xi} \right)^\ell \frac{\sin \xi}{\xi}$$

the spherical Bessel functions j_0, j_1, j_2 and j_3. Normalise j_0 and j_1.

6.5   Use tables to find for half-integer Bessel functions $j_\ell(\xi)$ all roots corresponding to $\xi \leq 10$ for $\ell \leq 5$. Check that, for this condition on ξ, no higher ℓ-values are of interest.

6.6   Show that for a particle with mass μ, charge e and magnetic moment $\mathbf{m} = (e/\mu)\mathbf{S}$, moving in a circular orbit in an electrostatic central field $\phi(r)$, the energy is given by

$$\zeta_{magn} = \frac{\varepsilon_0 \mu_0}{\mu^2} \frac{1}{r} \frac{\partial V}{\partial r} \mathbf{L} \cdot \mathbf{S}$$

where $V = e\phi$.

6.7   Show that the 'ℓ^2-term' of the modified oscillator potential does not change the centre of gravity for the orbitals of one N-shell, i.e. show that the average value of $\langle \ell^2 \rangle$ within one shell is equal to $N(N+3)/2$.

6.8   Show that, for harmonic oscillator wave functions, one has

$$\langle r_i^2 \rangle = (N_i + 3/2) \frac{\hbar}{M\omega_0}$$

for any orbital i.
Hint: the problem can for example be solved by use of the virial theorem, which states that for any potential V

$$2 \langle T \rangle = \langle \mathbf{r} \cdot \nabla V \rangle$$

6.9   Discuss for the case of filled oscillator shells the accuracy of the expressions

$$A = \frac{2}{3} (N + 2)^3$$

$$A \langle r^2 \rangle = \frac{1}{2} \frac{\hbar}{M\omega_0} (N + 2)^4$$

by performing the sums exactly.

6.10   How are the relations of the previous problem modified for the case of the last neutron and proton shells being only half filled?

6.11   Calculate the Coulomb potential within and outside a homogeneously charged sphere of total charge Ze and radius $R = r_0 A^{1/3}$.

6.12   Show that the condition of equality for the neutron and proton 'radii',

$$\left\langle r^2 \right\rangle_N = \left\langle r^2 \right\rangle_Z$$

leads to the following expressions for the neutron and proton frequencies,

$$\omega_0^N = \omega_0(x = 0)\left(1 + \tfrac{1}{3}x\right)$$
$$\omega_0^Z = \omega_0(x = 0)\left(1 - \tfrac{1}{3}x\right)$$

The small expansion parameter, x, is defined by

$$N = \frac{A}{2}(1 + x)\,;\; Z = \frac{A}{2}(1 - x)$$

i.e. $x = (N - Z)/A$. Use furthermore the approximate relations of problems 6.9 and 6.10.

6.13   Show that the wave function $Y_{\ell\ell}\alpha$ is an eigenfunction of $\mathbf{j}^2$ with the eigenvalue $(\ell + 1/2)(\ell + 3/2)$ and of j_z with the eigenvalue $(\ell + 1/2)$.

6.14   Show the relation

$$\left|\ell\,\tfrac{1}{2}\,j = \ell + \tfrac{1}{2}\,m\right\rangle = \left(\frac{\ell + m + \tfrac{1}{2}}{2\ell + 1}\right)^{1/2} Y_{\ell m - \frac{1}{2}}\alpha + \left(\frac{\ell - m + \tfrac{1}{2}}{2\ell + 1}\right)^{1/2} Y_{\ell m + \frac{1}{2}}\beta$$

by operating $j - m$ times with the operator j_- on the function

$$\left|\ell\,\tfrac{1}{2}\,j = \ell + \tfrac{1}{2}\,\ell + \tfrac{1}{2}\right\rangle = Y_{\ell\ell}\alpha$$

6.15   Show that

$$\sum_{jm} C^{j_1 j_2 j}_{m_1 m_2 m} C^{j_1 j_2 j}_{m'_1 m'_2 m} = \delta_{m_1 m'_1}\delta_{m_2 m'_2}$$

6.16   Use the explicit formula for Clebsch–Gordan coefficients to verify that C^{101}_{101} and C^{112}_{112} both take the value 1.

6.17   Use the same formula to calculate

$$C^{\ell\,\frac{1}{2}\,\ell + \frac{1}{2}}_{m - \frac{1}{2}\,\frac{1}{2}\,m}$$

Compare with table 6B.1.

Appendix 6A

The harmonic oscillator and infinite square well potentials in three dimensions

In many situations, it is useful to have some schematic nuclear one-body potential for which analytic solutions exist. Two such potentials are the harmonic oscillator and the infinite square well. In this appendix we will briefly consider the corresponding eigenvectors and eigenvalues.

The Schrödinger equation for the radial wave function, $R(r)$, of any potential, $V(r)$, having spherical symmetry, was given in the main text,

$$\left(-\frac{\hbar^2}{2M} \frac{1}{r} \frac{\mathrm{d}^2}{\mathrm{d}r^2} r + \frac{\hbar^2 \ell(\ell+1)}{2Mr^2} + V(r) - E \right) R(r) = 0$$

We first limit our considerations to the spherical oscillator case $V = \frac{1}{2} M \omega_0^2 r^2$. The transformation to the radial coordinate

$$\rho = \left(\frac{M\omega_0}{\hbar} \right)^{1/2} r$$

where a useful unit of length $(\hbar/M\omega_0)^{1/2}$ has been defined, leads to the equation†

$$\frac{1}{2} \hbar \omega_0 \left(-\frac{1}{\rho} \frac{\mathrm{d}^2}{\mathrm{d}\rho^2} \rho + \frac{\ell(\ell+1)}{\rho^2} + \rho^2 \right) R(\rho) = E \cdot R(\rho)$$

With the substitution $R(\rho) = u(\rho)/\rho$ we obtain

$$\frac{\mathrm{d}^2 u}{\mathrm{d}\rho^2} - \left(\frac{\ell(\ell+1)}{\rho^2} + \rho^2 - \frac{2E}{\hbar\omega_0} \right) u = 0$$

It is now easy to find that in the limits $\rho \to 0$ and $\rho \to \infty, u(\rho)$ becomes proportional to $\rho^{\ell+1}$ and $e^{-\rho^2/2}$, respectively. Thus, we make the 'ansatz'

$$u = \rho^{\ell+1} e^{-\rho^2/2} f(\rho)$$

† Here and in a few other places, the same symbol is used for two functions; in this case the symbol R for different functions of r and ρ, respectively.

and obtain for f

$$\rho\frac{d^2f}{d\rho^2} + \left(2\ell + 2 - 2\rho^2\right)\frac{df}{d\rho} + \rho\left(\frac{2E}{\hbar\omega_0} - 2\ell - 3\right)f = 0$$

Finally the substitution

$$z = \rho^2$$

leads to

$$zf'' + \left(\ell + \frac{3}{2} - z\right)f' + \frac{1}{2}\left(\frac{E}{\hbar\omega_0} - \ell - \frac{3}{2}\right)f = 0$$

The latter equation is recognisable as the differential equation for the confluent hypergeometric function, or Kummer function, $F(a, c; z)$ which fulfills the Kummer differential equation

$$zF'' + (c - z)F' - aF = 0$$

The eigenvectors of the Kummer differential equation are given in standard handbooks but are also easy to derive. Assume a power series expansion

$$F = a_0 + a_1 z + a_2 z^2 + \dots$$

Substitute into the differential equation, which leads to a simple recursion relation:

$$\frac{a_{k+1}}{a_k} = \frac{(k + a)}{(k + 1)(k + c)}$$

Thus for large k, $a_{k+1}/a_k \to 1/k$. This corresponds to a divergence at large distances, which cannot be allowed. Therefore, it must be required that the series terminates at some finite $k = n$, i.e. $a_n \neq 0$ but $a_{n+1} = 0$. This condition, which according to the recursion relation corresponds to $a = -n$, leads to an energy quantisation in the original wave function:

$$\frac{1}{2}\left(\frac{E}{\hbar\omega_0} - \ell - \frac{3}{2}\right) = n, \qquad n = 0, 1, 2, \dots$$

or

$$E = \hbar\omega_0\left(2n + \ell + \frac{3}{2}\right) = \hbar\omega_0\left(N + \frac{3}{2}\right)$$

with the principal and radial quantum numbers, N and n, related through

$$N = 2n + \ell$$

The energy-values are degenerate, as to each N there correspond wave functions of different ℓ (and n) such that $\ell = N, N - 2, \dots 0$ or 1.

Table 6A.1. *Closed shells of the spherical harmonic oscillator potential. The last column refers to whether the magic number is consistent with the closed shells observed in nuclei.*

N	$E/\hbar\omega_0$	States $(n+1)\ell$	Degeneracy	Magic number	Comment
0	3/2	1s	$1 \cdot 2 = 2$	2	yes
1	1 + 3/2	1p	$2 \cdot 3 = 6$	8	yes
2	2 + 3/2	2s,1d	$3 \cdot 4 = 12$	20	yes
3	3 + 3/2	2p,1f	$4 \cdot 5 = 20$	40	(no)
4	4 + 3/2	3s,2d,1g	$5 \cdot 6 = 30$	70	no
5	5 + 3/2	3p,2f,1h	$6 \cdot 7 = 42$	112	no

With C being a normalisation constant, the total wave function is

$$\psi = C\rho^\ell F\left(-n, \ell + \frac{3}{2}; \rho^2\right) e^{-\rho^2/2} Y_{\ell m}(\theta, \varphi)$$

where the expression for the Kummer function† follows directly from the recursion relation

$$F(a, c; z) \equiv 1 + \frac{a}{c}\frac{z}{1!} + \frac{a(a+1)}{c(c+1)}\frac{z^2}{2!} + \cdots$$

or

$$F\left(-n, \ell + \frac{3}{2}; \rho^2\right) \equiv 1 + \frac{(-n)}{\ell + 3/2}\rho^2 + \frac{(-n)(-n+1)}{\left(\ell + \frac{3}{2}\right)\left(\ell + \frac{5}{2}\right)}\frac{\rho^4}{2} + \cdots$$

The energy degeneracy of each N-value equals $\sum(2\ell + 1)$, where the sum is taken over the odd or even ℓ-values up to N. In both cases this sum comes out as $\frac{1}{2}(N + 1)(N + 2)$ (see problems). With spin included, the number of states $\mathcal{N}$ for each energy $(N + 3/2)\hbar\omega_0$ equals

$$\mathcal{N} = (N + 1)(N + 2)$$

The properties of the energy eigenvalues of the spherical oscillator are summarised in table 6A.1 (cf. tables 5.1 and 5.2) and the energy spectrum is given on the left in fig. 6.1. It turns out that the first magic numbers, 2, 8 and 20, are consistent with those observed experimentally in nuclei. However, for higher particle numbers, the magic numbers predicted by the harmonic oscillator model are not consistent with data.

† Note that the choice of phases is here such that the radial wave function always becomes positive when $\rho \to 0$.

We now consider a square well potential, $V = -V_0$ for $r \leq R$. With the substitution $u(r) = rR(r)$, the radial wave function valid inside the well reads

$$-\frac{\hbar^2}{2M}u'' + \frac{\hbar^2 \ell(\ell+1)}{2Mr^2}u = (E + V_0)u$$

The transformations

$$k^2 = 2M(E + V_0)/\hbar^2$$
$$\xi = kr$$

lead to

$$-\frac{d^2u}{d\xi^2} + \frac{\ell(\ell+1)}{\xi^2}u = u$$

With the substitution

$$u = \sqrt{\xi} \cdot J(\xi)$$

this goes over into

$$\xi^2 \frac{d^2 J}{d\xi^2} + \xi \frac{dJ}{d\xi} + \left[\xi^2 - \left(\ell + \frac{1}{2}\right)^2\right] J = 0$$

which is the half-integer case, $v = \ell + \frac{1}{2}$, of the differential equation for the Bessel function

$$\xi^2 \frac{\partial^2 J_v}{\partial \xi^2} + \xi \frac{\partial J_v}{\partial \xi} + \left(\xi^2 - v^2\right) J_v = 0$$

The total radial function in this case

$$R_\ell = \frac{u}{r}$$

can be written

$$R_\ell = N_\ell j_\ell(kr)$$

where

$$j_\ell(kr) = \left(\frac{\pi}{2kr}\right)^{1/2} J_{\ell+\frac{1}{2}}(kr)$$

The functions $j_\ell(kr)$ are the so-called spherical Bessel functions and can be generated from the operator relation

$$j_\ell(\xi) = \xi^\ell \left(-\frac{1}{\xi}\frac{d}{d\xi}\right)^\ell \frac{\sin \xi}{\xi}$$

It is easy to work out the first few R_ℓ-functions

$$R_0 = N_0 \frac{\sin kr}{kr}$$

$$R_1 = N_1 \left(\frac{\sin kr}{(kr)^2} - \frac{\cos kr}{kr} \right)$$

and so on. For an infinite square well ($V(r) = \infty, r > R$), the corresponding eigenvalues are found from the condition that

$$R_\ell(kr) = 0$$

In the case of $\ell = 0$, where the solution is proportional simply to $\sin kr$, the eigenvalues are roots to the equation

$$\sin kR = 0$$

which corresponds to $kR = n\pi$, $n = 1, 2, 3 \ldots$ In terms of E we have (in the following we put $V_0 = 0$, i.e. we count the energy eigenvalues relative to the bottom of the potential)

$$E(\ell = 0, n = 1) = \frac{\hbar^2}{2MR^2} \pi^2 \qquad \text{(corresponding to 1s)}$$

$$E(\ell = 0, n = 2) = \frac{\hbar^2}{2MR^2} 4\pi^2 \qquad \text{(corresponding to 2s)}$$

etc. For the other ℓ-values we may write correspondingly

$$E(\ell, n) = \frac{\hbar^2}{2MR^2} \cdot \xi_{\ell n}^2$$

where $\xi_{\ell n}$ is the root of

$$j_\ell(\xi) = 0$$

A table of the solutions is given in table 6A.2. The corresponding energy graph is given on the right in fig. 6.1. As for the harmonic oscillator, only the first few magic numbers are in agreement with those observed in nuclei.

Table 6A.2. *Single-particle energies of the infinite square well potential. The roots $\xi_{n\ell}$ of the spherical Bessel functions, $j_\ell(\xi_{n\ell})$, are listed. The squared quantity $\xi_{n\ell}^2$ is the energy eigenvalue in units of $\hbar^2/2MR^2$. The first root for given ℓ is denoted $n = 1$, etc. Also listed are the degeneracies and the corresponding closed shells.*

ℓ	n	State	$\xi_{n\ell} = \left(\dfrac{E}{\hbar^2/2MR^2}\right)^{1/2}$	Degeneracy	Total
0	1	1s	3.14	2	2
1	1	1p	4.49	6	8
2	1	1d	5.76	10	18
0	2	2s	6.28	2	20
3	1	1f	6.99	14	34
1	2	2p	7.73	6	40
4	1	1g	8.18	18	58
2	2	2d	9.10	10	68
5	1	1h	9.35	22	90
0	3	3s	9.43	2	92
3	2	2f	10.42	14	106
6	1	1i	10.51	26	132
1	3	3p	10.90	6	138

Appendix 6B

Coupling of spin and orbital angular momentum. Clebsch–Gordan coefficients

6B.1 The case of $s = \frac{1}{2}$

We have now to go back and make things a little tidier mathematically. What did the wave function really look like with ℓ and $\mathbf{s}$ appearing 'coupled'? Earlier we have considered angular-momentum wave functions of the type $|\ell m_\ell\rangle \cdot |sm_s\rangle$ or $Y_{\ell m_\ell}\chi_{sm_s}$. This product is an eigenvector of ℓ^2 and $\mathbf{s}^2$ with the eigenvalues $\hbar^2\ell(\ell+1)$ and $\hbar^2 s(s+1)$, i.e. $(3/4)\hbar^2$, furthermore of $\ell_z(\hbar m_\ell)$ and $s_z(\hbar m_s)$ and, if we like, $j_z(\hbar m_\ell + \hbar m_s)$. However, this product is not an eigenvector of $\mathbf{j}^2$ or $\ell \cdot \mathbf{s}$. Between the latter quantities there is the relation

$$\mathbf{j}^2 = \ell^2 + \mathbf{s}^2 + 2\ell \cdot \mathbf{s}$$

Let us form linear combinations

$$|\ell sjm\rangle = \sum_{m_\ell m_s} C^{\ell\,s\,j}_{m_\ell m_s m} |\ell m_\ell\rangle \, |sm_s\rangle$$

where the numbers $C^{\ell\,s\,j}_{m_\ell m_s m}$ are so-called Clebsch–Gordan coefficients (C–G coefficients).

We now exploit the freedom in the choice of the coefficients $C^{\ell\,s\,j}_{m_\ell m_s m}$ to satisfy the following desired relations:

(1) $\mathbf{j}^2|\ell sjm\rangle = \hbar^2 j(j+1)|\ell sjm\rangle$
(2) $j_z|\ell sjm\rangle = \hbar m|\ell sjm\rangle$
(3) $\langle \ell sj'm'|\ell sjm\rangle = \delta_{jj'}\delta_{mm'}$

The second relation combined with the equality $j_z = \ell_z + s_z$ leads to $C^{\ell\,s\,j}_{m_\ell m_s m} = 0$ if $m_\ell + m_s \neq m$ or equivalently that the sum over m_ℓ and m_s can be limited by the condition $m = m_\ell + m_s$. In our simple case under consideration, with $s = \frac{1}{2}$, one has

$$|\ell sjm\rangle = C^{\ell\,s\,j}_{m-\frac{1}{2}\,\frac{1}{2}\,m} Y_{\ell m-\frac{1}{2}}\alpha + C^{\ell\,s\,j}_{m+\frac{1}{2}\,-\frac{1}{2}\,m} Y_{\ell m+\frac{1}{2}}\beta$$

where α and β denote spin states with $s_z = \frac{1}{2}\hbar$ and $s_z = -\frac{1}{2}\hbar$, respectively.

To shorten the formulae we shall from now on suppress the factor $\hbar$. This is achieved by assuming that $\mathbf{j}$ be 'expressed' in units of $\hbar$ etc. Let us study the product $Y_{\ell\ell}\alpha$. Apparently this is 'by luck' already an eigenfunction of $\mathbf{j}^2$ with the eigenvalue $(\ell + \frac{1}{2})(\ell + \frac{3}{2})$ and of j_z with the eigenvalues $j_z = \ell + 1/2$ (see problem 6.13). In this case one of the Clebsch–Gordan coefficients is one and the other zero. We thus have

$$\left| \ell \frac{1}{2} j = \ell + \frac{1}{2}\, m = \ell + \frac{1}{2} \right\rangle = Y_{\ell\ell}\alpha$$

The wave function corresponding to the same $\mathbf{j}^2$ eigenvalue but with $m = j-1$ can be generated through the application of $j_- = \ell_- + s_-$. The operator ℓ_- should be well known from elementary courses in quantum mechanics. It is defined as $\ell_- = \ell_x - i\ell_y$ and has the property

$$\ell_- |\ell\, m\rangle = [(\ell + m)(\ell - m + 1)]^{1/2} |\ell\, m - 1\rangle$$

In a similar way, $\ell_+ = \ell_x + i\ell_y$ is introduced and one finds

$$\ell_+ |\ell\, m\rangle = [(\ell - m)(\ell + m + 1)]^{1/2} |\ell\, m + 1\rangle$$

Such lowering and raising operators can be analogously defined for $\mathbf{s}$ and also for $\mathbf{j} = \ell + \mathbf{s}$. This leads to

$$j_- \left| \ell \frac{1}{2} j = \ell + \frac{1}{2}\, m = \ell + \frac{1}{2} \right\rangle = (2j)^{1/2} \left| \ell \frac{1}{2} j\, j - 1 \right\rangle$$

or

$$\left| \ell \frac{1}{2} j\, j - 1 \right\rangle = \frac{1}{(2j)^{1/2}} (\ell_- + s_-)\, Y_{\ell\ell}\alpha = \left(\frac{2\ell}{2\ell+1} \right)^{1/2} Y_{\ell\ell-1}\alpha + \left(\frac{1}{2\ell+1} \right)^{1/2} Y_{\ell\ell}\beta$$

The arbitrary m-state $\left| \ell \frac{1}{2} j = \ell + \frac{1}{2}\, m \right\rangle$ can now be reached through the $(j - m)$ times repeated application of j_-. After some manipulations (see problem 6.14) one obtains

$$\left| \ell \frac{1}{2} j = \ell + \frac{1}{2}\, m \right\rangle = \left(\frac{\ell + m + \frac{1}{2}}{2\ell + 1} \right)^{1/2} Y_{\ell m-\frac{1}{2}}\alpha + \left(\frac{\ell - m + \frac{1}{2}}{2\ell + 1} \right)^{1/2} Y_{\ell m+\frac{1}{2}}\beta$$

To obtain the wave function with $j_z = m$ but with $j = \ell - \frac{1}{2}$ it turns out that the condition of orthogonality with the wave function above and normalisation immediately determines the form as

$$\left| \ell \frac{1}{2} j = \ell - \frac{1}{2}\, m \right\rangle = -\left(\frac{\ell - m + \frac{1}{2}}{2\ell + 1} \right)^{1/2} Y_{\ell m-\frac{1}{2}}\alpha + \left(\frac{\ell + m + \frac{1}{2}}{2\ell + 1} \right)^{1/2} Y_{\ell m+\frac{1}{2}}\beta$$

You should verify for yourself that this is really another eigenfunction of $\mathbf{j}^2$, now with the eigenvalue $(\ell - \frac{1}{2})(\ell + \frac{1}{2})$. One should note, however, that the wave function so conveniently derived is undetermined with respect to an overall phase. This phase is thus determined through a convention, which according to Condon and Shortley is the following, namely that $\langle \ell\, s\, j\, m | \ell_z | \ell\, s\, j - 1\, m \rangle$ should be real and positive for all j and m. In our case apparently

$$\left(\frac{\left(\ell + \frac{1}{2}\right)^2 - m^2}{(2\ell + 1)^2} \right)^{1/2} \left[-\left(m - \frac{1}{2}\right) + \left(m + \frac{1}{2}\right) \right] > 0$$

which thus confirms our choice of phase to be properly made.

We have thus derived explicit formulae for the coefficients $C^{\ell\, s\, j}_{m_\ell m_s m}$. They are called vector coupling coefficients, Wigner coefficients or Clebsch–Gordan coefficients (the last is the most common).

6B.2 The general case

We can now easily generalise the problem of construction of common eigenfunctions of the commuting operations ℓ and $\mathbf{s}$ (which both are angular-momentum operators but which refer to different Hilbert spaces) to the case of two general commuting angular momenta $\mathbf{j}_1$ and $\mathbf{j}_2$, which e.g. may be angular momenta of two different particles or alternatively spin and orbital angular momenta of the same particle. The problem is treated in most books on quantum mechanics and we shall therefore be moderately brief in our presentation. Consider thus

$$\mathbf{j} = \mathbf{j}_1 + \mathbf{j}_2$$

where $\mathbf{j}_1$ and $\mathbf{j}_2$ refer to two different spaces. Hence one should more properly write

$$\mathbf{j} = \mathbf{j}_1 \otimes 1 + 1 \otimes \mathbf{j}_2$$

The following commutation rules apply

$$[j_{1\kappa}, j_{2\lambda}] = 0$$
$$[j_{1\kappa}, j_{1\lambda}] = i j_{1\kappa\times\lambda}$$

There is a corresponding relation for the components of $\mathbf{j}_2$. The problem is to find eigenvalues and eigenvectors of j_z and $\mathbf{j}^2$. The eigenvectors are to be expressed in terms of the eigenvectors of j_{1z}, $\mathbf{j}_1^2$, j_{2z} and $\mathbf{j}_2^2$:

$$j_{1z} |j_1 m_1\rangle = m_1 |j_1 m_1\rangle$$

$$\mathbf{j}_1^2 \, |j_1 m_1\rangle = j_1 \, (j_1 + 1) \, |j_1 m_1\rangle$$

The following ket is a simultaneous eigenket of $j_{1z}, j_{2z}, \mathbf{j}_1^2$ and $\mathbf{j}_2^2$ namely

$$|j_1 j_2 m_1 m_2\rangle = |j_1 m_1\rangle \, |j_2 m_2\rangle$$

From this base one thus wants to construct eigenvectors of j_z and $\mathbf{j}^2$, and furthermore $\mathbf{j}_1^2$ and $\mathbf{j}_2^2$. This is possible as all four of the listed operators commute, as is easily proved. Thus $\mathbf{j}^2$, j_z, $\mathbf{j}_1^2$ and $\mathbf{j}_2^2$ can simultaneously be made good quantum numbers. The new eigenvectors remain eigenvectors of $\mathbf{j}_1^2$ and $\mathbf{j}_2^2$ but *not* of j_{1z} and j_{2z}. Formally we can write the relation between the new and the old system of eigenvectors in the following manner based on the notation of Dirac:

$$|j_1 j_2 jm\rangle = \sum_{m_1 m_2} \underbrace{|j_1 j_2 m_1 m_2\rangle}_{\text{vector}} \underbrace{\langle j_1 j_2 m_1 m_2| \, j_1 j_2 jm\rangle}_{\text{number}}$$

The transformation matrix element or scalar product $\langle j_1 j_2 m_1 m_2 | j_1 j_2 jm \rangle$ is identical with the earlier introduced coefficient $C^{j_1 j_2 j}_{m_1 m_2 m}$. Operate with $j_z = j_{1z} + j_{2z}$ on $|j_1 j_2 jm\rangle$:

$$j_z \, |j_1 j_2 jm\rangle = m \, |j_1 j_2 jm\rangle$$

$$(j_{1z} + j_{2z}) \, |j_1 j_2 jm\rangle = \sum_{m_1 m_2} (m_1 + m_2) \, C^{j_1 j_2 j}_{m_1 m_2 m} \, |j_1 m_1\rangle \, |j_2 m_2\rangle$$

$$= m \, |j_1 j_2 jm\rangle$$

The last equality requires that

$$C^{j_1 j_2 j}_{m_1 m_2 m} = 0 \qquad \text{if} \quad m \neq m_1 + m_2$$

6B.3 Recursion relations of Clebsch–Gordan coefficients

Operate now with j_+ and $j_{1+} + j_{2+}$ on $|j_1 j_2 jm\rangle$:

$$\sum_{m_1 m_2} [(j_1 - m_1)(j_1 + m_1 + 1)]^{1/2} \, |j_1 \, j_2 \, m_1 + 1 \, m_2\rangle \, C^{j_1 j_2 j}_{m_1 m_2 m}$$

$$+ \sum_{m_1 m_2} [(j_2 - m_2)(j_2 + m_2 + 1)]^{1/2} \, |j_1 \, j_2 \, m_1 \, m_2 + 1\rangle \, C^{j_1 j_2 j}_{m_1 m_2 m}$$

$$= [(j - m)(j + m + 1)]^{1/2} \sum_{m_1 m_2} |j_1 \, j_2 \, m_1 \, m_2\rangle \, C^{j_1 j_2 j}_{m_1 m_2 m+1}$$

The subsequent operation with j_- gives similar relations with a few changes of sign. Identification of the cofficients gives the recursion relations

$$[(j \mp m)(j \pm m + 1)]^{1/2} C^{j_1 j_2 j}_{m_1 m_2 m \pm 1} = [(j_1 \mp m_1 + 1)(j_1 \pm m_1)]^{1/2} C^{j_1 j_2 j}_{m_1 \mp 1 \, m_2 \, m}$$

$$+ [(j_2 \mp m_2 + 1)(j_2 \pm m_2)]^{1/2} C^{j_1 j_2 j}_{m_1 m_2 \mp 1 \, m}$$

For practical use of the formula, it is necessary to anchor the process at one or two points. This is done the following way.

(1) Set $m_1 = j_1$, $m = j$ and use the lower signs. It then follows that $m_2 = j - j_1 - 1$ in order to obtain Clebsch–Gordon coefficients different from zero. This leads to

$$(2j)^{1/2} C^{j_1 j_2 j}_{j_1 \, j - j_1 - 1 \, j - 1} = 0 + [(j_2 + j - j_1)(j_2 - j + j_1 + 1)]^{1/2} C^{j_1 j_2 j}_{j_1 j - j_1 \, j}$$

(2) Analogously, set $m_1 = j_1$, $m = j - 1$. This, together with the upper signs, gives

$$(2j)^{1/2} C^{j_1 \, j_2 \, j}_{j_1 \, j - j_1 \, j} = (2j_1)^{1/2} C^{j_1 \, j_2 \, j}_{j_1 - 1 \, j - j_1 \, j - 1}$$

$$+ [(j_2 + j - j_1)(j_2 - j + j_1 + 1)]^{1/2} C^{j_1 \, j_2 \, j}_{j_1 \, j - j_1 - 1 \, j - 1}$$

If $C^{j_1 j_2 j}_{j_1 \, j - j_1 \, j}$ is known, the relation 1 serves to determine $C^{j_1 j_2 j}_{j_1 \, j - j_1 - 1 \, j - 1}$. From the relation 2 subsequently, the coefficient $C^{j_1 \, j_2 \, j}_{j_1 - 1 \, j - j_1 \, j - 1}$ can be determined. One can thus determine all coefficients relative to one, which

(a) is chosen real and positive
(b) is determined in magnitude through the normalisation condition.

It is easy to see that $C^{j_1 j_2 j}_{j_1 \, j - j_1 \, j}$ is $\neq 0$ only if

$$|j_1 - j_2| \le j \le j_1 + j_2$$

In the representation $|j_1 j_2 m_1 m_2\rangle$ there are $(2j_1 + 1)(2j_2 + 1)$ basis vectors. In the new representation (assuming $j_1 \ge j_2$), there are

$$\sum_{j_1}^{j_1 + j_2} (2j +) = (2j_1 + 1)(2j_2 + 1)$$

basis vectors. Both sets of basis vectors are orthogonal because the corresponding sets of operators are Hermitian. Furthermore they are normalised. Thus the Clebsch–Gordon coefficients together build up a unitary matrix.

6B.4 Orthogonality relations

For the unitary transformation matrices the following orthogonality relations must hold:

$$\sum_{m_1 m_2} C_{m_1 m_2 m}^{j_1 j_2 j} C_{m_1 m_2 m'}^{j_1 j_2 j'} = \delta_{jj'} \delta_{mm'}$$

which is straightforward to prove. Thus, noting that j^2 and j_z are Hermitian operators:

$$\delta_{jj'} \delta_{mm'} = \langle j_1 j_2 j'm' | \, j_1 j_2 jm \rangle$$

$$\sum_{m_1 m_2 m_1' m_2'} \langle j_1 j_2 j'm' | \, j_1 j_2 m_1' m_2' \rangle \langle j_1 j_2 m_1' m_2' | \cdot | j_1 j_2 m_1 m_2 \rangle \langle j_1 j_2 m_1 m_2 | \, j_1 j_2 jm \rangle$$

$$= \sum_{m_1 m_2} \langle j_1 j_2 j'm' | \, j_1 j_2 m_1 m_2 \rangle \langle j_1 j_2 m_1 m_2 | \, j_1 j_2 jm \rangle$$

The last step is due to the orthogonality condition assumed

$$\langle j_1 j_2 m_1' m_2' | \, j_1 j_2 m_1 m_2 \rangle = \delta_{m_1 m_1'} \delta_{m_2 m_2'}$$

The second orthogonality condition

$$\sum_{jm} C_{m_1 m_2 m}^{j_1 j_2 j} C_{m_1' m_2' m}^{j_1 j_2 j} = \delta_{m_1 m_1'} \delta_{m_2 m_2'}$$

is entirely analogous and is proven as an exercise.

We now want to prove that the transformation coefficients for the inverse transformation are also Clebsch–Gordan coefficients. Thus we want to show that

$$\psi_{j_1 m_1}(1) \psi_{j_2 m_2}(2) = \sum_{jm} C_{m_1 m_2 m}^{j_1 j_2 j} \phi_{jm}^{j_1 j_2}(1, 2)$$

(This really follows directly from the unitarity of the transformation matrices.) We start from

$$\phi_{jm}^{j_1 j_2} = \sum_{m_1' m_2'} C_{m_1' m_2' m}^{j_1 j_2 j} \psi_{j_1 m_1'} \psi_{j_2 m_2'}$$

Multiply both sides with $C_{m_1 m_2 m}^{j_1 j_2 j}$ and sum over j and m

$$\sum_{jm} C_{m_1 m_2 m}^{j_1 j_2 j} \phi_{jm}^{j_1 j_2} = \sum_{m_1' m_2'} \sum_{jm} C_{m_1 m_2 m}^{j_1 j_2 j} C_{m_1' m_2' m}^{j_1 j_2 j} \psi_{j_1 m_1'} \psi_{j_2 m_2'}$$

$$= \sum_{m_1' m_2'} \delta_{m_1 m_1'} \delta_{m_2 m_2'} \psi_{j_1 m_1'} \psi_{j_2 m_2'} = \psi_{j_1 m_1} \cdot \psi_{j_2 m_2}$$

which is the desired relation.

6B.5 Practical calculations with Clebsch–Gordan coefficients

For practical calculation of Clebsch–Gordan coefficients three lines of approach are possible.

(1) Direct use of the recursion formulae. This was the most common method used a few years ago for the calculation of non-tabulated coefficients. However, today the recursion formulae have lost their importance for practical calculations.

(2) Use of available tabulations. Such tabulations are either in the form of factorials as exemplified in table 6B.1 or in the form of explicit numbers.

(3) Use of a closed expression involving factorials. This is the most common method nowadays on which computer programs are based. Without proof, we give the following explicit formula satisfying the recursion relations and the normalisation conditions:

$$
C^{j_1 j_2 j}_{m_1 m_2 m} = \left(\frac{(j_1 + j - j_2)!\,(j - j_1 + j_2)!\,(j_1 + j_2 - j)!}{(j + j_1 + j_2 + 1)!\,(j_1 - m_1)!\,(j_1 + m_1)!} \right.
$$

$$
\times \left. \frac{(j + m_1 + m_2)!\,(j - m_1 - m_2)!}{(j_2 - m_2)!\,(j_2 + m_2)!} \right)^{1/2}
$$

$$
\times \sum_H \left(\frac{(-1)^{H + j_2 + m_2}(2j + 1)^{1/2}}{(j - j_1 + j_2 - H)!\,(j + m_1 + m_2 - H)!} \right.
$$

$$
\times \left. \frac{(j + j_2 + m_1 - H)!\,(j_1 - m_1 + H)!}{H!\,(H + j_1 - j_2 - m_1 - m_2)!} \right)
$$

The sum over H is limited by the fact that the factorials are defined to equal infinity for negative integer arguments.

Some different symmetry properties of the Clebsch–Gordon coefficients are also very useful. If j_1 and j_2 are interchanged, it seems obvious that the absolute value does not change. However, a phase factor enters:

$$
\langle j_1 m_1 j_2 m_2 |\, j_1 j_2 j_3 m \rangle = (-1)^{j_1 + j_2 - j_3} \langle j_2 m_2 j_1 m_1 |\, j_2 j_1 j_3 m \rangle
$$

The same phase factor is obtained if the sign of all the m quantum numbers is changed

$$
\langle j_1 m_1 j_2 m_2 |\, j_1 j_2 j_3 m_3 \rangle = (-1)^{j_1 + j_2 - j_3} \langle j_1 - m_1\, j_2 - m_2 |\, j_1 j_2 j_3 - m_3 \rangle
$$

Finally, it is also possible to derive the relations

$$\langle j_1 m_1 j_2 m_2 | j_1 j_2 j_3 m_3 \rangle = (-1)^{j_2 + m_2} \left(\frac{2j_3 + 1}{2j_1 + 1} \right)^{1/2} \langle j_2 - m_2 j_3 m_3 | j_2 j_3 j_1 m_1 \rangle$$

$$\langle j_1 m_1 j_2 m_2 | j_1 j_2 j_3 m_3 \rangle = (-1)^{j_1 - m_1} \left(\frac{2j_3 + 1}{2j_2 + 1} \right)^{1/2} \langle j_3 m_3 j_1 - m_1 | j_3 j_1 j_2 m_2 \rangle$$

where the last one could be said to be superfluous because it can be obtained from the other three relations. Instead of Clebsch–Gordan coefficients, so-called 3j-symbols are sometimes used. The main advantage of the latter is that j_1, j_2 and j_3 enter in a more symmetric way.

6B.6 The addition theorem

We now give without proof what is called the addition theorem valid for the spherical harmonics. For the special case $j_1 = \ell$, $j_2 = \ell'$ one can obviously write

$$Y_{\ell m} (\theta_1, \varphi_1) Y_{\ell' m'} (\theta_2, \varphi_2) = \sum_{LM} C_{mm'M}^{\ell \ell' L} \phi_{LM}^{\ell \ell'} (\theta_1, \varphi_1, \theta_2, \varphi_2)$$

Assume now $\theta_1 = \theta_2 = \theta$, $\varphi_1 = \varphi_2 = \varphi$. The formula then evolves into

$$Y_{\ell m}(\theta, \varphi) Y_{\ell' m'}(\theta, \varphi) = \sum_{LM} C_{mm'M}^{\ell \ell' L} \phi_{LM}^{\ell \ell'}(\theta, \varphi)$$

The function $\phi_{LM}^{\ell \ell'}(\theta, \varphi)$ must be of the form

$$\phi_{LM}^{\ell \ell'}(\theta, \varphi) = F(\ell \ell' LM) \cdot Y_{LM}(\theta, \varphi)$$

where F is a constant. This is so because every function of (θ, φ), which is an eigenfunction of L_z with the eigenvalue $\hbar M$ and to $\mathbf{L}^2$ with the eigenvalues $\hbar^2 L(L + 1)$, must be proportional to $Y_{LM}(\theta, \varphi)$. The determination of F is most simply done from the case $\theta = 0$. One arrives at

$$Y_{\ell m}(\theta, \varphi) Y_{\ell' m'}(\theta, \varphi) = \sum_{LM} \left(\frac{(2\ell + 1)(2\ell' + 1)}{4\pi(2L + 1)} \right)^{1/2} C_{mm'M}^{\ell \ell' L} C_{000}^{\ell \ell' L} Y_{LM}(\theta, \varphi)$$

Alternatively we can couple the two first spherical harmonics to a total angular momentum L', M'. From the formula above, it is thus easy to derive

$$[Y_\ell(\theta, \varphi) Y_{\ell'}(\theta, \varphi)]_{M'}^{L'} = C_{000}^{\ell \ell' L'} \left(\frac{(2\ell + 1)(2\ell' + 1)}{4\pi(2L' + 1)} \right)^{1/2} Y_{L'M'}(\theta, \varphi)$$

where we have used the shorthand notation

$$[Y_\ell(\theta, \varphi) Y_{\ell'}(\theta, \varphi)]_{M'}^{L'} = \sum_{mm'} C_{mm'M'}^{\ell \ell' L'} Y_{\ell m}(\theta, \varphi) Y_{\ell' m'}(\theta, \varphi)$$

Table 6B.1 *Clebsch–Gordan coefficients* $\langle j_1 j_2 m_1 m_2 | j_1 j_2 jm \rangle \equiv \langle j_1 j_2 m_1 m_2 | jm \rangle$ *for* $j_1 = 1/2,\ 1,\ 3/2$ *and* 2.

$$\left\langle \tfrac{1}{2} j_2 m_1 m_2 | jm \right\rangle$$

j	$m_1 = \tfrac{1}{2}$	$m_1 = -\tfrac{1}{2}$
$j_2 + \tfrac{1}{2}$	$\left(\dfrac{j+m}{2j}\right)^{1/2}$	$\left(\dfrac{j-m}{2j}\right)^{1/2}$
$j_2 - \tfrac{1}{2}$	$\left(\dfrac{j-m+1}{2j+2}\right)^{1/2}$	$-\left(\dfrac{j+m+1}{2j+2}\right)^{1/2}$

$$\langle 1 j_2 m_1 m_2 | jm \rangle$$

j	$m_1 = 1$	$m_1 = 0$	$m_1 = -1$
$j_2 + 1$	$\left(\dfrac{(j+m-1)(j+m)}{(2j-1)(2j)}\right)^{1/2}$	$\left(\dfrac{2(j-m)(j+m)}{(2j-1)(2j)}\right)^{1/2}$	$\left(\dfrac{(j-m-1)(j-m)}{(2j-1)(2j)}\right)^{1/2}$
j_2	$\left(\dfrac{2(j+m)(j-m+1)}{2j(2j+2)}\right)^{1/2}$	$-\dfrac{2m}{(2j(2j+2))^{1/2}}$	$-\left(\dfrac{2(j-m)(j+m+1)}{2j(2j+2)}\right)^{1/2}$
$j_2 - 1$	$\left(\dfrac{(j-m+1)(j-m+2)}{(2j+2)(2j+3)}\right)^{1/2}$	$-\left(\dfrac{2(j-m+1)(j+m+1)}{(2j+2)(2j+3)}\right)^{1/2}$	$\left(\dfrac{(j+m+2)(j+m+1)}{(2j+2)(2j+3)}\right)^{1/2}$

$$\left\langle \tfrac{3}{2} j_2 m_1 m_2 | jm \right\rangle$$

j	$m_1 = \tfrac{3}{2}$	$m_1 = \tfrac{1}{2}$
$j_2 + \tfrac{3}{2}$	$\left(\dfrac{(j+m-2)(j+m-1)(j+m)}{(2j-2)(2j-1)(2j)}\right)^{1/2}$	$\left(\dfrac{3(j+m-1)(j+m)(j-m)}{(2j-2)(2j-1)(2j)}\right)^{1/2}$
$j_2 + \tfrac{1}{2}$	$\left(\dfrac{3(j+m-1)(j+m)(j-m+1)}{(2j-1)(2j)(2j+2)}\right)^{1/2}$	$(j - 3m + 1)\left(\dfrac{(j+m)}{(2j-1)(2j)(2j+2)}\right)^{1/2}$
$j_2 - \tfrac{1}{2}$	$\left(\dfrac{3(j+m)(j-m+1)(j-m+2)}{2j(2j+2)(2j+3)}\right)^{1/2}$	$-(j + 3m)\left(\dfrac{(j-m+1)}{2j(2j+2)(2j+3)}\right)^{1/2}$
$j_2 - \tfrac{3}{2}$	$\left(\dfrac{(j-m+1)(j-m+2)(j-m+3)}{(2j+2)(2j+3)(2j+4)}\right)^{1/2}$	$-\left(\dfrac{3(j+m+1)(j-m+1)(j-m+2)}{(2j+2)(2j+3)(2j+4)}\right)^{1/2}$

j	$m_1 = -\frac{1}{2}$	$m_1 = -\frac{3}{2}$
$j_2 + \frac{3}{2}$	$\left(\frac{3(j+m)(j-m-1)(j-m)}{(2j-2)(2j-1)(2j)}\right)^{1/2}$	$\left(\frac{(j-m-2)(j-m-1)(j-m)}{(2j-2)(2j-1)(2j)}\right)^{1/2}$
$j_2 + \frac{1}{2}$	$-(j+3m+1)\left(\frac{(j-m)}{(2j-1)(2j)(2j+2)}\right)^{1/2}$	$-\left(\frac{3(j+m+1)(j-m-1)(j-m)}{(2j-1)(2j)(2j+2)}\right)^{1/2}$
$j_2 - \frac{1}{2}$	$-(j-3m)\left(\frac{(j+m+1)}{2j(2j+2)(2j+3)}\right)^{1/2}$	$\left(\frac{3(j+m+1)(j+m+2)(j-m)}{2j(2j+2)(2j+3)}\right)^{1/2}$
$j_2 - \frac{3}{2}$	$\left(\frac{3(j+m+1)(j+m+2)(j-m+1)}{(2j+2)(2j+3)(2j+4)}\right)^{1/2}$	$-\left(\frac{(j+m+1)(j+m+2)(j+m+3)}{(2j+2)(2j+3)(2j+4)}\right)^{1/2}$

$$\langle 2j_2 m_1 m_2 | jm \rangle$$

j	$m_1 = 2$	$m_1 = 1$
$j_2 + 2$	$\left(\frac{(j+m-3)(j+m-2)(j+m-1)(j+m)}{(2j-3)(2j-2)(2j-1)(2j)}\right)^{1/2}$	$\left(\frac{4(j-m)(j+m)(j+m-1)(j+m-2)}{(2j-3)(2j-2)(2j-1)(2j)}\right)^{1/2}$
$j_2 + 1$	$\left(\frac{4(j+m-2)(j+m-1)(j+m)(j-m+1)}{(2j-2)(2j-1)(2j)(2j+2)}\right)^{1/2}$	$2(j-2m+1)\left(\frac{(j+m)(j+m-1)}{(2j-2)(2j-1)(2j)(2j+2)}\right)^{1/2}$
j_2	$\left(\frac{6(j+m-1)(j+m)(j-m+1)(j-m+2)}{(2j-1)(2j)(2j+2)(2j+3)}\right)^{1/2}$	$(1-2m)\left(\frac{6(j-m+1)(j+m)}{(2j-1)(2j)(2j+2)(2j+3)}\right)^{1/2}$
$j_2 - 1$	$\left(\frac{4(j+m)(j-m+1)(j-m+2)(j-m+3)}{2j(2j+2)(2j+3)(2j+4)}\right)^{1/2}$	$-2(j+2m)\left(\frac{(j-m+2)(j-m+1)}{2j(2j+2)(2j+3)(2j+4)}\right)^{1/2}$
$j_2 - 2$	$\left(\frac{(j-m+1)(j-m+2)(j-m+3)(j-m+4)}{(2j+2)(2j+3)(2j+4)(2j+5)}\right)^{1/2}$	$-\left(\frac{4(j-m+3)(j-m+2)(j-m+1)(j+m+1)}{(2j+2)(2j+3)(2j+4)(2j+5)}\right)^{1/2}$

j	$m_1 = 0$	$m_1 = -1$
$j_2 + 2$	$\left(\frac{6(j-m)(j-m-1)(j+m)(j+m-1)}{(2j-3)(2j-2)(2j-1)(2j)}\right)^{1/2}$	$\left(\frac{4(j-m)(j-m-1)(j-m-2)(j+m)}{(2j-3)(2j-2)(2j-1)(2j)}\right)^{1/2}$
$j_2 + 1$	$-2m\left(\frac{6(j-m)(j+m)}{(2j-2)(2j-1)(2j)(2j+2)}\right)^{1/2}$	$-2(j+2m+1)\left(\frac{(j-m)(j-m-1)}{(2j-2)(2j-1)(2j)(2j+2)}\right)^{1/}$
j_2	$\frac{2\left(3m^2-j(j+1)\right)}{[(2j-1)(2j)(2j+2)(2j+3)]^{1/2}}$	$(2m+1)\left(\frac{6(j-m)(j+m+1)}{(2j-1)(2j)(2j+2)(2j+3)}\right)^{1/2}$
$j_2 - 1$	$2m\left(\frac{6(j-m+1)(j+m+1)}{2j(2j+2)(2j+3)(2j+4)}\right)^{1/2}$	$2(j-2m)\left(\frac{(j+m+2)(j+m+1)}{2j(2j+2)(2j+3)(2j+4)}\right)^{1/2}$
$j_2 - 2$	$\left(\frac{6(j-m+2)(j-m+1)(j+m+2)(j+m+1)}{(2j+2)(2j+3)(2j+4)(2j+5)}\right)^{1/2}$	$-\left(\frac{4(j-m+1)(j+m+3)(j+m+2)(j+m+1)}{(2j+2)(2j+3)(2j+4)(2j+5)}\right)^{1/2}$

j	$m_1 = -2$
$j_2 + 2$	$\left(\frac{(j-m-3)(j-m-2)(j-m-1)(j-m)}{(2j-3)(2j-2)(2j-1)(2j)}\right)^{1/2}$
$j_2 + 1$	$-\left(\frac{4(j-m-2)(j-m-1)(j-m)(j+m+1)}{(2j-2)(2j-1)(2j)(2j+2)}\right)^{1/2}$
j_2	$\left(\frac{6(j-m-1)(j-m)(j+m+1)(j+m+2)}{(2j-1)(2j)(2j+2)(2j+3)}\right)^{1/2}$
$j_2 - 1$	$-\left(\frac{4(j-m)(j+m+1)(j+m+2)(j+m+3)}{2j(2j+2)(2j+3)(2j+4)}\right)^{1/2}$
$j_2 - 2$	$\left(\frac{(j+m+1)(j+m+2)(j+m+3)(j+m+4)}{(2j+2)(2j+3)(2j+4)(2j+5)}\right)^{1/2}$

7

The magnetic dipole moment and electric quadrupole moment for nuclei with closed shells ±1 nucleon

Two nuclear quantities are easily accessible to measurements, the magnetic dipole moment and the nuclear quadrupole moment. These represent the lowest non-trivial electric and magnetic multipoles, as the electric dipole moment and the magnetic monopole probably vanish exactly while the electric monopole is the nuclear charge.

7.1 The magnetic dipole moment. Schmidt diagram

The first established method of measurement of the magnetic dipole moment μ_I was to study the hyperfine splitting of the electron spectrum (see e.g. Kopfermann, 1958). The interaction energy between the nuclear magnetic moment directed approximately along the nuclear spin $\mathbf{I}$ and the magnetic field, $\mathbf{B}$, generated by the electrons and on the average directed along the electron angular momentum $\mathbf{J}$, can be written

$$W = -\frac{\mu_I B(0)}{IJ}\mathbf{I}\cdot\mathbf{J}$$

By this interaction $\mathbf{I}$ and $\mathbf{J}$ are coupled to a total angular momentum $\mathbf{F}$. This leads to a spectrum of the following type (the analogy with the $\boldsymbol{\ell}\cdot\mathbf{s}$ term is obvious).

$$\Delta W_{I,J} = -\frac{1}{2}\frac{\mu_I B(0)}{IJ}\left[F(F+1) - I(I+1) - J(J+1)\right]$$

This relation is called Lande's interval rule.

Thus, provided the field strength B in the nucleus can be calculated, μ_I can be determined from the magnitude of the splitting. The calculation of the field strength is, however, often a complicated problem, as it requires detailed information on the electron wave functions. If possible, one therefore uses more direct methods, so called 'brute-force' methods, based on a directly

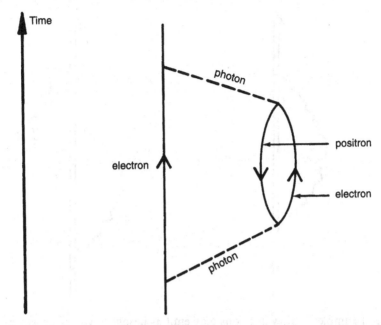

Fig. 7.1. Diagram illustrating the formation of a virtual electron–positron pair from the interaction of an electron with the electromagnetic field.

measureable *external* field overcoming the field generated by the electronic motion.

It turns out that the measured magnetic moment, based on the formula above, is calculated as the expectation value of the z-component of the *nuclear* magnetic moment. The direction of the z-axis is defined as the direction of the **B** field at the origin.

Let us first consider the electron case, which should be well known. For an electron the magnetic moment derives in part from the orbital motion, in part from the electron spin. We have thus

$$\boldsymbol{\mu}_{\mathrm{el}} = -\frac{e\hbar}{2m_{\mathrm{e}}}(\boldsymbol{\ell} + 2\mathbf{s}) = \frac{e\hbar}{2m_{\mathrm{e}}}\left(g_{\ell}^{\mathrm{e}}\boldsymbol{\ell} + g_{s}^{\mathrm{e}}\mathbf{s}\right)$$

where

$$g_{\ell}^{\mathrm{e}} = -1 \quad \text{and} \quad g_{s}^{\mathrm{e}} = -2$$

and where $e\hbar/2m_{\mathrm{e}}$ is the Bohr magneton.

From non-relativistic Schrödinger theory one might have expected both g-factors to equal −1 (the classical result). On the other hand first relativistic quantum mechanics (Dirac theory) properly connects spin with the associated magnetic moment. The surprising result is that $g_{s}^{\mathrm{e}} = -2$. Precise

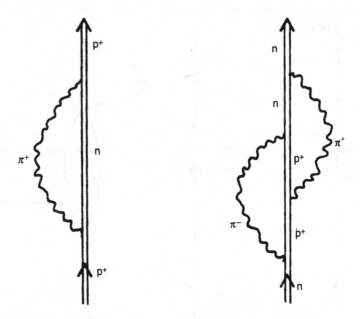

Fig. 7.2. Examples of how nucleons may emit π-mesons in virtual processes confined within the limits set by the indeterminacy relation.

measurements show that $|g_s^e|$ is in fact somewhat greater than 2. This has to do with the fact that the electron cannot be considered a 'naked' particle. When it interacts with the electromagnetic field an electron–positron pair is formed virtually†, as illustrated in fig. 7.1.

In analogy with the electron case, we shall assume for the nucleons that

$$\mu = \frac{e\hbar}{2M}\,(g_\ell\boldsymbol{\ell} + g_s\mathbf{s})$$

where $e\hbar/2M$ is the nuclear magneton, M being the nucleon mass. The nucleons are Dirac particles, and one expects therefore

$$g_\ell^p = 1 \qquad g_s^p = 2$$
$$g_\ell^n = 0 \qquad g_s^n = 0$$

Empirically one finds that for *free* nucleons (cf. problem 7.6)

$$g_s^p = 5.59$$
$$g_s^n = -3.83$$

This 'anomalous' g_s-factor depends on the fact that the nucleon emits π-

---

† 'Virtually' stands here for something like 'in isolation from energy conservation'; the pairs therefore only exist for ranges of time compatible with the indeterminacy relation.

mesons as illustrated in fig. 7.2 (cf. also chapter 13). As a first approximation we shall use the free nucleon g_s-values also for the nucleons in the nucleus, as these are also surrounded by virtual meson clouds.

According to the relations above we have

$$\mu_I = \langle jm = j | \mu_z | jm = j \rangle$$

i.e., the magnetic moment is defined for the state with $m = j$. The operator μ_z is given by

$$\mu_z = \frac{e\hbar}{2M}(g_\ell \ell_z + g_s s_z) = \frac{e\hbar}{2M}(g_\ell j_z + (g_s - g_\ell)s_z)$$

To calculate $\langle s_z \rangle$, we have to expand the wave functions $|j\,m = j\rangle \equiv |\ell\,s\,j\,m = j\rangle$ into components with m_s as a good quantum number:

$$\mu_I = \frac{e\hbar}{2M} \int R^*_{n\ell} R_{n\ell} r^2 \, dr \cdot \left\{ g_\ell j + (g_s - g_\ell) \right.$$

$$\left. \times \int d\Omega \sum_{m_\ell m_s} C^{\ell s j}_{m_\ell m_s j} Y^*_{\ell m_\ell} \chi^+_{s m_s} \cdot s_z \sum_{m'_\ell m'_s} C^{\ell s j}_{m'_\ell m'_s j} Y_{\ell m'_\ell} \cdot \chi_{s m'_s} \right\}$$

Now exploit the normalisation of the radial functions

$$\int R^*_{n\ell} R_{n\ell} r^2 \, dr = 1$$

and the orthogonality relations

$$\int d\Omega\, Y^*_{\ell m_\ell} Y_{\ell m'_\ell} = \delta_{m_\ell m'_\ell}; \quad \chi^+_{s m_s} \cdot \chi_{s m'_s} = \delta_{m_s m'_s}$$

to obtain

$$\mu_I = \frac{e\hbar}{2M}\left(g_\ell j + (g_s - g_\ell) \sum_{m_\ell m_s} \left(C^{\ell s j}_{m_\ell m_s j} \right)^2 \cdot m_s \right)$$

Set $j = \ell + \frac{1}{2}$, which leads to

$$\mu_{\ell + \frac{1}{2}} = \frac{e\hbar}{2M}\left\{ g_\ell j + (g_s - g_\ell)\left[\frac{1}{2}\left(C^{\ell\,\frac{1}{2}\,\ell + \frac{1}{2}}_{\ell\,\frac{1}{2}\,\ell + \frac{1}{2}} \right)^2 - \frac{1}{2}\left(C^{\ell\,\frac{1}{2}\,\ell + \frac{1}{2}}_{(\ell+1)\,-\frac{1}{2}\,\ell + \frac{1}{2}} \right)^2 \right] \right\}$$

$$= \frac{e\hbar}{2M} j \left\{ g_\ell + (g_s - g_\ell)\frac{1}{2\ell + 1} \right\}$$

where the last equality follows because the first Clebsch–Gordan coefficient is equal to one while the second one (with $m_\ell > \ell$) is zero.

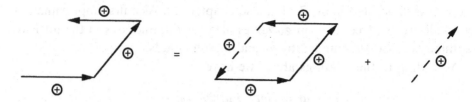

Fig. 7.3. Schematic vector addition of magnetic moments showing that a missing particle in a shell (a hole) gives the same magnetic moment as a particle in the shell.

The Clebsch–Gordan coefficients to be evaluated in the case of $j = \ell - \frac{1}{2}$ are somewhat more difficult but can be found in table 6B.1. One then obtains

$$\mu_{\ell-1/2} = \frac{e\hbar}{2M} j \left(g_\ell - (g_s - g_\ell) \frac{1}{2\ell + 1} \right)$$

and thus the general formula

$$\mu_{\ell \pm 1/2} = \frac{e\hbar}{2M} j \left(g_\ell \pm (g_s - g_\ell) \frac{1}{2\ell + 1} \right)$$

Based on the empirical g_s-values, valid for free nucleons, we obtain finally the following simplified expressions, representing the 'extreme' single-particle model values:

(1) for odd-Z nuclei (i.e. protons)

$$\mu_{\ell+1/2} = 2.29 + j$$
$$\mu_{\ell-1/2} = j \left(1 - 4.59 \cdot \frac{1}{2(j+1)} \right)$$

(2) for odd-N nuclei (i.e. neutrons)

$$\mu_{\ell+1/2} = -1.91$$
$$\mu_{\ell-1/2} = 1.91 \frac{j}{j+1}$$

These values are expressed in nuclear magnetons, $e\hbar/2M$.

We shall now consider the magnetic moment, μ, for nuclei with closed shells ± 1 nucleon. For a closed shell, the particles are assumed to be coupled to spin 0. Consider then fig. 7.3. There is shown on the left a missing leg, representing a hole, in the polygon. On the right the dashed cancelling angular momenta are added. Thus a shell where a particle is missing behaves as a closed shell plus one particle of the same charge and angular momentum as an ordinary one-particle state. This results in the relation $\mu(\text{hole}) = \mu(\text{particle})$.

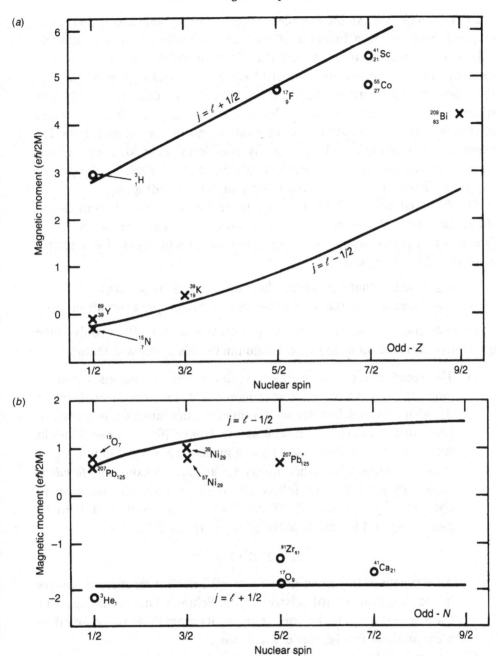

Fig. 7.4. a) Measured magnetic moments of some 'closed-shell ± 1 proton' nuclei compared with the corresponding free proton values (Schmidt lines). Nuclei with ℓ and s coupled in parallel are denoted by open circles, those with ℓ and s in antiparallel by crosses. b) Same as part a) but for 'closed-shell ± 1 neutron' nuclei instead.

In figs. 7.4a,b, we exhibit so-called Schmidt (1937) diagrams showing the magnetic moment as a function of angular momentum I. The theoretical values of the single-particle model (the Schmidt lines) are compared with measured magnetic moments for nuclei having closed shells ±1 nucleon. Note the tendency that, with increasing mass number, the discrepancy between theory and experiment increases. Note also that the experimental values always fall inside the Schmidt lines or more or less on the Schmidt lines. No experimental value is found significantly outside the Schmidt lines. All the nuclei also fall closest to the one Schmidt line characterised by the proper coupling scheme, $j = \ell + \frac{1}{2}$ or $j = \ell - \frac{1}{2}$ (correct Schmidt group).

The fact that nearly all nuclei, even those further removed from closed shells, fall inside the Schmidt lines, indicates a systematic reduction of g_s. One may analyse the empirical magnetic moments on the basis of an effective g_s-factor, $g_s^{\text{eff}} = x \cdot g_s \equiv x \cdot g_s^{\text{free}}$:

> $x \leq 1$ means that the nucleus falls inside the Schmidt lines
> $x > 0$ means that the nucleus belongs to the proper Schmidt group

Experience shows that, with few exceptions, $0 < x < 1$. Here, only some qualitative considerations are used to illuminate this systematic feature:

(1) **The recoil effect.** As mentioned above, the anomalous magnetic moment derives from emitted π-mesons. If now several nucleons in the vicinity of each other emit π-mesons, the latter do not perturb each other because the π-mesons are *bosons* (follow Bose–Einstein statistics), while on the other hand a nucleon cannot recoil into states already occupied by other nucleons, as the nucleons are *fermions* (which means that they follow Fermi–Dirac statistics and have to obey the Pauli principle). This effect should result in a reduced π-emission and for example for protons we would expect

$$g_s^{\text{free}} \geq g_s^{\text{eff}} \geq g_s^{\text{Dirac}}$$

An estimate of this effect gives $\delta\mu \approx \pm 0.2$ nuclear magnetons, always in the direction inward relative to the Schmidt lines. Although this effect is certainly of the correct trend, its magnitude is too small to account alone for the entire discrepancy.

(2) **The spin polarisation effect.** As already stated, one may analyse *empirical* moments from

$$\mu = j\left(g_\ell \pm \left(g_s^{\text{eff}} - g_\ell \right) \frac{1}{2\ell + 1} \right); \quad j = \ell \pm 1/2$$

leading to values of the ratio $g_s^{\text{eff}}/g_s^{\text{free}} = x$. The tendency here is

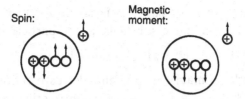

Fig. 7.5. Schematic illustration of the spin polarisation effect.

that the ratio falls from about 1 for light nuclei to about 0.6 for heavy nuclei. These systematically reduced values of x are indicative of a negative spin polarisation, which can be ascribed to a specific component of the residual interaction (i.e. the interaction not included in the nuclear potential). The component in question is $(\boldsymbol{\sigma}_1 \cdot \boldsymbol{\sigma}_2)(\boldsymbol{\tau}_1 \cdot \boldsymbol{\tau}_2) \cdot V_{\tau\sigma}$ (cf. chapter 13), where $\boldsymbol{\sigma}$ and $\boldsymbol{\tau}$ refer to the nucleon spin and isospin. The isospin factor, $\boldsymbol{\tau}_1 \cdot \boldsymbol{\tau}_2$ is positive for like particles (e.g. a proton interacting with a proton) and negative for unlike particles (proton–neutron interaction) just as the spin factor, $\boldsymbol{\sigma}_1 \cdot \boldsymbol{\sigma}_2$ is positive for parallel spins and negative for antiparallel spins (the isospin formalism will be introduced in chapter 13).

The $(\boldsymbol{\sigma}_1 \cdot \boldsymbol{\sigma}_2)(\boldsymbol{\tau}_1 \cdot \boldsymbol{\tau}_2)$ interaction attempts to align in parallel the spins of unlike particles and consequently antiparallel the spins of like particles. Now consider a proton outside the nucleus as exhibited in fig. 7.5. There is thus a trend for this proton to set the spins of the internal protons antiparallel to the spin of the valence proton. For internal neutrons on the other hand, the trend will be to put the spins in parallel. Owing to the sign of g_s, the magnetic moment and spin are parallel for a proton, while the magnetic moment and spin are antiparallel for a neutron. Thus, the trend for the magnetic moments of both internal protons and neutrons will be to put them antiparallel to a proton outside the nucleus (fig. 7.5). In an analogous way, it is found that, for an external neutron, its magnetic moment tends to be antiparallel to those of the internal nucleons. Apparently the trend towards polarisation of the nucleons of the closed shells can be realised only a fraction of the time. The result is a systematic suppression of the magnetic moment associated with the spin. Thus we have

$$\left| g_s^{\text{eff}} \right| < \left| g_s^{\text{free}} \right|$$

In fact the subshell pulled down by the spin–orbit force is most easily polarisable. For the doubly magic ± 1 nucleus ^{209}Bi the subshells $h_{11/2}$

and $i_{13/2}$ are the most easily polarisable for protons and neutrons, respectively. From these subshells, a one-particle–one-hole state can easily be excited as $\left(h_{11/2}\right)^{-1} h_{9/2}$ and $\left(i_{13/2}\right)^{-1} i_{11/2}$ respectively.

As a final remark on magnetic moments, let us mention that not only g_s but also g_ℓ might be subject to a modification relative to the 'free'-nucleon value. However, both experiment and theory point to a $|\delta g_\ell|$ not exceeding about 0.1 nuclear magnetons (see e.g. Häusser *et al.*, 1977, and references therein).

7.2 The electric quadrupole moment

Another nuclear quantity first observed on the basis of optical spectra is the electric quadrupole moment. One found certain deviations from Lande's interval rule for the hyperfine structure. These could be ascribed to the coupling between the electric quadrupole moment of the nucleus and the wave function of the electrons (Kopfermann, 1958).

Classically the electrostatic interaction energy due to the coupling between the electron cloud and the nucleus is given by the expression

$$W = e \int \rho_Z (\mathbf{r}) \, \phi (\mathbf{r}) \, d\tau$$

where ρ_Z is the electric charge density of the nucleus and ϕ the electrostatic potential due to the electrons. By expansion of $\phi(\mathbf{r})$ around the nuclear centre of gravity point one obtains

$$W = e \int \rho_Z \phi(0) \, d\tau + e \int \rho_Z \nabla\phi \cdot \mathbf{r} \, d\tau + \frac{e}{2} \int \rho_Z \left(\phi_{xx} x^2 + 2\phi_{xy} xy + \dots\right) d\tau + \dots$$

where $\phi_{xy} = \partial^2\phi/\partial x \, \partial y$ etc., all derivatives assumed evaluated at the origin. The first integral corresponds to the case of a point charge at the origin. In the absence of a nuclear dipole moment, $\int \rho_Z \mathbf{r} \, d\tau = 0$, the second term vanishes identically. Thus, the third integral gives the lowest order effect of the finite extension of the nucleus. With the coordinate system having its z-axis along the electronic axis of rotational symmetry (then excluding spherically symmetric $s_{1/2}$ and $p_{1/2}$ electrons for which such an axis cannot be defined, cf. fig. 7.6 below), the non-diagonal terms ϕ_{xy} etc. disappear leaving an energy contribution from the third term as

$$W_Q = \frac{e}{2}\left[\phi_{xx} \int \rho_Z x^2 \, d\tau + \phi_{yy} \int \rho_Z y^2 \, d\tau + \phi_{zz} \int \rho_Z z^2 \, d\tau\right]$$

The relation

$$\nabla^2\phi = 0$$

combined with the assumption of axial symmetry around the z-axis leads to

$$\phi_{zz} = -2\phi_{xx} = -2\phi_{yy}$$

We then obtain the third integral simplified to the following form

$$W_Q = -\frac{\phi_{xx}}{2} e \int \left(3z^2 - r^2\right) \rho z \, d\tau$$

Assuming that the nucleus is rotationally symmetric, a new coordinate system (x', y', z') with the z'-axis along the nuclear symmetry axis is introduced. Now, if the angle between the z- and z'-axes is β, it is straightforward to make a coordinate transformation (cf. problem 7.2) to obtain:

$$W_Q = -\frac{\phi_{xx}}{2} e \int \left(3\, (z')^2 - r^2\right) \rho z \, d\tau \left(\frac{3}{2}\cos^2 \beta - \frac{1}{2}\right)$$

We now define the quadrupole moment (Q or Q_2) as the integral

$$Q = \int \left(3\, (z')^2 - r^2\right) \rho z \, d\tau$$

or its quantum mechanical generalisation (dropping the 'primes' on the nuclear oriented coordinate system)

$$Q = \int \psi^* \sum_p \left(3z_p^2 - r_p^2\right) \psi \, d\tau$$

where the sum is to be taken over the charge-carrying particles, i.e. in our case the protons in the nucleus. Note that the quadrupole moment, according to the standard definition, has the dimension of an area. The unit of charge thus does not enter. In terms of spherical coordinates we write for a single proton

$$Q_p = 3z_p^2 - r_p^2 = r_p^2 \left(\frac{16\pi}{5}\right)^{1/2} Y_{20}\left(\theta_p, \phi_p\right)$$

For the extreme single-particle model, only the 'valence'-particle contributes to the quadrupole moment. Furthermore, by definition the quadrupole moment is calculated in the $m = j$ state and we thus obtain

$$Q = \begin{bmatrix} e_n \\ e_p \end{bmatrix} \left(\frac{16\pi}{5}\right)^{1/2} \langle r^2 \rangle_{N\ell} \left\langle \ell \tfrac{1}{2} jj \left| Y_{20} \right| \ell \tfrac{1}{2} jj \right\rangle = - \begin{bmatrix} e_n \\ e_p \end{bmatrix} \cdot \frac{2j-1}{2j+2} \langle r^2 \rangle_{N\ell}$$

where for the neutron and the proton we have $e_n = 0$ and $e_p = 1$, respectively. We shall derive this result for the case $j = \ell + \tfrac{1}{2}$ and will leave the proof that it holds also for $j = \ell - \tfrac{1}{2}$ to the exercises. We consider the matrix element

$$\left(\ell \tfrac{1}{2} jj \left| Y_{20} \right| \ell \tfrac{1}{2} jj \right)$$

Let us here introduce the notation that bra- and ket-vectors, written with *rounded parentheses* imply that *only the angular integral* enters. The wave function in the $j = \ell + \frac{1}{2}$ case (corresponding to the case of parallel spin and orbital angular momentum),

$$\left| \ell\, s\, j = \ell + \frac{1}{2}\, m = j \right) = \sum_{m_\ell m_s} Y_{\ell m_\ell} \cdot \chi_{sm_s} C^{\ell \, s \, j = \ell + \frac{1}{2}}_{m_\ell \, m_s \, m = \ell + \frac{1}{2}} = Y_{\ell\ell} \chi_{\frac{1}{2}\frac{1}{2}}$$

comes out very simple because there is only one possibility of having $m_\ell + m_s = \ell + \frac{1}{2}$, namely $m_\ell = \ell, m_s = \frac{1}{2}$ (in the derivation of the magnetic moment above, this same fact was obtained from a direct evaluation of the Clebsch–Gordan coefficients obtained as 1 and 0, respectively). We now obtain

$$\left(\ell\, \frac{1}{2}\, j = \ell + \frac{1}{2}\, j \,\middle|\, Y_{20} \,\middle|\, \ell\, \frac{1}{2}\, j = \ell + \frac{1}{2}\, j \right) = \int d\Omega \; Y^*_{\ell\ell} \chi^+_{\frac{1}{2}\frac{1}{2}} Y_{20} \chi_{\frac{1}{2}\frac{1}{2}} Y_{\ell\ell} \equiv \mathcal{I}$$

Note that

$$\chi^+_{\frac{1}{2}\frac{1}{2}} \chi_{\frac{1}{2}\frac{1}{2}} = (1 \quad 0) \begin{pmatrix} 1 \\ 0 \end{pmatrix} = 1$$

and therefore

$$\mathcal{I} = \int Y^*_{\ell\ell}(\theta, \phi)\, Y_{20}(\theta, \phi)\, Y_{\ell\ell}(\theta, \phi)\, d\Omega$$

In the preceding chapter, we quoted the addition formula for spherical harmonics

$$Y_{\ell m}(\theta, \phi)\, Y_{\ell' m'}(\theta, \phi) = \sum_{LM} \left(\frac{(2\ell + 1)(2\ell' + 1)}{4\pi(2L + 1)} \right)^{1/2} \cdot C^{\ell \ell' L}_{m m' M} \cdot C^{\ell \ell' L}_{000} \, Y_{LM}(\theta, \phi)$$

The sum over M obviously contains one term, $M = m + m'$ and consequently

$$Y_{20} Y_{\ell\ell} = \sum_L C^{\ell 2 L}_{\ell 0 \ell} Y_{L\ell} C^{\ell 2 L}_{000} \left(\frac{(2\ell + 1)(2 \cdot 2 + 1)}{4\pi(2L + 1)} \right)^{1/2}$$

Owing to orthogonality it follows that the only non-vanishing contributions to the angular integral above are those with $L = \ell$. We have thus

$$\mathcal{I} = \left(\frac{5}{4\pi} \right)^{1/2} C^{\ell 2 \ell}_{\ell 0 \ell} C^{\ell 2 \ell}_{000}$$

and if the values of the Clebsch–Gordan coefficients (available e.g. from table 6B.1) are inserted, one obtains

$$Q = - \begin{bmatrix} e_n \\ e_p \end{bmatrix} \frac{2\ell}{2\ell + 3} \left\langle r^2 \right\rangle_{N\ell}$$

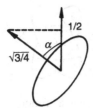

Fig. 7.6. Vector model of a spin $\frac{1}{2}$ particle. The angle α is quite large. This leads to a wave function that is smeared over all angles in such a way that the quadrupole moment Q vanishes.

which can be transformed to

$$Q = - \begin{bmatrix} e_n \\ e_p \end{bmatrix} \frac{2j-1}{2j+2} \langle r^2 \rangle_{N\ell}$$

This relation has thus been proven for $j = \ell + \frac{1}{2}$. In the exercises it is shown that this expression, with Q expressed in terms of j, also holds for $j = \ell - \frac{1}{2}$. This means that the quadrupole moment depends only on j and not on ℓ.

Note that directly from the expression for Q in terms of j it is apparent that $Q = 0$ for $j = \frac{1}{2}$. The quantum fluctuations are then so large that no deviation from sphericity can be observed. The picture provided by the vector model for these relations is illustrated in fig. 7.6. The length of the vector is $[j(j+1)]^{1/2} = \sqrt{\frac{3}{4}}$ (in units of $\hbar$) and the length of the z-component is $\frac{1}{2}$. In terms of this model the orbit precesses around the z-axis. This precession smears the charge distribution in space relative to that of the orbital by the smearing factor $(2j-1)/(2j+2)$. For $j = \frac{1}{2}$ the angle α is so large and the smearing as a consequence so severe that Q vanishes exactly.

We have derived an expression for the quadrupole moment valid for closed shells plus 1 particle. For the case of closed shells plus 1 hole the corresponding formula holds but with the opposite sign. We have thus

$$Q(\text{hole}) = -Q(\text{particle})$$

The situation is thus opposite to that for the magnetic moment. The reason for the change of sign for the quadrupole moment can be understood from an inspection of fig. 7.7. We have thus replaced the hole with cancelling $+$ and $-$ charge clouds. The $+$ charge completes the closed shell. The $-$ charge is subsequently responsible for the shift in sign of the quadrupole moment for a hole state relative to a particle state.

Let us estimate the magnitude of Q. We assume that the orbital of the

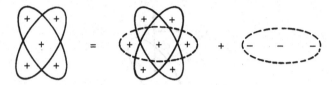

Fig. 7.7. Schematic addition of wave functions showing that a hole gives a quadrupole moment with a different sign than that of a particle (cf. fig. 7.3).

Table 7.1. *Experimental quadrupole moments compared with theoretical 'one-proton values' for nuclei with one proton or one neutron outside closed shells. Also given are the corresponding effective charges and polarisation factors.*

	Nucleus (doubly magic +1)		
	$^{17}_{8}O_9$	$^{17}_{9}F_8$	$^{209}_{83}Bi_{126}$
Orbital	$d_{5/2}$	$d_{5/2}$	$h_{9/2}$
Q^{exp} (barn)	−0.026	−0.10	−0.46
$\langle Q^{\text{one proton}} \rangle$ (barn)	−0.051	−0.051	−0.30
$e^{\text{eff}} = Q^{exp}/\langle Q^{\text{one proton}} \rangle$	0.51	2.0	1.5
$\alpha = \left(e^{\text{eff}} - e_{(n,p)}\right)/\left(Ze_p/A\right)$	1.1	1.8	1.3

valence nucleon lies near the surface of the nucleus. This leads to

$$|Q| \approx \langle r^2 \rangle \approx R^2 \approx 1.22 \cdot A^{2/3} \cdot 10^{-26} \, \text{cm}^2 = 1.44 \cdot A^{2/3} \cdot 10^{-2} \quad \text{barns}$$
$$(1 \, \text{barn} = 10^{-24} \, \text{cm}^2)$$

Thus for light nuclei we expect in terms of the 'extreme' one-particle model a quadrupole moment of a few hundredths of a barn, and for heavier nuclei some tenths of a barn. A comparison with a few selected experimental values is shown in table 7.1.

Because $e = 1$ for protons and $e = 0$ for neutrons, one should expect to have $Q = 0$ for the odd-N case. Instead, the quadrupole moment turns out to be approximately equal to that expected for an odd proton. For the odd-proton case, Q comes out larger than the single-proton value by about a factor of 2. These deviations in the quadrupole moment are considered to represent matter polarisation as illustrated in fig. 7.8 (one often refers to the tidal waves caused on the surface of the earth by the orbiting moon).

Let us introduce the factor of polarisation α. This implies a charge $\alpha \cdot (Z/A)$

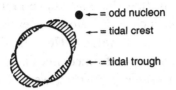

Fig. 7.8. Illustration of how a particle outside a spherical core polarises the core.

Table 7.2. *Quadrupole moments of nuclei removed from doubly closed shells. Subshells for protons and neutrons are denoted by π and ν, respectively. In the last five cases of the table, the polarisation factor α has no apparent meaning. These nuclei are deformed, permanently or momentarily through vibrations. The two In nuclei are so-called vibrational nuclei. The last three nuclei are permanently deformed. The quadrupole moment is 20–30 times larger than that of a single proton.*

Nucleus	State	Q^{exp} (barn)	$Q^{one\ proton}$ (barn)	Polarisation factor	Comment
$^{7}_{3}Li_4$	$(\pi p_{3/2})$	−0.040	−0.020	$\alpha > 1$	
$^{9}_{4}Be_5$	$(\nu p_{3/2})^{-1}$	0.053	0.020	$\alpha > 1$	
$^{35}_{17}Cl_{18}$	$\pi d_{3/2}$	−0.082	−0.047	$\alpha > 1$	
$^{37}_{17}Cl_{20}$	$\pi d_{3/2}$	−0.065	−0.049	$\alpha > 1$	
$^{113}_{49}In_{64}$	$(\pi g_{9/2})^{-1}$	0.8	0.22	$(\alpha \gg 1)$	Vibrational
$^{115}_{49}In_{66}$	$(\pi g_{9/2})^{-1}$	0.86	0.22	$(\alpha \gg 1)$	nuclei $\alpha \simeq 10$
$^{165}_{67}Ho_{98}$	7/2	2.73	$\simeq 0.2$	$(\alpha \ggg 1)$	Nuclei with
$^{175}_{71}Lu_{104}$	7/2	5.68	$\simeq 0.2$	$(\alpha \ggg 1)$	stable deformations
$^{181}_{73}Ta_{108}$	7/2	3.9	$\simeq 0.2$	$(\alpha \ggg 1)$	$\alpha \simeq 100$

from the polarisation. One thus obtains

$$e_n^{eff} = e_n + \alpha_n \frac{Z}{A} e_n \approx 0 + \alpha_n \cdot \frac{1}{2}$$

$$e_p^{eff} = e_p + \alpha_p \frac{Z}{A} e_p \approx 1 + \alpha_p \cdot \frac{1}{2}$$

As a very rough estimate, $\alpha_n^{exp} \approx 1$ and $\alpha_p^{exp} \approx 1 - 2$ (cf. table 7.1). The theoretical value obtained for the harmonic oscillator is $\alpha = 1$ and for the infinite square well potential $\alpha \approx 2$–4.

Some examples of other types of nuclei than closed-shells ±1 nuclei are given in table 7.2. It is apparent that, in some cases, α becomes much larger than one. This must mean that the quadrupole moment is built from many nucleons and the whole nucleus is deformed, either vibrationally or more permanently (Bohr, 1976; Rainwater, 1976). In the coming chapters, we will consider the permanently deformed nuclei. As a preparation for the description of their measurable properties, we will study the single-particle orbitals of a deformed potential in chapter 8.

Exercises

7.1   Use the formula

$$\hbar^2 j(j+1)\langle jm|t|jm'\rangle = \langle jm|\mathbf{j}|jm'\rangle\langle jm'|\mathbf{t}\cdot\mathbf{j}|jm'\rangle$$

where **t** represents a vector of type ℓ, **s** etc., to derive the formula for the magnetic moment:

$$\mu = \langle jj |g_s s_z + g_\ell \ell_z| jj\rangle = j\left(g_\ell \pm (g_s - g_\ell)\frac{1}{(2\ell+1)}\right)$$

The upper and lower signs are valid for $j = \ell \pm \frac{1}{2}$, respectively. Interpret the given formula in the case $m = m'$.

7.2   (a) Show that the quadrupole moment Q with regard to the symmetry axis of a homogeneously charged spheroid is $\frac{2}{5}Z\left(b^2 - a^2\right)$. The half-axes are given by a and b with b referring to the symmetry axis.

           (b) If the symmetry axis of the spheroid is rotated by an angle β, show that the quadrupole moment with regard to the rotated axis is given by

$$Q' = Q\left(\frac{3}{2}\cos^2\beta - \frac{1}{2}\right)$$

7.3   Calculate the magnetic moment for the following closed shell ± 1 nuclei (experimental values are given in square brackets)

$$^{15}\text{N} \left(p_{1/2}\right)^{-1} \quad [-0.28]$$
$$^{15}\text{O} \left(p_{1/2}\right)^{-1} \quad [0.719]$$
$$^{17}\text{O} \quad d_{5/2} \quad [-1.894]$$
$$^{39}\text{K} \left(d_{3/2}\right)^{-1} \quad [0.391]$$
$$^{41}\text{Ca} \quad f_{7/2} \quad [-1.595]$$
$$^{207}\text{Pb} \left(p_{1/2}\right)^{-1} \quad [0.590]$$
$$^{209}\text{Bi} \quad h_{9/2} \quad [4.080]$$

Which g_s^{eff} is needed for the different nuclei to reproduce the experimental values? Use the formula $\mu^{\text{exp}} = j\left[g_\ell \pm \left(g_s^{\text{eff}} - g_\ell\right)/(2\ell+1)\right]$.

7.4 Because of the Coulomb field, the average potential for protons and neutrons becomes somewhat different. Estimate this difference due to the Coulomb energy for a $1g_{7/2}$ and a $2d_{5/2}$ proton, respectively in the following case.

(a) The wave function of the proton is of harmonic oscillator type
(b) The charge distribution in the nucleus is assumed homogeneous out to the radius $R = 1.2 \cdot A^{1/3}$ fm.

7.5 Calculate the quadrupole moment

$$Q = \left\langle \ell \frac{1}{2} jj \,|\, Q^{\text{op}} \,|\, \ell \frac{1}{2} jj \right\rangle$$

where $Q^{\text{op}} = e(16\pi/5)\langle r^2 \rangle_{N\ell} Y_{20}$ is the quadrupole moment operator.

7.6 In the quark picture, the proton is described as being built from two u quarks and one d quark while the neutron is built from one u quark and two d quarks. The u quark has charge $(2/3)e$ and the d quark $(-1/3)e$. They both have an internal spin, $s = 1/2$. The magnetic moment operator for a quark with charge Q is now assumed to be given as $\mu_q = Q\mu_0 s_z$ where μ_0 is a constant. Furthermore, to fulfill the necessary symmetry relations, the two equal quarks in a proton or neutron must couple to spin 1. Through the coupling of the third quark, a total spin of $1/2$ is then obtained. Show that this leads to the relation $\mu_n = (-2/3)\mu_p$ for the neutron and proton magnetic moments. How does this relation compare with the experimental values?

8

Single-particle orbitals in deformed nuclei

For the spherical case we have discussed the isotropic harmonic-oscillator field. To amend for the radial deficiencies we have seen that the addition of a term proportional to $-\ell^2$ has the desired properties of giving rise to an effective interpolation between a harmonic oscillator and a square well. One thus obtains a fair reproduction of the spherical single-particle levels by the following Hamiltonian – the modified-oscillator (MO) Hamiltonian:

$$H_{\text{sph}} = -\frac{\hbar^2}{2M}\Delta + \frac{1}{2}M\omega_0^2 r^2 - C\boldsymbol{\ell}\cdot\mathbf{s} - D\left(\ell^2 - \left\langle\ell^2\right\rangle_N\right)$$

As a secondary and undesirable effect of ℓ^2 alone, there is a general compression of the distance between the shells below $\hbar\omega_0$. This basic energy spacing is restored by the subtraction of the term $\left\langle\ell^2\right\rangle_N = N(N+3)/2$ (see problem 6.7), which thus assumes a constant value within each shell. One argument for the introduction of the $\left\langle\ell^2\right\rangle_N$ term is the following. Only the terms proportional to r^2 are conveniently included in the volume conservation condition (see below). In order not to upset volume conservation by the effective widening of the radial shape by the ℓ^2 term, it appears reasonable to subtract from this term the average value appropriate to each shell. A resulting level scheme is exhibited in fig. 6.3. In that figure, different strength parameters, κ and μ' are used. The relations $C = 2\kappa\hbar\omega_0$ and $D = \mu'\hbar\omega_0$ are straightforward to derive. The parameters κ and μ' (or μ where $\mu' = \kappa\mu$) are the standard parameters used together with dimensionless oscillator units.

This potential easily lends itself to a generalisation so as to be applicable to the deformed case. If we allow for the potential extension along the nuclear z-axis (3-axis) being different from the extension along the x- and

110

y-axes, we may write the single-particle Hamiltonian in the form

$$H = -\frac{\hbar^2}{2M}\left(\frac{\partial^2}{\partial x^2} + \frac{\partial^2}{\partial y^2} + \frac{\partial^2}{\partial z^2}\right) + \frac{M}{2}\left[\omega_\perp^2\left(x^2 + y^2\right) + \omega_z^2 z^2\right] - C\boldsymbol{\ell}\cdot\mathbf{s}$$

$$-D\left(\ell^2 - \left\langle\ell^2\right\rangle_N\right)$$

The anisotropy corresponds to the difference introduced between $\omega_\perp$ and ω_z. It is convenient to introduce an elongation parameter ε (Nilsson, 1955):

$$\omega_z = \omega_0(\varepsilon)\left(1 - \frac{2}{3}\varepsilon\right)$$

$$\omega_\perp = \omega_0(\varepsilon)\left(1 + \frac{1}{3}\varepsilon\right)$$

where $\omega_0(\varepsilon)$ is weakly ε-dependent, enough to conserve the nuclear volume (see below). The distortion parameter ε is obtained as $\varepsilon = (\omega_\perp - \omega_z)/\omega_0$. It is defined so that $\varepsilon > 0$ and $\varepsilon < 0$ correspond to so-called prolate and oblate shapes, respectively.

8.1 Perturbation treatment for small ε

Let us first study the situation for very small ε-values. Expanding in ε we may write

$$H = H_0 + \varepsilon h' + 0\left(\varepsilon^2\right) + \cdots$$

where H_0 is the spherical shell model Hamiltonian. Furthermore $\varepsilon h'$ is given as

$$\varepsilon h' = \varepsilon\frac{M}{2}\omega_0^2\frac{2}{3}\left(x^2 + y^2 - 2z^2\right) = -\frac{M}{2}\omega_0^2\frac{4}{3}\varepsilon r^2 P_2(\cos\theta)$$

As shown in preceding chapters we may write the eigenfunctions in the spherical case as

$$\phi(N\ell sj\Omega) = R_{N\ell}(r)\sum_{\Lambda\Sigma}\langle\ell s\Lambda\Sigma|\ell sj\Omega\rangle\, Y_{\ell\Lambda}\chi_{s\Sigma}$$

where the constants of the motion are $\mathbf{j}^2$ and Ω, the total angular momentum and its z-component, and furthermore ℓ^2 and s^2, the orbital and spin angular momenta. The projections of the orbital and spin angular momenta are denoted by Λ and Σ, respectively.

In the spherical case each j state is $(2j+1)$-fold degenerate. This degeneracy is removed by the perturbation h' to first order as (see problem 8.1)

$$\langle N\ell sj\Omega|\varepsilon h'|N\ell sj\Omega\rangle = \frac{1}{6}\varepsilon M\omega_0^2\left\langle r^2\right\rangle\frac{3\Omega^2 - j(j+1)}{j(j+1)}$$

This result of the deformation of the field is easily understood qualitatively. For a so-called prolate distortion ($\varepsilon > 0$) of the field, matter is removed from the 'waistline' and placed at the 'poles'. This corresponds to a softer potential in the z-direction and a steeper potential in the perpendicular x- and y-directions where the equatorial orbitals with $\Omega \simeq j$ are mainly located. Classically, this is understood from the fact that the $\Omega \simeq j$ angular momentum vector is almost parallel to the z-axis and the particle moves in a plane perpendicular to this vector. Consequently, the $\Omega \simeq j$ orbitals move up in energy. On the other hand, the polar orbitals with $\Omega \ll j$ are associated with a negative energy contribution for $\varepsilon > 0$. They are thus favoured by a prolate deformation, i.e. they move down in energy. For an oblate distortion the opposite is true, i.e. the large Ω-values are suppressed energywise (cf. the splitting of the j-shells for small distortions in fig. 8.3 below).

8.2 Asymptotic wave functions

Before we discuss the case of moderate deformations of $\varepsilon \simeq 0.2-0.3$, acquired by most deformed nuclei, we shall now consider the limit of very large ε-values. Beyond very small ε-values, say $\varepsilon \simeq 0.1$, the exhibited perturbation treatment of the ε-term is no longer applicable. Instead one may at large ε introduce a representation that exactly diagonalises the harmonic oscillator field while instead the ℓ^2 and $\ell \cdot \mathbf{s}$ terms are treated as perturbations.

Let us write

$$H = H_{\text{osc}} + H'$$

where

$$H_{\text{osc}} = -\frac{\hbar^2}{2M}\Delta + \frac{M}{2}\left[\omega_\perp^2\left(x^2 + y^2\right) + \omega_z^2 z^2\right]$$

It is now convenient to introduce what one may call 'stretched' coordinates (Nilsson, 1955)

$$\xi = x\left(\frac{M\omega_\perp}{\hbar}\right)^{1/2} , \quad \eta = y\left(\frac{M\omega_\perp}{\hbar}\right)^{1/2} , \quad \zeta = z\left(\frac{M\omega_z}{\hbar}\right)^{1/2}$$

Thus

$$H_{\text{osc}} = \frac{1}{2}\hbar\omega_\perp\left[-\left(\frac{\partial^2}{\partial\xi^2} + \frac{\partial^2}{\partial\eta^2}\right) + \left(\xi^2 + \eta^2\right)\right] + \frac{1}{2}\hbar\omega_z\left(-\frac{\partial^2}{\partial\zeta^2} + \zeta^2\right)$$

In the spherical case we added correction terms of the type

$$H'_{\text{sph}} = -2\kappa\hbar\omega_0\ell \cdot \mathbf{s} - \mu'\hbar\omega_0\left(\ell^2 - \left\langle\ell^2\right\rangle_N\right)$$

Introducing an ℓ_t corresponding to the stretched coordinates ξ, η and ζ, where e.g.

$$(\ell_t)_x = -i\hbar \left(\eta \frac{\partial}{\partial \zeta} - \zeta \frac{\partial}{\partial \eta} \right)$$

it seems natural to generalise H' into

$$H'_{\text{def}} = -2\kappa\hbar\omega_0 \ell_t \cdot \mathbf{s} - \mu'\hbar\omega_0 \left(\ell_t^2 - \left\langle \ell_t^2 \right\rangle_N \right)$$

One may show (Nilsson, 1955) that the difference terms between H'_{sph} and H'_{def} have no matrix elements within each N-shell. There exist alternatives but no really definite prescription for how the terms ℓ^2 and $\ell \cdot \mathbf{s}$ are to be generalised in going from the spherical to the distorted case (see e.g. Bengtsson, 1975). The presently employed recipe of simply replacing ℓ by ℓ_t is thus associated with a high degree of arbitrariness.

Consider first the oscillator part of H given above as H_{osc}. To proceed with the case of cylinder symmetry only, H_{osc} is transformed by going over to cylindrical coordinates (ρ, φ, ζ) where

$$\xi = \rho \cos \varphi$$

$$\eta = \rho \sin \varphi$$

(contrary to above, ρ is here the 'cylinder radius' $\rho = (\xi^2 + \eta^2)^{1/2}$). We can write the Schrödinger equation in terms of these coordinates:

$$\left[\frac{1}{2}\hbar\omega_\perp \left(-\frac{1}{\rho}\frac{\partial}{\partial\rho}\rho\frac{\partial}{\partial\rho} - \frac{1}{\rho^2}\frac{\partial^2}{\partial\varphi^2} + \rho^2 \right) + \frac{1}{2}\hbar\omega_z \left(-\frac{\partial^2}{\partial\zeta^2} + \zeta^2 \right) - E \right] \psi = 0$$

We separate off the φ-dependence by the assumption

$$\psi = U(\rho)Z(\zeta)\phi(\varphi)$$

where

$$-\frac{\partial^2}{\partial\varphi^2}\phi = \Lambda^2\phi$$

with the solution

$$\phi = e^{i\Lambda\varphi}$$

corresponding to the fact that $[L_z, H] = 0$ and thus $L_z = \Lambda$ a constant of the motion. We can also separate off the ζ-dependence and obtain

$$\hbar\omega_z \left(-\frac{\partial^2}{\partial\zeta^2} + \zeta^2 \right) Z(\zeta) = E_z Z(\zeta)$$

with the usual one-dimensional harmonic oscillator solution where $E_z =$

$(n_z + 1/2)\,\hbar\omega_z$. Finally, with $E = E_\perp + E_z$, one ends up with the following equation for $U(\rho)$:

$$\frac{1}{2}\hbar\omega_\perp \left(-\frac{1}{\rho}\frac{\partial}{\partial\rho}\rho\frac{\partial}{\partial\rho} + \frac{\Lambda^2}{\rho^2} + \rho^2 \right) U(\rho) = E_\perp U(\rho)$$

or

$$\frac{d^2 U}{d\rho^2} + \frac{1}{\rho}\frac{dU}{d\rho} - \frac{\Lambda^2}{\rho^2}U - \rho^2 U = -\frac{2E_\perp}{\hbar\omega_\perp} \cdot U$$

From the behaviour in the limiting cases $\rho \to 0$ and $\rho \to \infty$ it is natural to assume

$$U = \rho^{|\Lambda|}e^{-\rho^2/2}W(\rho)$$

Hence W fulfils

$$W'' + \left(\frac{2|\Lambda|+1}{\rho} - 2\rho \right) W' - 2\left(|\Lambda| + 1 - \frac{E_\perp}{\hbar\omega_\perp} \right) W = 0$$

After the substitution

$$z = \rho^2$$

we obtain

$$zW'' + (|\Lambda| + 1 - z)\,W' - \frac{1}{2}\left(|\Lambda| + 1 - \frac{E_\perp}{\hbar\omega_\perp} \right) W = 0$$

This should be compared with the equation obtained in the spherical oscillator case (chapter 6) with the 'ansatz' $R_\ell(\rho) = \rho^\ell e^{-\rho^2/2}f(\rho)$, where

$$\rho f'' + (\ell + 3/2 - z)\,f' + \frac{1}{2}\left(\frac{E}{\hbar\omega} - (\ell + 3/2) \right) f = 0$$

The differential equation for W is, similarly to the equation for f, a Kummer equation and the solution is a confluent hypergeometric function

$$W = F\left(\frac{1}{2}\left(|\Lambda| + 1 - E_\perp/\hbar\omega_\perp \right),\ |\Lambda| + 1\ ;\ z \right)$$

with the (finiteness) condition

$$\frac{1}{2}\left(|\Lambda| + 1 - E_\perp/\hbar\omega_\perp \right) = -n_\rho\ ;\quad n_\rho = 0, 1, 2, \ldots$$

or

$$E_\perp = \hbar\omega_\perp \left(2n_\rho + |\Lambda| + 1 \right) = \hbar\omega_\perp \left(n_\perp + 1 \right)$$

The quantity n_ρ is thus the number of (cylinder) radial nodes and $n_\perp$ is the total number of oscillator quanta in the x- and y-directions, $n_\perp = n_x + n_y$. Note that

$$2n_\rho + |\Lambda| = n_\perp$$

Thus

$$|\Lambda| = n_\perp, \ n_\perp - 2, \ n_\perp - 4 \ldots 0 \text{ or } 1$$

We may now sum up the final results for the cylindrical oscillator case

$$E(n_z, n_\perp) = \hbar\omega_z \left(n_z + \frac{1}{2}\right) + \hbar\omega_\perp (n_\perp + 1) = \hbar\omega_0 \left(N + \frac{3}{2} + (n_\perp - 2n_z)\frac{\varepsilon}{3}\right)$$

$$\psi(n_z, n_\perp, \Lambda) = C \cdot \mathrm{e}^{-\zeta^2/2} H_{n_z}(\zeta) \rho^{|\Lambda|} \mathrm{e}^{-\rho^2/2} F\left(-\frac{n_\perp - |\Lambda|}{2}, |\Lambda| + 1; \rho^2\right) \cdot \mathrm{e}^{\mathrm{i}\Lambda\varphi}$$

Let us see how our oscillator levels look with distortion (fig. 8.1). Obviously each N shell, $N = n_\perp + n_z$, is split into $N + 1$ levels corresponding to $n_\perp = 0, 1, \ldots, N$. The degeneracy of each level is $2 \times (n_\perp + 1)$ corresponding to the two spin values $\Sigma = \pm 1/2$ and the $n_\perp + 1$ different Λ-values for each $n_\perp$ (for example $\Lambda = 2, 0$ and -2 for $n_\perp = 2$).

We may note in passing that, although the spherical shell structure is lost when $\varepsilon \neq 0$, new shells are re-appearing e.g. at $\varepsilon = 0.6$, $\varepsilon = 1$ and $\varepsilon = -0.75$ etc. Thus for $\varepsilon = 0.6$ we have $\hbar\omega_\perp = 2\hbar\omega_z$, i.e. we can replace one quantum $\hbar\omega_\perp$ with two quanta $\hbar\omega_z$ without change of energy just as for spherical symmetry $\hbar\omega_x$, $\hbar\omega_y$ and $\hbar\omega_z$ are interchangeable (cf. problem 5.1).

This covers the pure oscillator Hamiltonian. There remains, however, the $\ell_t \cdot \mathbf{s}$ and ℓ_t^2 terms. One can actually treat these by first-order perturbation theory. The matrix elements of these terms are evaluated by operator methods in the last section of this chapter. For the perturbation treatment, we only need the diagonal terms:

$$\langle N n_z \Lambda\Sigma | \ell_t \cdot \mathbf{s} | N n_z \Lambda\Sigma \rangle = \Lambda\Sigma$$

$$\langle N n_z \Lambda\Sigma | \ell_t^2 | N n_z \Lambda\Sigma \rangle = \Lambda^2 + 2n_\perp n_z + 2n_z + n_\perp$$

The diagonal value of the total Hamiltonian then becomes

$$\langle N n_z \Lambda\Sigma | H_{\mathrm{osc}} - 2\kappa\hbar\omega_0\ell_t \cdot \mathbf{s} - \mu'\hbar\omega_0 \left(\ell_t^2 - \langle \ell_t^2 \rangle_N\right) | N n_z \Lambda\Sigma \rangle$$

$$= \left(N + \frac{3}{2}\right)\hbar\omega_0 + \frac{1}{3}\varepsilon\hbar\omega_0 (N - 3n_z) - 2\kappa\hbar\omega_0\Lambda\Sigma$$

$$- \mu'\hbar\omega_0 \left(\Lambda^2 + 2n_\perp n_z + 2n_z + n_\perp - \frac{N(N+3)}{2}\right)$$

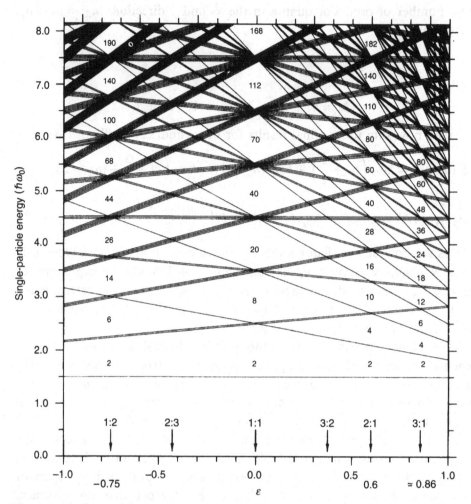

Fig. 8.1. Single-particle energies of the axially symmetric harmonic oscillator potential drawn as functions of the deformation parameter, $\varepsilon = (\omega_\perp - \omega_z)/\omega_0$. The orbitals that stay degenerate for all ε-values are drawn with a small spacing to indicate the total level density more clearly. The high degeneracy for spherical shape is partly broken for $\varepsilon \neq 0$ but is then largely regained for $\omega_\perp : \omega_z$ corresponding to small integer numbers where the most important ones are indicated in the lower part of the figure. For $\omega_\perp : \omega_z = $ 1:2, 1:1, 2:1, and 3:1, the corresponding magic numbers are given.

The effect of the inclusion of the $\ell_t \cdot \mathbf{s}$ and ℓ_t^2 terms in perturbation approximation is thus a lifting of the $2 \times (n_\perp + 1)$-fold degeneracy. After the inclusion of these terms only a two-fold degeneracy (the time-reversal degeneracy) remains. The splitting can be seen from fig. 8.2, where first the $\ell_t \cdot \mathbf{s}$ term and subsequently the ℓ_t^2 term have been added to an $N = 5$

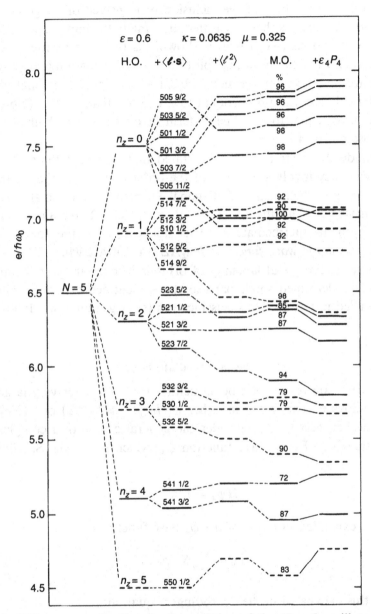

Fig. 8.2. Different contributions to the splitting of the $N = 5$ oscillator shell are illustrated. First the pure oscillator is deformed to $\varepsilon = 0.6$, then diagonal matrix elements of the $\ell \cdot \mathbf{s}$ and ℓ^2-terms are added to the $|Nn_z\Lambda\Omega\rangle$ basis. Finally, to the right, an exact diagonalisation was carried out, with $\varepsilon_4 = 0$ and 0.07 respectively. The latter value of ε_4 corresponds approximately to the liquid drop valley in the actinide region. The κ and μ parameters are chosen for neutrons in this same region. In the case of $\varepsilon_4 = 0$, the pureness of the asymptotic representation is indicated (from Ragnarsson *et al.*, 1978).

shell of orbitals. The effect of the inclusion of neglected off-diagonal terms (exact diagonalisation) is then illustrated and finally the importance of hexa-decapole deformations (see below) is shown. As the $\ell_t \cdot \mathbf{s}$ term enters with a negative coefficient the resulting splitting is such that orbitals with Λ and Σ parallel are favoured. Furthermore the spin–orbit splitting is smaller than in the spherical case and proportional to 2Λ rather than $2\ell + 1$. (Note that Λ is at the most equal to ℓ_t.) The ℓ_t^2-term favours states of high Λ-values among those of given n_z and $n_\perp$.

To conclude this section, it may be stated that, although the splitting of the $n_\perp$ degeneracy involves energies of the order of $h\omega_0$, i.e. the separation between spherical shells, still off-diagonal elements are so small that the so called *asymptotic quantum numbers* N, n_z, Λ and Σ remain essentially preserved even at intermediate ε. This holds also for the spectroscopic data. The *asymptotic wave functions* $Nn_z\Lambda\Sigma$ (or $Nn_z\Lambda\Omega$ where $\Omega = \Lambda + \Sigma$) have become a very useful labelling of orbitals in nuclear spectroscopy. In fact nowadays deformed single-particle states identified experimentally are usually labelled in terms of this classification (see e.g. Jain *et al.*, 1990).

8.3 The intermediate region

In the intermediate region of ε being neither very small nor very large one usually employs an expansion in either $\phi(N\ell j\Omega)$, $\phi(N\ell\Lambda\Sigma)$ or $\phi(Nn_z\Lambda\Omega)$. Let us, however, first briefly consider the general case with a complete set of basis states ϕ_ν. Every wave function ψ_α, which solves the Schrödinger equation:

$$H\psi_\alpha = E_\alpha\psi_\alpha$$

can now be expanded in terms of the ϕ_ν base functions

$$\psi_\alpha = \sum_\nu S_{\alpha\nu}\phi_\nu$$

We insert this expansion in the Schrödinger equation:

$$\sum_\nu S_{\alpha\nu}H\phi_\nu = E_\alpha\sum_\nu S_{\alpha\nu}\phi_\nu$$

Multiplying from the left by ϕ_μ^* and integrating (taking the scalar product) we obtain

$$\sum_\nu S_{\alpha\nu}\int \phi_\mu^* H\phi_\nu \, \mathrm{d}\tau = E_\alpha\sum_\nu S_{\alpha\nu}\delta_{\mu\nu}$$

Denoting

$$\int \phi_\mu^* H \phi_\nu \, d\tau = \langle \mu | H | \nu \rangle = H_{\mu\nu}$$

we have thus a set of equations, one for each μ

$$\sum_\nu S_{\alpha\nu} \left(H_{\mu\nu} - E_\alpha \delta_{\mu\nu} \right) = 0$$

The condition that a solution exists is that the determinant vanishes:

$$\text{Det} \left(H_{\mu\nu} - E_\alpha \delta_{\mu\nu} \right) = 0$$

The problem is two-fold, first to calculate the matrix elements $H_{\mu\nu} = \int \phi_\mu^* H \phi_\nu \, d\tau$ and second to 'diagonalise' the matrix or in principle find the roots that make $\text{Det} \left(H_{\mu\nu} - E_\alpha \delta_{\mu\nu} \right)$ vanish. Technically the computer programs constructed find a transformation matrix $S_{\alpha\mu}$ such that the transformed H-matrix is diagonal.

The set of equations can be written:

$$E_\alpha S_{\alpha\mu} = \sum_\nu S_{\alpha\nu} H_{\mu\nu}$$

or in matrix form

$$E \, S = S \, H$$

where E is a diagonal matrix:

$$E = \begin{pmatrix} E_1 & 0 & 0 & \dots \\ 0 & E_2 & 0 & \\ 0 & 0 & E_3 & \\ & & & \vdots \end{pmatrix}$$

The eigenvalues E_α are now given as

$$E = S \, H \, S^{-1}$$

and the eigenvectors are the rows of the transformation matrix S.

Let us now return to the practical case. We assume a wave function

$$\psi_{\alpha N \Omega} = \sum_{\ell \Lambda \Sigma} a_{N\ell\Lambda\Sigma} R_{N\ell}(\rho) Y_{\ell\Lambda}(\theta_t, \varphi_t) \chi_{s\Sigma} = \sum_{\ell \Lambda \Sigma} a_{N\ell\Lambda\Sigma} |N\ell\Lambda\Sigma\rangle$$

where we assume that $R_{N\ell}$ is normalised. The spherical harmonics $Y_{\ell\Lambda}$ are already normalised. The spin wave function $\chi_{s\Sigma}$ is simply

$$\begin{pmatrix} 1 \\ 0 \end{pmatrix} \quad \text{or} \quad \begin{pmatrix} 0 \\ 1 \end{pmatrix}$$

The quantity ρ is now defined (in contradistinction to the cylindrical case) as the radius in the stretched coordinates,

$$\rho^2 = \xi^2 + \eta^2 + \zeta^2$$

while θ_t and φ_t are the polar and azimuthal angles in these coordinates. Note also that all the quantum numbers, $N, \ell, \ldots$ are defined in the stretched coordinates and should more appropriately be denoted $N_t, \ell_t, \ldots$. In these cases, we will however drop the index 't'.

The coefficients $a_{N\ell\Lambda\Sigma}$ are our unknowns. The assumption is that any wave function can be expressed in terms of these basis vectors $|N\ell\Lambda\Sigma\rangle$. These vectors form a complete basis but the problem is that our computer can only accept a finite number of basis states. Usually, we limit ourselves to the lowest N-values.

There are restrictions on the sum fortunately. First we note that $j_z = \ell_z + s_z = (\ell_t)_z + s_z$ (with $\omega_x = \omega_y = \omega_\perp$, it follows that $\ell_z = (\ell_t)_z$) is a constant of the motion as (cf. problem 8.4)

$$[j_z, H] = 0$$

Indeed $(\ell_t)_z$ and s_z both commute with the cylindrically symmetrical H_{osc} and also with ℓ_t^2. With $\ell_t \cdot s$ only the sum $(\ell_t)_z + s_z$ commutes. In the wave function expansion only two Λ values can occur as we have to fulfill $\Lambda + \Sigma = \Omega$. Thus we can have $\Lambda = \Omega - 1/2$ and $\Lambda = \Omega + 1/2$ corresponding to $\Sigma = +1/2$ and $\Sigma = -1/2$.

It is convenient to rewrite the Hamiltonian in the following form

$$H = H_d + H_\varepsilon$$

where H_d is a part of the Hamiltonian that is rotationally symmetric, in the stretched coordinates, (ξ, η, ζ);

$$H_d = \frac{1}{2}\hbar\omega_0 \left(-\Delta_\xi + \rho^2\right) - 2\kappa\hbar\omega_0 \ell_t \cdot s - \mu'\hbar\omega_0 \left(\ell_t^2 - \left\langle \ell_t^2 \right\rangle_N\right)$$

We have furthermore

$$H_\varepsilon = \frac{1}{6}\varepsilon\hbar\omega_0 \left(-\frac{\partial^2}{\partial\xi^2} - \frac{\partial^2}{\partial\eta^2} + 2\frac{\partial^2}{\partial\zeta^2} + \xi^2 + \eta^2 - 2\zeta^2\right)$$

The quantity H_ε now has the remarkable property that its matrix elements vanish identically between base states with different N quantum numbers. Between *states belonging to one N-shell* (or rather one N_t-shell) one may prove (see appendix 8) that the matrix elements are identical to those of

$$H_\varepsilon' = \frac{1}{3}\varepsilon\hbar\omega_0 \left(\xi^2 + \eta^2 - 2\zeta^2\right) = -\frac{2}{3}\varepsilon\hbar\omega_0\rho^2 P_2(\cos\theta_t)$$

In the expansion of $\psi_{\alpha N\Omega}$, we used the fact that there are no matrix elements between states of different N or different Ω. The sum thus runs only over ℓ and Λ (with Σ determined from $\Sigma = \Omega - \Lambda$) and we can limit our diagonalisation to one N-shell and one Ω-value at a time. The index α is used to distinguish states, $\psi_{\alpha N\Omega}$ having the same N- and Ω-values.

The matrix elements are simple to generate within the computer or to calculate by hand. We have thus

$$\langle N\ell\Lambda\Sigma | H_d | N\ell\Lambda\Sigma \rangle = \frac{1}{2}\hbar\omega_0 \left(N + \frac{3}{2} \right)$$

$$- 2\kappa\hbar\omega_0\Lambda\Sigma - \mu'\hbar\omega_0 \left(\ell(\ell+1) - \frac{N(N+3)}{2} \right)$$

For the $\ell \cdot$ s-term, we note that $\ell \cdot \mathbf{s} = \frac{1}{2}(\ell_+ s_- + \ell_- s_+) + \ell_z s_z$ and obtain

$$\langle N\,\ell\,\Lambda+1\,\Sigma-1 \,| \ell_t \cdot \mathbf{s} | N\,\ell\,\Lambda\,\Sigma \rangle = \frac{1}{2}[(\ell-\Lambda)(\ell+\Lambda+1)]^{1/2}$$

$$\langle N\,\ell\,\Lambda-1\,\Sigma+1 \,| \ell_t \cdot \mathbf{s} | N\,\ell\,\Lambda\,\Sigma \rangle = \frac{1}{2}[(\ell+\Lambda)(\ell-\Lambda+1)]^{1/2}$$

Noting that H'_ε may be written $H'_\varepsilon = -\hbar\omega_0\varepsilon\frac{4}{3}(\pi/5)^{1/2}\rho^2 Y_{20}$:

$$\langle N\ell'\Lambda'\Sigma' | H_\varepsilon | N\ell\Lambda\Sigma \rangle = \langle N\ell'\Lambda'\Sigma' | H'_\varepsilon | N\ell\Lambda\Sigma \rangle$$

$$= -\frac{2}{3}\varepsilon\hbar\omega_0\delta_{\Lambda\Lambda'}\delta_{\Sigma\Sigma'} \int R_{N\ell'}R_{N\ell}\rho^2\,\mathrm{d}\rho \cdot \left(\frac{2\ell+1}{2\ell'+1} \right)^{1/2} C^{\ell 2\ell'}_{\Lambda 0\Lambda'} C^{\ell 2\ell'}_{000}$$

We can thus easily construct the matrix and use the computer to diagonalise it. As already stated, the eigenvalues and eigenvectors both come out as a result. An example of a calculated single-particle diagram is provided in fig. 8.3.

If $\psi_{\alpha N\Omega}$ is expanded in normal coordinates x, y, z instead of the stretched set ξ, η, ζ, there enter components with $N\pm 2$, $N\pm 4$. Their amplitude is not negligible and it is certainly only together with stretched coordinates that the use of pure N-shells (N_t-shells) can be justified.

8.4 The volume conservation condition

In the oscillator calculations the absolute energy scale is given by $\hbar\omega_0$. The energy scale is as well as its ε-dependence, however, irrelevant in establishing the single-particle level order. On the other hand, for a determination of the equilibrium distortion (chapter 9) the volume conservation condition is of primary importance. This condition is based on the empirical finding

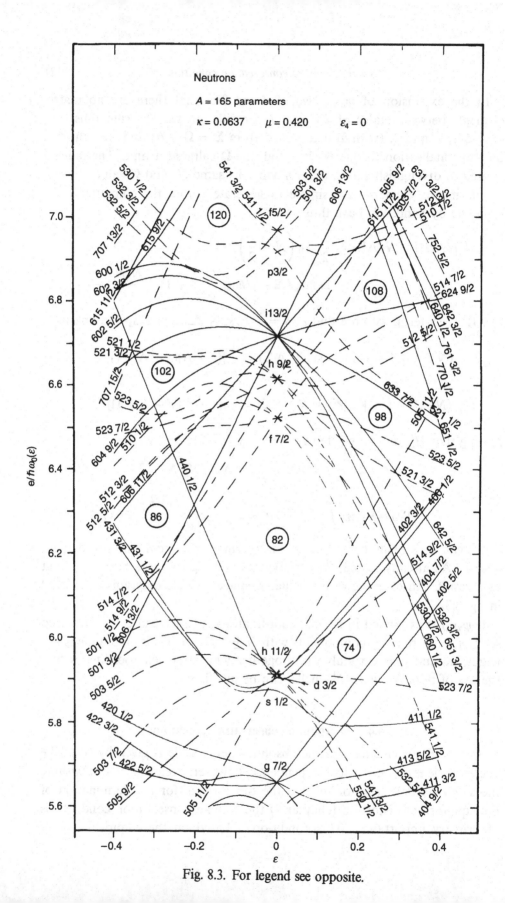

Fig. 8.3. For legend see opposite.

that nuclear matter is incompressible and the nuclear volume proportional to A. The short-range character of the nuclear force then appears to justify the condition that the volume enclosed by a given equipotential surface (the locus of points of the same nuclear potential) is preserved. For the simple harmonic oscillator potential an equipotential surface is obviously an ellipsoid.

$$V_{osc} = \frac{1}{2}M\left(\frac{x^2+y^2}{1/\omega_\perp^2} + \frac{z^2}{1/\omega_z^2}\right) = V_0 = \text{constant}$$

The half-axes in this ellipsoid are given as proportional to $\omega_\perp^{-1}$ and ω_z^{-1} and the volume enclosed is thus proportional to $\omega_\perp^{-2}\omega_z^{-1}$. The remarkable feature for the pure oscillator is that for any equipotential surface (any value of V_0) we obtain the same volume conservation condition

$$\omega_\perp^2 \omega_z = \text{constant} = (\omega_0(\varepsilon=0))^3 = \left(\overset{0}{\omega}_0\right)^3$$

Using the definitions cited for ω_z and $\omega_\perp$ we obtain

$$\omega_0(\varepsilon) = \overset{0}{\omega}_0\left(1 - \frac{1}{3}\varepsilon^2 - \frac{2}{27}\varepsilon^3\right)^{-1/3} = \overset{0}{\omega}_0\left(1 + \frac{1}{9}\varepsilon^2 + \ldots\right)$$

All the energies in a level diagram such as fig. 8.3 are conveniently expressed in the ε-dependent energy unit $\hbar\omega_0(\varepsilon)$. Even though some energy levels appear wildly down-sloping as a function of ε in the $\hbar\omega_0$ scale, when expressed in the ε-independent unit $\hbar\overset{0}{\omega}_0$ they eventually bend upwards for large enough ε. Thus the ellipsoidal shape at very large distortions ultimately becomes unfavourable for any combination of nuclear orbitals filled.

Fig. 8.3. (*opposite*) Single-neutron orbitals calculated from the modified oscillator potential for nuclei in the mass region $150 < A < 190$ as a function of the ellipsoidal deformation parameter ε. Solid and dashed lines are used to distinguish between orbitals having even parity (even N) and odd parity (odd N), respectively. For large ε-values, the orbitals are labelled by the asymptotic quantum numbers $Nn_z\Lambda\Omega$, which, of course, are only approximate. N refers to the oscillator shell quantum number, n_z to the number of nodes along the z-axis, Λ to the value of the orbital angular momentum along the intrinsic z-axis and Ω to the value of the total angular momentum along the same axis. The spin projection along this axis is implied as $\Sigma = \Omega - \Lambda$.

8.5 Single-particle orbitals in an axially symmetric modified oscillator potential

The nuclear potential we have worked with so far can be written

$$V = V_{\text{osc}} + V'$$

$$V_{\text{osc}} = \frac{1}{2}\hbar\omega_0\rho^2\left(1 - \frac{2}{3}\varepsilon P_2\left(\cos\theta_t\right)\right)$$

$$V' = -2\kappa\hbar\overset{0}{\omega}_0\,\ell_t\cdot\mathbf{s} - \mu'\hbar\overset{0}{\omega}_0\left(\ell_t^2 - \left\langle\ell_t^2\right\rangle_N\right)$$

Here, we have by convention (Nilsson, 1955) used the volume conserved frequency, $\overset{0}{\omega}_0$, together with the $\ell_t\cdot$ s- and ℓ_t^2-terms. To be able to describe nuclear shapes other than ellipsoidal ones, the potential must be generalised in some way. A natural generalisation is (Nilsson *et al.*, 1969)†

$$V_{\text{osc}} = \frac{1}{2}\hbar\omega_0\rho^2\left(1 - \frac{2}{3}\varepsilon P_2\left(\cos\theta_t\right) + 2\sum_{\lambda=3,4,\ldots}\varepsilon_\lambda P_\lambda\left(\cos\theta_t\right)\right)$$

With this potential, we preserve the property that for any equipotential surface, $V_{\text{osc}} = V_0$, the same volume conservation condition is obtained. However, the ratio $\omega_0\left(\varepsilon, \varepsilon_3, \varepsilon_4, \ldots\right)/\overset{0}{\omega}_0$ must be calculated numerically (see e.g. Bengtsson, Ragnarsson and Åberg, 1991).

The shape of an equipotential surface is obtained as

$$\rho^2 = \frac{\text{constant}}{\left(1 - \frac{2}{3}\varepsilon P_2\left(\cos\theta_t\right) + 2\sum_\lambda\varepsilon_\lambda P_\lambda\left(\cos\theta_t\right)\right)}$$

Using the definition of the stretched coordinates, we obtain a relation between the stretched and unstretched radii:

$$\rho^2 = \frac{\omega_\perp M}{\hbar}\left(x^2 + y^2\right) + \frac{\omega_z M}{\hbar}z^2 = \frac{M\omega_0}{\hbar}r^2\left(1 - \frac{2\varepsilon}{3}P_2\left(\cos\theta\right)\right)$$

Here we will only calculate r to first order in the distortion coordinates. It is then possible to neglect the difference between $\cos\theta$ and $\cos\theta_t$ ($\cos\theta_t = (\xi^2 + \eta^2)^{1/2}/\rho$) to obtain

$$r = r_0\left(1 + \frac{2\varepsilon}{3}P_2 - \sum_\lambda\varepsilon_\lambda P_\lambda + \ldots\right)$$

This expression, which should only be used for small values of ε, ε_3 and ε_4,

---

† The quadrupole deformation parameter is sometimes referred to as $\varepsilon_2 \equiv \varepsilon$. Here we have used ε to indicate that the constant of the εP_2 term is different from that of the $\varepsilon_\lambda P_\lambda$, $\lambda \geq 3$, terms. Furthermore, the 'Cartesian coordinates' in the (ε, γ) plane are sometimes denoted $(\varepsilon_2, \varepsilon_{22})$ see e.g. Larsson (1973).

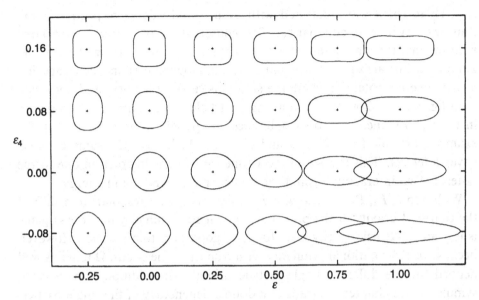

Fig. 8.4. The equipotential shapes generated by the parameters ε and ε_4 of the modified oscillator potential (from Nilsson *et al.*, 1969).

..., describes the same shapes as the coordinates β_λ (or $\alpha_{\lambda 0}$) used in chapter 4. Thus, to first order, one obtains for example

$$\varepsilon \approx \frac{3}{2}\left(\frac{5}{4\pi}\right)^{1/2}\beta_2 \approx 0.95\beta_2 \quad \text{(quadrupole deformation)}$$

$$\varepsilon_3 \approx (-)\left(\frac{7}{4\pi}\right)^{1/2}\beta_3 \approx (-)0.75\beta_3 \quad \text{(octupole deformation)}$$

$$\varepsilon_4 \approx -\left(\frac{9}{4\pi}\right)^{1/2}\beta_4 \approx -0.85\beta_4 \quad \text{(hexadecapole deformation)}$$

Higher order expressions have been published e.g. by Bengtsson *et al.* (1989). At large distortions, the two parametrisations become rather different. The shapes generated in the $(\varepsilon, \varepsilon_4)$ parametrisation are illustrated in fig. 8.4. Observe that, if only $\varepsilon \neq 0$, one obtains pure spheroids, in contrast to the β_4-parametrisation with only $\beta_2 \neq 0$ (see problem 4.2).

In the Legendre polynomials, the coordinates ξ and η (or x and y) enter in a symmetric way (independence of φ_t or φ) and it follows that $[H, j_z] = 0$. Thus, Ω remains as a good quantum number. On the other hand, the different (stretched) N_t-shells are no longer uncoupled. One finds instead that, with λ even, P_4, P_6, ..., only the shells with $\Delta N_t = 2$ couple. This means that the even shells, $N_t = 0, 2, 4, \ldots$, do not mix with the odd shells, $N_t = 1, 3, 5,$

.... These shells are associated with even ℓ and odd ℓ, respectively, and thus with even and odd parity. The conservation of parity is associated with the fact that the potential is mirror symmetric, not only in the z–x and z–y-plane, but also in the x–y-plane. The analogous case in one dimension is a symmetric potential where the solutions separate in even and odd ones.

In fig. 8.5, we show the proton single-particle orbitals drawn along a path in the $(\varepsilon, \varepsilon_4)$-plane. A distinct difference compared with fig. 8.3 is that, for example, orbitals from $N = 5$ and $N = 7$, which have the same Ω-value, never intersect. This is a simple consequence of their interaction due to the ε_4-term. Two interacting orbitals never become degenerate in energy.

With odd λ, P_3, P_5, ... (and also P_1 which mainly corresponds to a shift of the centre of gravity of the potential), the mirror symmetry in the x–y-plane is lost and (intrinsic) parity is no longer a good quantum number. However, because of the rotational symmetry with respect to the z-axis, Ω remains well defined for the different single-particle orbitals. In addition, time-reversal symmetry (see chapter 13) leads to a double degeneracy of the single-particle energies. Thus, all orbitals are filled by two particles 'moving in different directions' (corresponding to $j_z = \Omega$ and $j_z = -\Omega$, respectively).

The calculation of the matrix elements of the $r^2 P_\lambda(\cos\theta_t)$ terms in a $|N\ell\Lambda\Sigma\rangle$ (or $|N\ell j\Omega\rangle$) basis involves one radial and one angular integral. The latter is solved exactly analogous to the $P_2(\cos\theta_t)$ term and results in a sum over Clebsch–Gordan coefficients. The matrix elements of ρ^2 can be given in closed form if the properties of the confluent hypergeometric functions are explored. The formula has been given for example by B. Nilsson (1969). When using such formulae, care must be taken that the same phase convention is used througout.

8.6 Triaxial nuclear shapes – the anisotropic harmonic oscillator potential

For description of the properties of really well-deformed nuclei, it seems to be a good approximation to consider only axially symmetric nuclear shapes. Thus, most of the early calculations (e.g. Mottelson and Nilsson, 1959a; Nilsson *et al.*, 1969; Brack *et al.*, 1972) on deformed nuclei were restricted to such shapes. However, for description of nuclei that are not so strongly deformed (transitional nuclei) it is in many cases necessary to consider also axially asymmetric (triaxial) shapes. Furthermore, it was pointed out in chapter 4 that the (inner) fission barrier of a large number of actinide nuclei is lowered by 1–2 MeV for a fission path involving triaxial shapes. It could also be mentioned that, even if a nucleus is axially symmetric in the ground state, it might change its deformation if it begins to rotate and thus

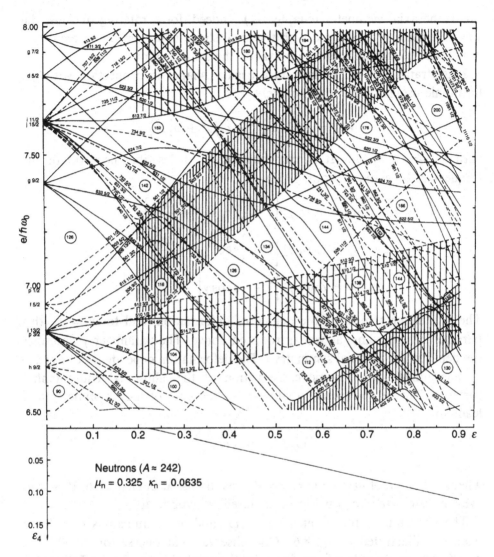

Fig. 8.5. Single-neutron levels in the actinide region drawn as functions of the set of prolate shapes defined in terms of ε and ε_4 as noted below the figure. The $\varepsilon_4\rho^2 Y_{40}$ term couples the different N-shells, $N' = N \pm 2$, which means that orbitals having the same parity and the same Ω-value never cross. Regions of almost degenerate orbitals with $n_z = 0$ and $n_z = 1$ are indicated. These orbitals lead to a high level density for special deformations and particle numbers and are largely responsible for the octupole ε_3-deformations at large ε-values as discussed in chapter 9 (from Ragnarsson *et al.*, 1978).

break the axial symmetry (chapter 12). Indeed, for a rotating nucleus, the special symmetry associated with the axial shape is broken. Thus, axial and non-axial deformations enter on exactly the same footing.

It is straightforward to introduce triaxial shapes in connection with the harmonic oscillator potential, namely

$$V_{\text{osc}} = \frac{1}{2}M\left(\omega_x^2 x^2 + \omega_y^2 y^2 + \omega_z^2 z^2\right)$$

with $\omega_x \neq \omega_y \neq \omega_z$. It is then customary to describe the ratio of the frequencies by ε and γ (Bohr, 1952):

$$\omega_x = \omega_0(\varepsilon, \gamma)\left[1 - \frac{2}{3}\varepsilon \cos\left(\gamma + \frac{2\pi}{3}\right)\right]$$

$$\omega_y = \omega_0(\varepsilon, \gamma)\left[1 - \frac{2}{3}\varepsilon \cos\left(\gamma - \frac{2\pi}{3}\right)\right]$$

$$\omega_z = \omega_0(\varepsilon, \gamma)\left[1 - \frac{2}{3}\varepsilon \cos\gamma\right]$$

The parameter ε specifies the degree of deformation, and for $\gamma = 0°$, this parameter is the same as was introduced above. The parameter γ describes the departure from axial symmetry.† The deformation dependence of $\omega_0(\varepsilon, \gamma)$ is determined from volume conservation of the ellipsoidal equipotential surfaces. This is completely analogous to the case of axial symmetry and leads to

$$\omega_x\omega_y\omega_z = \left(\overset{0}{\omega}_0\right)^3$$

where the value of $\overset{0}{\omega}_0 \equiv \omega_0(\varepsilon = \gamma = 0)$ was discussed in chapter 6. It is thus evident that $\omega_0(\varepsilon, \gamma)$ can be given in closed analytic form.

The variation of the frequencies ω_x, ω_y and ω_z as functions of γ for a fixed ε is illustrated in fig. 8.6. One observes that one sector of 60°, e.g. $\gamma = 0\text{–}60°$, is enough to describe all ellipsoidal shapes. The different 60° sectors then only correspond to a different labelling of the three principal axes. Thus, for description of static nuclei we need only consider one sector, $\gamma = 0\text{–}60°$. The nucleus is prolate for $\gamma = 0°$, oblate for $\gamma = 60°$ and triaxial for intermediate γ-values. For a rotating nucleus, one often considers the case where the axis of rotation coincides with the x-axis. Then, to describe all possible situations, it is necessary to consider three sectors with the x-axis being the smaller ($\gamma = 0°$ to 60°) the intermediate ($\gamma = 0°$ to $-60°$) and the larger ($\gamma = -60°$ to $-120°$) principal axis, respectively.

† Different conventions for the sign of γ are used in the literature. This has no importance for the shape of the nucleus but corresponds to a different labelling of the principal axes, see fig. 8.6.

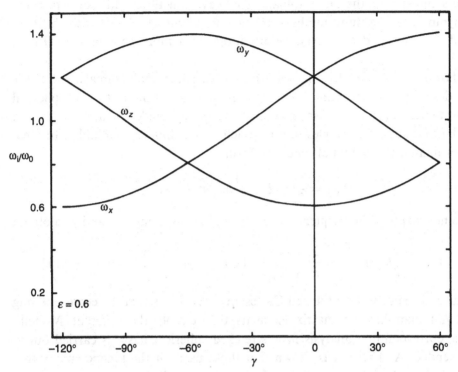

Fig. 8.6. The harmonic oscillator frequencies, ω_x, ω_y and ω_z, versus the asymmetry parameter γ for fixed value of ε, $\varepsilon = 0.6$.

With the introduction of stretched coordinates, $\xi = x(M\omega_x/\hbar)^{1/2}$ etc., the three-dimensional oscillator potential separates into three uncoupled oscillators

$$H_{\text{osc}} = H_x + H_y + H_z$$

where

$$H_x = \frac{1}{2}\hbar\omega_x\left(-\frac{\partial^2}{\partial\xi^2} + \xi^2\right)$$

and analogous expressions for H_y and H_z. The single-particle energies are thus obtained as

$$e_i = \hbar\omega_x\left(n_x + \frac{1}{2}\right) + \hbar\omega_y\left(n_y + \frac{1}{2}\right) + \hbar\omega_z\left(n_z + \frac{1}{2}\right)$$

In the case of $\omega_x = \omega_y = \omega_\perp$ ($\gamma = 0°$), these energies are exhibited in fig. 8.1. The present calculations thus show that the energies of fig. 8.1 are much easier to get out in Cartesian than in cylindrical coordinates. With the introduction of cylindrical coordinates, however, one takes advantage of the symmetry of the Hamiltonian. The wave functions with corresponding (asymptotic) quantum numbers are thus more suited to deal with the $\ell \cdot$ s- and ℓ^2-terms and also axial deformations of a more general type ($P_3, P_4, \ldots$).

For axially symmetric shapes, we used Legendre polynomials, $P_\lambda(\cos \theta_t)$, to describe the potential. An obvious generalisation is to use spherical harmonics, $Y_{\lambda\mu}(\theta_t, \varphi_t)$, for triaxial shapes. The angles θ_t and φ_t are the polar and azimuthal angles in the stretched coordinates, ξ, η and ζ. In these coordinates, the potential takes the form

$$V_{\text{osc}} = \frac{1}{2}\hbar\omega_x\xi^2 + \frac{1}{2}\hbar\omega_y\eta^2 + \frac{1}{2}\hbar\omega_z\zeta^2$$

We now express the frequencies, ω_x, ω_y and ω_z in terms of ε and γ to obtain

$$V_{\text{osc}} = \frac{1}{2}\hbar\omega_0\rho^2 \left[1 - \frac{2}{3}\varepsilon \left(\frac{4\pi}{5}\right)^{1/2} \left(\cos (\gamma Y_{20}) - \frac{\sin \gamma}{\sqrt{2}} (Y_{22} + Y_{2-2}) \right) \right]$$

Our treatment in the stretched Cartesian basis shows that the corresponding Hamiltonian has no matrix elements that couple the different N-shells. The corresponding analysis in a 'stretched spherical basis' is carried out in appendix 8A. There, it is shown that those parts of the kinetic energy that break the 'spherical symmetry' *in the stretched coordinates* can be transformed to a potential energy term. This term cancels the $\Delta N = 2$ matrix elements of the original potential while it increases the matrix elements within an N-shell by a factor of two.

The triaxial potential, V_{osc}, can now easily be generalised by the addition of $\ell \cdot$ s- and ℓ^2-terms and also by deformations of higher order, $\rho^2\varepsilon_4 P_4(\cos \theta_t)$, etc. In addition it is straightforward, and in principle also necessary to introduce terms like $\rho^2\varepsilon_{42} (Y_{42} + Y_{4-2})$ and $\rho^2\varepsilon_{44} (Y_{44} + Y_{4-4})$ (Larsson *et al.*, 1976; Rohozinski and Sobiczewski, 1981; Nazarewicz and Rozmej, 1981). The importance of such terms can be understood from the fact that for triaxial shape, it is not clear with respect to which axis the angle θ_t should be defined.

The γ-deformation does not alter the fact that each orbital is filled by two particles 'moving in different directions' (due to the time-reversal symmetry). Also the parity is preserved as a good quantum number as long as no odd P_λ are introduced. Thus, the odd and even N-shells remain uncoupled. There are however no other exact quantum numbers and it is also difficult to find any approximate quantum numbers to compare with the asymptotic ones, $|Nn_z\Lambda\Omega\rangle$, which are used for axial symmetry.

The single-particle energies of the triaxial modified oscillator Hamiltonian

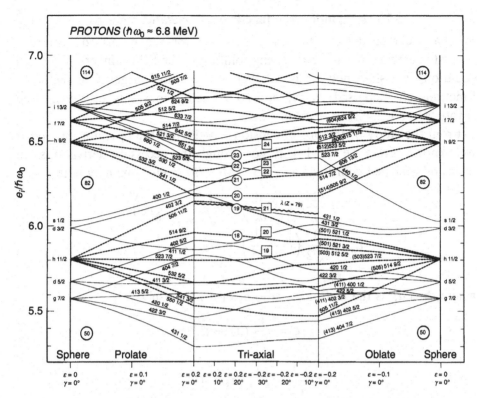

Fig. 8.7. Single-proton orbitals in the $Z \simeq 80$ region. The orbitals at the left edge and the right edge are identical and correspond to spherical shape. In between, ε and γ are varied in such a way that the left part of the figure is drawn for prolate shapes and ε-values up to 0.2, the middle part shows the orbitals for triaxial shape and in the right part of the figure, the orbitals corresponding to oblate shape are exhibited. The asymptotic quantum numbers for prolate and oblate shape, respectively, are given. At the small deformation of $\varepsilon = 0.2$, they are rather impure and in some cases, two alternatives are indicated. For γ-deformations, it is difficult to find any approximate quantum numbers, but the orbitals of different parity can be numbered from 'the bottom' as indicated in the figure. Furthermore, a calculated Fermi level, λ, (chapter 14) for $Z = 79$ is indicated by a wavy line (from C. Ekström *et al.*, 1989, *Nucl. Phys.* **A348**, 25).

are obtained from diagonalisation. An example of such energies, shown along a closed path in the (ε, γ)-plane, is provided in fig. 8.7. The orbitals are labelled by the asymptotic quantum numbers for oblate and prolate shape.

We will come back to the question of triaxial deformations in chapter 12. Now, we will, however, finish this chapter by discussing an operator method, which provides a concise method to calculate single-particle matrix elements in an oscillator basis.

8.7 Asymptotic wave functions by operator methods

We have earlier derived the wave function for the deformed harmonic oscillator with axial symmetry by the solution of the Schrödinger differential equation. We shall now utilise an operator method to generate the wave functions characterised by N, n_z and Λ being constants of the motion (Mottelson and Nilsson, 1959b).

From the study of the one-dimensional oscillator, the operator method is well known. Here we will briefly go through the necessary derivations. In terms of stretched coordinates, $\xi = x \cdot (M\omega_x/\hbar)^{1/2}$ etc., one may write

$$a_x = \frac{1}{\sqrt{2}}\left(\xi + \frac{\partial}{\partial \xi}\right) = \frac{1}{\sqrt{2}}\left(\xi + \frac{i}{\hbar}p_\xi\right)$$

$$a_x^+ = \frac{1}{\sqrt{2}}\left(\xi - \frac{\partial}{\partial \xi}\right) = \frac{1}{\sqrt{2}}\left(\xi - \frac{i}{\hbar}p_\xi\right)$$

where thus a_x^+ is the Hermitian adjoint of a_x. Inversely

$$\xi = \frac{1}{\sqrt{2}}\left(a_x + a_x^+\right)$$

$$\frac{\partial}{\partial \xi} = \frac{1}{\sqrt{2}}\left(a_x - a_x^+\right)$$

Similar definitions hold for a_y and a_z. It is easy to verify that the a_x operators fulfill the commutation relations

$$[a_x, a_x^+] = 1$$

Expressing

$$H_x = \frac{1}{2}\hbar\omega_x\left(-\frac{\partial^2}{\partial \xi^2} + \xi^2\right)$$

in terms of a_x^+ and a_x we obtain

$$H_x = \hbar\omega_x\left(a_x^+a_x + \frac{1}{2}\right)$$

Thus H_x commutes with $a_x^+a_x$, which means that these two operators can be assumed to have the same eigenvectors. We introduce eigenvectors ψ_k and eigenvalues λ_k

$$a_x^+a_x\psi_k = \lambda_k\psi_k$$

Scalar multiplication with ψ_k leads to

$$\langle \psi_k | a_x^+a_x\psi_k \rangle = \langle a_x\psi_k | a_x\psi_k \rangle = \lambda_k \langle \psi_k | \psi_k \rangle$$

Since $\langle a_x \psi_k | a_x \psi_k \rangle$ and $\langle \psi_k | \psi_k \rangle$ are non-negative, we conclude that

$$\lambda_k \geq 0$$

From the commutation relations we furthermore get

$$(a_x^+ a_x) \, a_x^+ \psi_k = a_x^+ (a_x^+ a_x + 1) \, \psi_k = (\lambda_k + 1) \, a_x^+ \psi_k$$

which means that $a_x^+ \psi_k$ is also an eigenvector of $a_x^+ a_x$, the corresponding eigenvalue being $\lambda_k + 1$. Similarly, one can show that $a_x \psi_k$ is an eigenvector with the eigenvalue $\lambda_k - 1$. The operator a_x^+ is thus called a raising operator, while a_x is a lowering operator. Since $\lambda_k \geq 0$, there must be a lowest value for λ_k:

$$a_x^+ a_x \psi_0 = \lambda_0 \psi_0$$

The eigenvalue cannot become smaller and consequently

$$a_x \psi_0 = 0$$

This leads to

$$\lambda_0 = 0$$

and consequently $\lambda_1 = 1$, $\lambda_2 = 2$, etc., i.e.

$$a_x^+ a_x \psi_n \equiv a_x^+ a_x |n_x\rangle = n_x |n_x\rangle$$

We thus find for the oscillator Hamiltonian

$$H_x |n_x\rangle = \hbar \omega_x \left(a_x^+ a_x + \frac{1}{2} \right) |n_x\rangle = \hbar \omega_x \left(n_x + \frac{1}{2} \right) |n_x\rangle$$

The eigenvalue, n_x, of $a_x^+ a_x$ is referred to as the number of oscillator quanta. This justifies the designation of $a_x^+ a_x$ as a number operator, $\mathcal{N}_x$, with respect to the x quanta. The eigenvector Ψ_0 is of course the no-quantum ground state wave function

$$\psi_0 = |0\rangle = \pi^{-1/4} e^{-\xi^2/2}$$

The one-quantum state is now easily calculated

$$a_x^+ |0\rangle = \pi^{-1/4} \frac{1}{\sqrt{2}} \left(\xi - \frac{\partial}{\partial \xi} \right) e^{-\xi^2/2} = \pi^{-1/4} \sqrt{2} \cdot \xi e^{-\xi^2/2}$$

Generally the normalised state with n_x quanta can be written

$$|n_x\rangle = \frac{1}{(n_x!)^{1/2}} \left(a_x^+ \right)^{n_x} |0\rangle$$

The normalisation constant is obtained from the commutation relations. These lead to (see problem 8.7)

$$a_x^+ |n_x\rangle = (n_x + 1)^{1/2} |n_x + 1\rangle$$
$$a_x |n_x\rangle = \sqrt{n_x} |n_x - 1\rangle$$

After these general considerations in one dimension, we go back to the three-dimensional oscillator with cylinder symmetry. To exploit this symmetry between x and y we define

$$R^+ = \frac{1}{\sqrt{2}} \left(a_x^+ + i a_y^+ \right)$$

$$R = \frac{1}{\sqrt{2}} \left(a_x - i a_y \right)$$

$$S^+ = \frac{1}{\sqrt{2}} \left(a_x^+ - i a_y^+ \right)$$

$$S = \frac{1}{\sqrt{2}} \left(a_x + i a_y \right)$$

One can now verify the commutation rules

$$[R, R^+] = [S, S^+] = 1$$

while all other commutations vanish. It is apparent that one can subsequently construct eigenfunctions with the help of the R^+ and S^+ operators analogous to what is done in the a_x^+ case.

The operators R^+ and S^+ increase while R and S lower the quantum number $n_\perp$ by one unit. This corresponds to the fact that the number operator with respect to the quanta perpendicular to the z-axis can be written

$$\mathcal{N}_\perp = a_x^+ a_x + a_y^+ a_y = R^+ R + S^+ S$$

In terms of these new operators, the Hamiltonian of the cylindrically symmetrical harmonic oscillator can be written

$$H_{\text{cyl. osc}} = \hbar\omega_z \left(a_z^+ a_z + \frac{1}{2} \right) + \hbar\omega_\perp \left(a_x^+ a_x + a_y^+ a_y + 1 \right)$$

$$= \hbar\omega_z \left(\mathcal{N}_z + \frac{1}{2} \right) + \hbar\omega_\perp (\mathcal{N}_\perp + 1)$$

$$= \hbar\omega_z \left(\mathcal{N}_z + \frac{1}{2} \right) + \hbar\omega_\perp (R^+ R + S^+ S + 1)$$

As we have seen in the earlier parts of this chapter, from the Schrödinger equation we can find a solution to this Hamiltonian, which in addition has a good Λ quantum number, and is of the type

$$\psi = Z(\zeta)U(\rho)\phi(\varphi) = |n_z\rangle\,|n_\perp\Lambda\rangle$$

Alternatively using the operators we can write a solution

$$\psi = |n_z rs\rangle = \frac{1}{(n_z!r!s!)^{1/2}}\,(a_z^+)^{n_z}\,(R^+)^r\,(S^+)^s\,|0\rangle$$

It is easily seen that

$$H_{\text{cyl. osc}}\,|n_z rs\rangle = \left[\left(n_z + \frac{1}{2}\right)\hbar\omega_z + (n_\perp + 1)\hbar\omega_\perp\right]|n_z rs\rangle$$

where

$$n_\perp = r + s$$

Our next step is to express ℓ_z in terms of the R and S operators. We shall use the simplified notation ℓ_x etc. to denote $(\ell_t)_x$ etc. and furthermore assume the angular momentum vector to be expressed in units of $\hbar$ (thus set $\hbar = 1$)

$$\ell_z = \frac{1}{i}\left(\xi\frac{\partial}{\partial\eta} - \eta\frac{\partial}{\partial\xi}\right) = \frac{1}{i}\left(a_x^+ a_y - a_y^+ a_x\right) = (R^+ R - S^+ S)$$

Thus,

$$\ell_z\,|n_z rs\rangle = (r - s)\,|n_z rs\rangle = \Lambda\,|n_z rs\rangle$$

where

$$\Lambda = r - s$$

Hence $|n_z rs\rangle$ is an eigenfunction both of $H_{\text{cyl. osc}}$ and ℓ_z.

The operators R^+ and S increase the quantum number $\ell_z = \Lambda$ by one unit while R and S^+ lower Λ by one unit. This corresponds to commutation rules

$$[\ell_z, R^+] = R^+, \quad [\ell_z, S] = S, \quad [\ell_z, R] = -R, \quad [\ell_z, S^+] = -S^+$$

which are easily proven if the expression of ℓ_z in terms of the R- and S-operators is used.

For the ℓ_x and ℓ_y operators, we obtain

$$i\ell_x = a_y^+ a_z - a_z^+ a_y = \frac{-i}{\sqrt{2}}\left[a_z^+(R - S) + a_z\,(R^+ - S^+)\right]$$

$$i\ell_y = a_z^+ a_x - a_x^+ a_z = \frac{-i}{\sqrt{2}}\left[a_z\,(R^+ + S^+) - a_z^+\,(R + S)\right]$$

and furthermore

$$\ell_+ = \sqrt{2}\,(a_z^+ S - a_z R^+)$$
$$\ell_- = \sqrt{2}\,(a_z S^+ - a_z^+ R)$$

which latter expressions will prove useful later on.

Starting from the wave function $|n_z, n_\perp = r + s, \Lambda = r - s\rangle$, we may generate a wave function with $n_\perp \to n_\perp + 1$ and $\Lambda \to \Lambda + 1$ or $|n_z, n_\perp + 1, \Lambda + 1\rangle$ by operation with R^+. We utilise $[H_{\text{cyl. osc}}, R^+] = \hbar\omega_\perp R^+$ to obtain

$$H_{\text{cyl. osc}}\{R^+ |n_z rs\rangle\} = R^+ (H_{\text{cyl. osc}} + \hbar\omega_\perp) |n_z rs\rangle$$
$$= \left[\left(n_z + \frac{1}{2}\right)\hbar\omega_z + (r + s + 2)\hbar\omega_\perp\right]\{R^+ |n_z rs\rangle\}$$

Similarly

$$\ell_z \{R^+ |n_z rs\rangle\} = R^+ (\ell_z + 1) |n_z rs\rangle = R^+ (r - s + 1) |n_z rs\rangle$$
$$= (\Lambda + 1)\{R^+ |n_z rs\rangle\}$$

Alternatively we may just exploit the fact that

$$R^+ |n_z rs\rangle = (r + 1)^{1/2} |n_z\, r + 1\, s\rangle$$

and the relations

$$n_\perp = r + s, \qquad \Lambda = r - s$$

With inclusion of spin the total wave function is

$$\psi = |n_z rs\rangle\, |\Sigma\rangle$$

We are now in a position to evaluate the matrix elements of $\ell_t \cdot \mathbf{s}$ and ℓ_t^2 in this asymptotic representation. Let us start with $\ell_t \cdot \mathbf{s}$. From the expansion

$$\ell_t \cdot \mathbf{s} = (\ell_t)_z\, s_z + \frac{1}{2}\,((\ell_t)_+ s_s + (\ell_t)_- s_+)$$

we obtain, simplifying the notation by leaving out t as is done above,

$$\ell \cdot \mathbf{s} = (R^+ R - S^+ S)\, s_z - \frac{1}{\sqrt{2}}\,(a_z R^+ - a_z^+ S)\, s_- - \frac{1}{\sqrt{2}}\,(a_z^+ R - a_z S^+)\, s_+$$

The first term is an eigenoperator of $|n_z rs\Sigma\rangle$

$$\ell_z s_z |n_z rs\Sigma\rangle = (r - s)\Sigma |n_z rs\Sigma\rangle = \Lambda\Sigma |n_z rs\Sigma\rangle$$

and

$$\langle n_z rs\Sigma | \ell \cdot \mathbf{s} | n_z rs\Sigma\rangle = \Lambda\Sigma$$

The next two terms of $(\ell \cdot s)$ give rise to non-diagonal coupling terms associated with selection rules

$$\Delta\Sigma = -1, \; \Delta\Lambda = 1, \; \Delta n_z = -\Delta n_\perp = \pm 1$$

and

$$\Delta\Sigma = 1, \; \Delta\Lambda = -1, \; \Delta n_z = -\Delta n_\perp = \pm 1$$

For the matrix elements, we get for example

$$\langle n_z - 1\, r + 1\, s\, \Sigma - 1 | \ell \cdot s | n_z\, r\, s\, \Sigma \rangle = -\frac{1}{\sqrt{2}}\, [n_z(r+1)]^{1/2}$$

and thus

$$\langle n_z - 1\, n_\perp + 1\, \Lambda + 1\, \Sigma - 1 | \ell \cdot s | n_z\, n_\perp\, \Lambda\, \Sigma \rangle = -\frac{1}{2}\, [n_z\,(n_\perp + \Lambda + 2)]^{1/2}$$

In a similar way, the other matrix elements are obtained as

$$\langle n_z + 1\, n_\perp - 1\, \Lambda + 1\, \Sigma - 1 | \ell \cdot s | n_z\, n_\perp\, \Lambda\, \Sigma \rangle = \frac{1}{2}\, [(n_z + 1)\,(n_\perp - \Lambda)]^{1/2}$$

$$\langle n_z + 1\, n_\perp - 1\, \Lambda - 1\, \Sigma + 1 | \ell \cdot s | n_z\, n_\perp\, \Lambda\, \Sigma \rangle = -\frac{1}{2}\, [(n_z + 1)\,(n_\perp + \Lambda)]^{1/2}$$

$$\langle n_z - 1\, n_\perp + 1\, \Lambda - 1\, \Sigma + 1 | \ell \cdot s | n_z\, n_\perp\, \Lambda\, \Sigma \rangle = \frac{1}{2}\, [n_z\,(n_\perp - \Lambda + 2)]^{1/2}$$

From the expressions for ℓ_+ and ℓ_- it is easy to find an expression for $\ell_\perp^2 = \ell_x^2 + \ell_y^2$

$$\ell_\perp^2 = \frac{1}{2}(\ell_+\ell_- + \ell_-\ell_+) = 2a_z^+ a_z\,(R^+R + S^+S + 1) + R^+R + S^+S$$
$$- 2\,(a_z^+)^2\, RS - 2\,(a_z)^2\, R^+S^+$$

The first three terms are an eigenoperator of $|n_z rs\rangle$ with the eigenvalue $2n_z\,(n_\perp + 1) + n_\perp$ while the last two terms have selection rules $\Delta\Lambda = 0$, $\Delta N = 0, \; \Delta n_z = -\Delta n_\perp = \pm 2$:

$$\left\langle n_z + 2\, r - 1\, s - 1 \left| \ell_\perp^2 \right| n_z\, r\, s \right\rangle = -2\,[(n_z + 2)\,(n_z + 1)\cdot r\cdot s]^{1/2}$$

and

$$\left\langle n_z - 2\, r + 1\, s + 1 \left| \ell_\perp^2 \right| n_z\, r\, s \right\rangle = -2\,[(n_z - 1)\, n_z\,(r + 1)\,(s + 1)]^{1/2}$$

Thus with $\ell^2 = \ell_\perp^2 + \ell_z^2$:

$$\left\langle n_z\, n_\perp\, \Lambda\, \Sigma \left| \ell^2 \right| n_z\, n_\perp\, \Lambda\, \Sigma \right\rangle = 2n_z\,(n_\perp + 1) + n_\perp + \Lambda^2$$

$$\langle n_z + 2\,n_\perp - 2\,\Lambda\,\Sigma \,|\, \ell^2 \,|\, n_z\,n_\perp\,\Lambda\,\Sigma \rangle$$

$$= -\,[(n_z + 2)\,(n_z + 1)\,(n_\perp + \Lambda)\,(n_\perp - \Lambda)]^{1/2}$$

$$\langle n_z - 2\,n_\perp + 2\,\Lambda\,\Sigma \,|\, \ell^2 \,|\, n_z n_\perp \Lambda \Sigma \rangle$$

$$= -\,[(n_z - 1)\,n_z\,(n_\perp + \Lambda + 2)\,(n_\perp - \Lambda + 2)]^{1/2}$$

Exercises

8.1   Show that

$$\left\langle N\ell j\Omega \left| -M\omega_0^2\frac{2}{3}\varepsilon r^2 P_2 \right| N\ell j\Omega \right\rangle = \frac{1}{6}\varepsilon M\omega_0^2 \left\langle r^2 \right\rangle \frac{3\Omega^2 - j(j+1)}{j(j+1)}$$

8.2   Find the solution in hypergeometrical functions of

$$\left(-\frac{1}{\rho}\frac{\partial}{\partial\rho}\rho\frac{\partial}{\partial\rho} + \frac{\Lambda^2}{\rho^2} + \rho^2 - \frac{2E_\perp}{\hbar\omega_\perp} \right) U(\rho) = 0$$

Use the substitution

$$U(\rho) = \rho^{|\Lambda|}e^{-\rho^2/2}f(\rho)$$

8.3   Start from the harmonic oscillator Hamiltonian

$$H = -\frac{\hbar^2}{2M}\Delta_x + \frac{M}{2}\left[\omega_\perp^2\left(x^2 + y^2\right) + \omega_z^2 z^2 \right]$$

Make the substitution

$$\xi = x\,(M\omega_\perp/\hbar)^{1/2} \quad \text{etc.;} \qquad \rho^2 = \xi^2 + \eta^2 + \zeta^2$$

and show that H can be expressed as

$$H = H_d + H_\varepsilon$$

where

$$H_d = \frac{1}{2}\hbar\omega_0\left(-\Delta_\xi + \rho^2 \right)$$

and

$$H_\varepsilon = \frac{1}{6}\varepsilon\hbar\omega_0\left(-\frac{\partial^2}{\partial\xi^2} - \frac{\partial^2}{\partial\eta^2} + 2\frac{\partial^2}{\partial\zeta^2} + \xi^2 + \eta^2 - 2\zeta^2 \right)$$

8.4   Show that an axially symmetric modified oscillator Hamiltonian, H_{MO}, commutes with the j_z operator, $[H_{MO}, j_z] = 0$. Calculate also $[\ell \cdot s, \ell_z]$. Comments!

8.5   For the modified oscillator potential at small deformations, the energy levels can be calculated in first-order perturbation theory. At large deformations, a good approximation is to treat the harmonic oscillator exactly but only to consider diagonal contributions from the $\ell \cdot$ s- and ℓ^2-terms. Carry through these calculations for the $N = 2$ levels and sketch them for $|\varepsilon| \leq 0.75$. Put $\kappa = 0.1$ and $\mu' = 0.02$.

8.6   For the $\Omega = 3/2$, $N = 2$ levels of the modified oscillator potential, it is easy to make an exact diagonalisation. Carry through the calculations and compare with the approximate energies of problem 8.5.

8.7   Use the properties $[a, a^+] = 1$ and $a^+ a \, |n\rangle = n \, |n\rangle$ of the step operator a^+ to find the normalisation constant C_n in the formula

$$a^+ \, |n\rangle = C_n \, |n+1\rangle$$

8.8   Prove that $[R, R^+] = 1$ and that $[R^+, S] = 0$.

8.9   Express ℓ_z in terms of R and S and their Hermitian conjugates.

8.10   Let $|A\rangle$ be an eigenvector of the operator A with the eigenvalue a, $A \, |A\rangle = a \, |A\rangle$. Show that, if $[A, B] = B$, then $B \, |A\rangle$ is an eigenvector of A with the eigenvalue $(a + 1)$.

8.11   Prove

$$\langle n_z + 1 \, n_\perp - 1 \, \Lambda - 1 \, \Sigma + 1 | \ell \cdot s | \, n_z n_\perp \Lambda \Sigma \rangle = \frac{-1}{2} \, [(n_z + 1)(n_\perp + \Lambda)]^{1/2}$$

8.12   Use operator methods to calculate the radius of a triaxial harmonic oscillator wave function,

$$\left\langle n_x n_y n_z \left| r^2 \right| n_x n_y n_z \right\rangle$$

Specialise to spherical shape!

Appendix 8A
The anisotropic harmonic oscillator in a 'spherical basis'

Consider a triaxial harmonic oscillator Hamiltonian

$$H_{\text{osc}} = -\frac{\hbar^2}{2M}\Delta + \frac{1}{2}M\left(\omega_1^2 x^2 + \omega_2^2 y^2 + \omega_3^2 z^2\right)$$

The solution of this Hamiltonian in a (stretched) Cartesian basis was given in the main text. It was found that the number of quanta n_1, n_2 and n_3 was preserved, i.e. this is the case also for the total number of quanta (the index t, which is sometimes suppressed, is used in the transformed basis, i.e. the stretched basis).

$$N_t = n_1 + n_2 + n_3$$

When working in the polar coordinates of the stretched system, $(\rho, \theta_t, \varphi_t)$, we have to go through a somewhat more tedious derivation to find the solution (Nilsson, 1955; Larsson, 1973). Let us introduce the deformation coordinates ε and γ in the standard way

$$\omega_i = \omega_0(\varepsilon, \gamma)\left[1 - \frac{2}{3}\varepsilon\cos\left(\gamma + i\frac{2\pi}{3}\right)\right]$$

and then split the Hamiltonian into two parts,

$$H_{\text{osc}} = H_0 + H_\varepsilon$$

where H_0 is isotropic in the stretched system;

$$\begin{aligned}
H_0 &= \frac{1}{2}\hbar\omega_0(\varepsilon)\left[-\left(\frac{\partial^2}{\partial\xi^2} + \frac{\partial^2}{\partial\eta^2} + \frac{\partial^2}{\partial\zeta^2}\right) + \left(\xi^2 + \eta^2 + \zeta^2\right)\right] \\
&= \frac{1}{2}\hbar\omega_0(\varepsilon)\left[-\Delta_\xi + \rho^2\right]
\end{aligned}$$

and

$$H_\varepsilon = \frac{1}{6}\varepsilon \cos\gamma\, \hbar\omega_0 \left[\left(-\frac{\partial^2}{\partial\xi^2}+\xi^2\right) + \left(-\frac{\partial^2}{\partial\eta^2}+\eta^2\right) - 2\left(-\frac{\partial^2}{\partial\zeta^2}+\zeta^2\right) \right]$$

$$+ \frac{\varepsilon\sin\gamma}{2\sqrt{3}}\hbar\omega_0 \left[\left(-\frac{\partial^2}{\partial\xi^2}+\xi^2\right) - \left(-\frac{\partial^2}{\partial\eta^2}+\eta^2\right) \right]$$

We now introduce a representation, $|N_t\alpha_t\rangle$, which makes H_0 diagonal

$$H_0\,|N_t\alpha_t\rangle = \left(N_t + \frac{3}{2}\right)\hbar\omega_0\,|N_t\alpha_t\rangle$$

The quantum numbers, which, in addition to N_t, are needed to specify the basis, are denoted by α_t, i.e. the natural choices are $\alpha_t = (\ell_t \Lambda_t s\Sigma)$ or $\alpha_t = (\ell_t s j_t \Omega_t)$.

For the evaluation of the matrix elements of H_ε it is convenient to rewrite the derivatives in terms of double commutators containing Laplacians because in an oscillator basis (in a stretched or non-stretched system)

$$\left\langle N'\alpha' \left| \left[\Delta, \left[\Delta, x^\lambda y^\mu z^\nu\right]\right] \right| N\alpha \right\rangle$$
$$= 4\left[(N'-N)^2 - (\lambda+\mu+\nu) \right] \left\langle N'\alpha' \left| x^\lambda y^\mu z^\nu \right| N\alpha \right\rangle$$

This equality is easily proven by use of the eigenvalue equation for H_0 and the commutator

$$\left[\left[\Delta, x^\lambda y^\mu z^\nu\right], \rho^2\right] = 4(\lambda+\mu+\nu)x^\lambda y^\mu z^\nu$$

The double commutators needed are

$$\left[\Delta_\xi, \left[\Delta_\xi, \xi^2+\eta^2-2\zeta^2\right]\right] = 8\left(\frac{\partial^2}{\partial\xi^2} + \frac{\partial^2}{\partial\eta^2} - 2\frac{\partial^2}{\partial\zeta^2}\right)$$

and

$$\left[\Delta_\xi, \left[\Delta_\xi, \xi^2-\eta^2\right]\right] = 8\left(\frac{\partial^2}{\partial\xi^2} - \frac{\partial^2}{\partial\eta^2}\right)$$

It is now straightforward to calculate

$$\left\langle N_t'\alpha_t' \left| -\frac{\partial^2}{\partial\xi^2} - \frac{\partial^2}{\partial\eta^2} + 2\frac{\partial^2}{\partial\zeta^2} \right| N_t\alpha_t \right\rangle$$

$$= -\frac{1}{8}\left\langle N_t'\alpha_t' \left| \left[\Delta_\xi, \left[\Delta_\xi, \left(\xi^2+\eta^2-2\zeta^2\right)\right]\right] \right| N_t\alpha_t \right\rangle$$

$$-\frac{1}{2}\left[(N_t'-N_t)^2 - 2 \right] \left\langle N_t'\alpha_t' \left| \left(\xi^2+\eta^2-2\zeta^2\right) \right| N_t\alpha_t \right\rangle$$

$$= \pm\left\langle N_t'\alpha_t' \left| \left(\xi^2+\eta^2-2\zeta^2\right) \right| N_t\alpha_t \right\rangle \quad \begin{cases} N_t' = N_t \\ N_t' = N_t \pm 2 \end{cases}$$

For $|N'_t - N_t| \neq 0$ or 2, the matrix elements of $\left(\xi^2 + \eta^2 - 2\zeta^2\right) \propto r^2 Y_{20}(\theta_t, \varphi_t)$ vanish. This is easily seen, e.g. by help of the operator technique of the last section in chapter 8 because at most two quanta can be created or destroyed by help of quadratic expressions in the coordinates.

The matrix elements of the derivatives in the second part of H_ε are now treated in an analogous manner leading to a similar formula as for the first part:

$$\left\langle N'_t \alpha'_t \left| -\left(\frac{\partial^2}{\partial \xi^2} - \frac{\partial^2}{\partial \eta^2} \right) \right| N_t \alpha_t \right\rangle$$

$$= \pm \left\langle N'_t \alpha'_t \left| \xi^2 - \eta^2 \right| N_t \alpha_t \right\rangle \quad \begin{cases} N'_t = N_t \\ N'_t = N_t \pm 2 \end{cases}$$

It is thus obvious that the matrix elements of the kinetic and potential energy parts of H_ε have the same magnitude and also the same sign for $\Delta N_t = 0$ but different signs for $\Delta N_t = 2$. Thus, the formula mentioned in the main text follows:

$$\langle N'_t \alpha'_t | H_\varepsilon | N_t \alpha_t \rangle =$$
$$\left\langle N'_t \alpha'_t \left| \frac{1}{3} \hbar \omega_0 \varepsilon \cos \gamma \left(\xi^2 + \eta^2 - 2\zeta^2 \right) + \frac{1}{\sqrt{3}} \hbar \omega_0 \varepsilon \sin \gamma \left(\xi^2 - \eta^2 \right) \right| N_t \alpha_t \right\rangle \delta_{N_t N'_t}$$

where the two terms in parentheses are proportional to $\rho^2 Y_{20}(\theta_t, \varphi_t)$ and $\rho^2 \left(Y_{22}(\theta_t, \varphi_t) + Y_{2-2}(\theta_t, \varphi_t) \right)$, respectively.

9

The shell correction method and the nuclear deformation energy

We have previously calculated the total energy of the atomic nucleus by use of the *'macroscopic' liquid-drop model*. In this model the energy is assumed to be a sum of a volume term, a surface term and a Coulomb term $(I = (N - Z)/A)$:

$$E = -a_v \left(1 - \kappa_v I^2\right) A + a_s \left(1 - \kappa_s I^2\right) A^{2/3} B_s(\text{def}) + a_C \frac{Z^2}{A^{1/3}} B_C(\text{def})$$

We have found that this model could reasonably well explain, in addition to the variation in nuclear mass, various phenomena associated with fission, e.g. why the heavier elements undergo spontaneous fission and, furthermore, the approximate heights of the fission barriers.

However, it is also apparent that many phenomena could not be understood in terms of this model. Thus it does not reproduce the detailed variation in fission barrier height with particle number or the two-peak character of the barriers in the actinide region. Neither could it explain why many nuclei are deformed and not spherical in their ground state. One might say that various nuclear properties are only explained *on the average* (where the average might be taken over particle number or alternatively over deformation) by the liquid-drop model.

To reproduce other aspects of nuclear structure such as ground state spins and energy spectra, it was found that a different description was necessary. In the preceding chapters, we have therefore introduced the *single-particle model*. In this connection we calculated single-particle energies e_ν as functions of the deformation parameters, $e_\nu = e_\nu (\varepsilon, \varepsilon_3, \varepsilon_4, \ldots)$. It is now tempting to consider the total energy of the nucleus (often referred to as the potential energy) obtained by the addition of the single-particle energies e_ν:

$$E_{\text{sp}} (\varepsilon, \varepsilon_3, \varepsilon_4 \ldots) = \sum_\nu e_\nu (\varepsilon, \varepsilon_3, \varepsilon_4, \ldots)$$

143

There are some problems connected with this procedure, however. First the single-particle energy e_ν is a sum of a kinetic-energy contribution $\langle T_\nu \rangle$ and a potential-energy contribution $\langle V_\nu \rangle$, the latter representing the expectation value of a sum of all the two-particle interactions

$$V_\nu = \sum_\mu U_{\nu\mu}$$

As all the terms $\langle V_\nu \rangle$ are added, the problem arises of whether or not the interactions are counted twice. A second problem concerns the volume conservation condition, which is difficult to generalise to include the effects of the $\ell \cdot$ s-term and, in the modified oscillator model, also the ℓ^2-term.

The recipe of single-particle energy summation has been tried, however, and found to have fair success. The energy surface (i.e. the energy considered as a function of two variables, e.g. ε and ε_4, see fig. 9.3 below) given by the single-particle sum is found to give a lowest minimum usually somewhat removed from spherical shape. The equilibrium shapes of well-deformed nuclei can be directly related to a quadrupole moment Q_2 and in some cases a hexadecapole moment Q_4, which can also be obtained from experiment (optical spectroscopy, Coulomb excitation cross sections etc.). It turns out that at least Q_2 is in good agreement with data. When extended to larger distortions the energy surface should then also account for the fission barrier. For this application, however, the single-particle sum recipe is found to be inadequate.

One may note that the restoring energy introduced by the volume conservation condition is a term of very large magnitude, being roughly proportional to $(\varepsilon^2/9)$ times the total nuclear energy, or for $\varepsilon = 0.9$ of the order of 1000 MeV. As, among other things, the entire nuclear potential is not included in the volume conservation condition, 'small' corrections to the gross trends of the total energy are not unexpected. On the other hand, the vicinity of the spherical shape appears to be correctly reproduced as long as the corrrect level order is reproduced. Indeed the entire topological character of the energy surface may be obtained although the entire surface appears to be tilted.

Indeed a renormalisation of the energy surface appears to be called for and it is brought about by the introduction of the Strutinsky (1967) procedure. The basic idea behind this is the following. The average, long-range behaviour of nuclear binding energy as a function of the nuclear charge and size is well reproduced by the liquid-drop model. One then surmises that on the average this model also adequately describes deformation†. The relative success of the liquid-drop theory of fission may be taken as a warrant for

† This is, of course, what is conjectured in the original application of the model to the theory of fission in the classical paper by N. Bohr and J.A. Wheeler (1939).

this. One should therefore require that on the average – the average taken over so many nuclei that shell effects are averaged out – the total energy has the same distortion dependence as that of a liquid drop. This requirement can be enforced by subtracting out of the total energy an *averaged energy* and replacing the latter by the liquid-drop energy. The main problem consists of forming this average in a satisfactorily unique way.

To this end, Strutinsky (1967) first defines a smoothed level density $\tilde{g}(e)$ by smearing the calculated single-particle levels e_ν over a range γ, where γ is an energy of the order of the shell spacing, $\hbar\omega_0$. Strutinsky thus considers a comparison between the actual discrete level structure g, where

$$g(e) = \sum_\nu \delta\,(e - e_\nu)$$

and a smeared level density $\tilde{g}(e)$ (see below). One then defines a shell energy as the difference

$$E_{\mathrm{sh}} = 2\Sigma e_\nu - 2\int e\tilde{g}(e)\,\mathrm{d}e$$

(the factor 2 coming from the double degeneracy of the deformed levels).

To obtain the smooth density $\tilde{g}(e)$ the discrete levels are associated with a smearing function,

$$\tilde{g}(e) = \frac{1}{\gamma\sqrt{\pi}} \sum_\nu \int \mathrm{d}e' f_{\mathrm{corr}} \left(\frac{e-e'}{\gamma}\right) \delta\,(e' - e_\nu) \exp\left(-\frac{(e-e')^2}{\gamma^2}\right)$$

or

$$\tilde{g}(e) = \frac{1}{\gamma\sqrt{\pi}} \sum_\nu f_{\mathrm{corr}} \left(\frac{e-e_\nu}{\gamma}\right) \exp\left(-\frac{(e-e_\nu)^2}{\gamma^2}\right)$$

The role of the exponential function is the obvious one of smearing. Through this we eliminate fluctuations of order $L \ll \hbar\omega_0$ by a choice of $\gamma \approx \hbar\omega_0$. In order not to disturb the long-range variations of order $L \gg \hbar\omega_0$, a correction function f_{corr} is introduced (cf. problem 9.3),

$$f_{\mathrm{corr}}(u) = 1 + \left(\frac{1}{2} - u^2\right) + \left(\frac{3}{8} - \frac{3}{2}u^2 + \frac{1}{2}u^4\right) + \ldots = \sum_{k=0}^{n} \frac{(-)^k}{(2k)!!2^k} H_{2k}(u)$$

where H_n are Hermite polynomials†.

† One can understand the occurrence of Hermite polynomials by the fact that the δ-function can be expanded as follows:

$$\delta(u) = \sum_{n=0}^{\infty} \frac{(-)^n}{\sqrt{\pi}(2n)!!2^n} H_{2n}(u)e^{-u^2}$$

If all the H_n polynomials up to $n = \infty$ were retained, then obviously the 'smeared' function $\tilde{g}(e)$ would be identical to the unsmeared one $g(e)$. It is thus by a proper break-off of the δ-function expansion that the desired $\tilde{g}(e)$ is obtained.

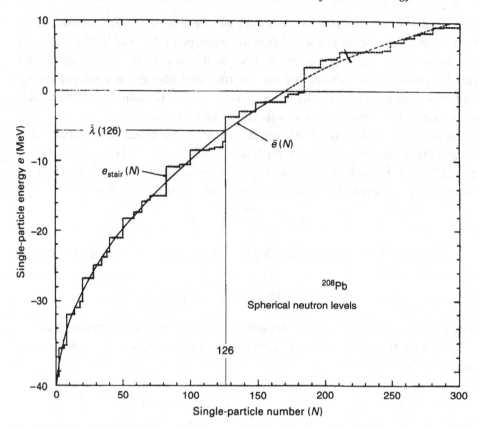

Fig. 9.1. Theoretical single-particle (neutron) energies in a Woods–Saxon type radial potential applicable to ^{208}Pb. The discrete energies define a 'staircase' function $e_{stair}(N)$. The smooth curve $\bar{e}(N)$ removes the local fluctuations but retains the long-range behaviour. The Fermi surface $\bar{\lambda}$ of the smooth distribution of levels is illustrated for 126 neutrons. The corresponding shell correction is given by the difference between the areas under the two curves up to $N = 126$. Note that a large section of the staircase curve is below the smooth one for N values just below $N = 126$ (from M. Bolsterli, E.O. Fiset, J.R. Nix and J.L. Norton, *Phys. Rev.* **C5** (1972) 1050).

The shell correction method is illustrated in fig. 9.1. One understands that the shell correction is negative when the sum of single-particle levels is below average, i.e. when there is a gap. It is positive and large where there is a high level density. To the shell energy, E_{sh}, which is defined independently for protons and neutrons, is to be added the liquid-drop energy. For an even nucleus the total energy is therefore given by the following expression:

$$E_{tot} = E_{L.D.} + E_{sh}(prot) + E_{sh}(neutr)$$

For the method to be well defined, the results should not be too sensitive to the value of the smearing range γ. In fig. 9.2 the shell correction energy is calculated for various distortions as a function of γ for the nucleus ^{208}Pb. It is evident that, if terms up to the order 6 are included in the correction function f_{corr}, the shell correction is insensitive to the choice of γ over a wide range of values.

To get the total nuclear energy, it is also necessary to add a pairing energy (chapter 14). The pairing Hamiltonian is introduced to account for the short range nuclear interactions, which are not taken care of in the mean field approximation (interaction terms not accounted for by the mean field are generally referred to as the residual interaction). In the present context, the main effect of the pairing interaction is to smoothen the fluctuations, considered as functions of deformation and particle number, which result from the shell energy.

Only after Strutinsky had suggested the above renormalisation procedure in 1966, did it become possible to calculate realistic potential-energy surfaces covering the whole of the fission process (see e.g. Nilsson *et al.*, 1969; Brack *et al.*, 1972). We shall briefly look at some of the main results. The potential energy is studied for various sets of shapes for the nucleus. One such set was illustrated in fig. 8.4. An actinide nucleus undergoing fission will assume a sequence of shapes corresponding to a line from the lower left hand corner to the upper right hand one as shown in fig. 9.3. In fig. 8.5 the single-neutron levels were plotted as functions of deformation corresponding to this line. One should note the gaps for $N = 152$ at $\varepsilon \approx 0.25$ and for $N = 144$ at $\varepsilon \approx 0.6$. These gaps and the corresponding low level densities for neighbouring particle numbers are associated with the first and second minimum in the potential-energy surface illustrated in fig. 9.3. This surface is typical for the actinide region. The second minimum at $\varepsilon \approx 0.6$, corresponding to a 2 : 1 ratio of the nuclear axes, gives rise to a two-humped fission barrier as discussed already in chapter 4 (cf. fig. 4.2).

One may also wish to study the effects of reflection asymmetric distortions of the nucleus, that is distortions that make one developing fragment of the nucleus larger and the other correspondingly smaller. To describe such pear-shaped nuclei, one introduces terms of the type $\rho^2 \varepsilon_3 Y_{30}$ (and $\rho^2 \varepsilon_5 Y_{50}$) in the nuclear potential (see chapter 8).

In fig. 9.4 the potential energy for ^{236}U is plotted in contour form as a function of an elongation-necking coordinate (approximately corresponding to the line in fig. 9.3) on the x-axis and a reflection asymmetry coordinate (pear-shape) on the y-axis. We note that the two minima and the first saddle are stable towards asymmetric distortions. The second saddle, however, is moved into the reflection asymmetry direction, and the latter degree of

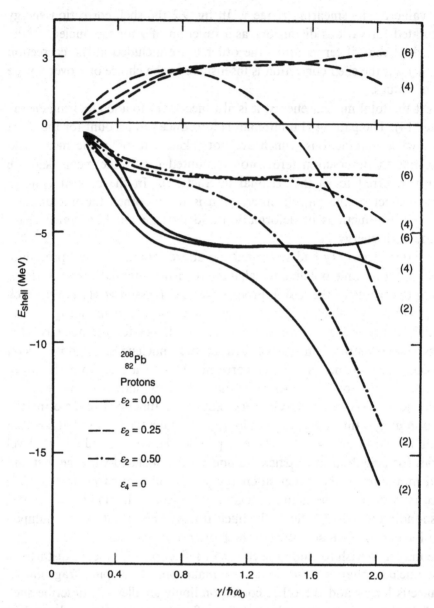

Fig. 9.2. Proton shell energy, E_{shell}, as a function of the smearing parameter γ for the nucleus ^{208}Pb. Three different orders of the correction function f_{corr} corresponding to second-, fourth- and sixth-order terms are included. Also three different shapes are considered. In all cases, a sixth-order correction function and $\gamma = 1.2\hbar\omega_0$ appears a reasonable combination (from Nilsson *et al.*, 1969).

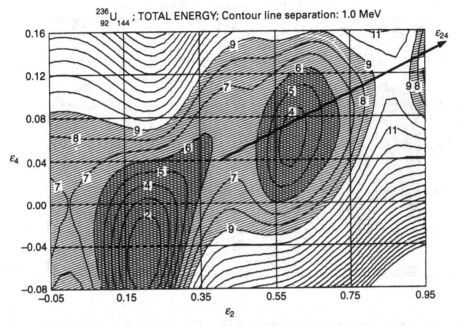

Fig. 9.3. A calculated potential-energy surface for ^{236}U in terms of the shape parameters ε and ε_4. The corresponding nuclear shapes are illustrated in fig. 8.4. Regions with a low energy are shaded and the contour line separation is 1 MeV. The solid line defines an average fission path and an associated coordinate ε_{24} (from P. Möller and S.G. Nilsson, *Phys. Lett.* **31B** (1970) 283).

freedom is found to lower the second peak of the fission barrier considerably. The effect also appears to explain the fission fragment asymmetry effect.

The instability of the second saddle to asymmetric distortions is obviously due to the fact that the shell energy decreases with increasing asymmetry. The liquid-drop energy is stable towards asymmetric distortions at these values of the fissility parameter, so the increase in liquid-drop energy must be overcome by a rather sharp decrease in the shell correction part of the energy. This in turns means that the level density at the Fermi surface decreases with asymmetry as can be studied in fig. 9.5. The asymmetry effect is caused mainly by levels of type $[Nn_z\Lambda] = [40\Lambda]$ (Johansson, 1961; Gustafsson *et al.*, 1971), which are strongly favoured (down-sloping) in the ε_3 direction due to their interaction with $[51\Lambda]$ levels (one would then have expected the latter to be strongly up-sloping but this is prevented by their interaction with $[62\Lambda]$ levels, cf. problem, 9.2). Furthermore, it has been found that the position of the interacting levels is rather insensitive to the exact radial shape of the single-particle potential.

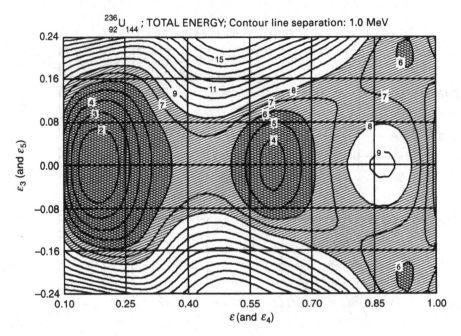

Fig. 9.4. The potential-energy surface of ^{236}U in terms of the combined ε and ε_4 parameter (in fig. 9.3 denoted ε_{24}) and a combined ε_3 and ε_5 parameter. Note that the second barrier is lowered approximately 2 Mev by reflection-asymmetric shapes (from P. Möller, *Proc. Int. Conf. on Nuclei far from Stability*, Leysin, Switzerland, 1970 (CERN 70-30, Genève, 1970) vol. 2, p. 689).

One could also imagine that the shell energy of the fragments that are formed could help to understand the outer part of the fission barrier. As the strongest shell effects are found for spherical nuclei, such an effect should be especially large if both fragments were doubly magic. The question is then of what happens if the two magic nuclei are placed side by side and allowed to overlap to some extent. If the shell energy remains low, it appears reasonable to consider a fission path leading to an approximate scission configuration of two touching spherical nuclei. This in turn means that the centres of mass of the two fragments are exceptionally close at scission so that the Coulomb energy leads to an unusually large kinetic energy of the two fragments.

It turns out to be difficult to find magic fragments for 'normal' nuclei undergoing fission. One interesting case is ^{264}Fm, which, if formed, could split into two magic ^{132}Sn nuclei. The nucleus ^{264}Fm is, however, too neutron-rich to be accessible with present techniques and we must go to somewhat lighter Fm isotopes, e.g. ^{258}Fm. Indeed, it turns out that, also for this isotope, the fission path leading to two essentially spherical fragments is very

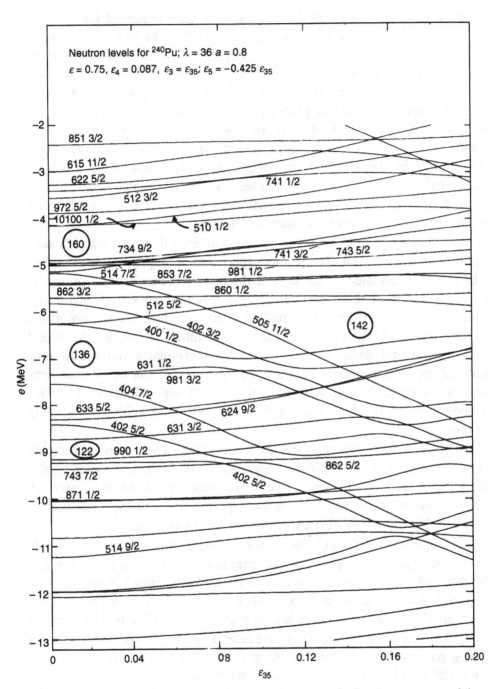

Fig. 9.5. Single-neutron energies calculated in a Woods–Saxon type potential as functions of the asymmetry coordinate ε_3 (and ε_5). It is striking how a few special orbitals profit from the inclusion of ε_3, such as orbitals [402 5/2], [404 7/2], [400 1/2], [402 3/2], [505 11/2] etc. These are all 'waist-line' orbitals (from P. Möller and J.R. Nix, *Nucl. Phys.* **A229** (1974) 269).

competitive. Calculations along this line have been presented for example by Mosel and Schmitt (1971) and Mustafa (1975). The idea was put on a more solid ground by the detailed experimental studies of fragment mass and kinetic energy distributions by Hulet *et al.* (1986). In particular, they concluded that the two paths of compact and more elongated scission shapes could be seen as competing processes in the same nucleus. Then, in more complete calculations (Möller, Nix and Swiatecki, 1989; Pashkevich, 1988; Ćwiok *et al.*, 1989) such phenomena have been described in more detail. We show in fig. 9.6 some different shapes close to scission as calculated by Pashkevich. We note the two possibilities discussed above but also a mixture of them. In this 'mixed' case, a large left–right shape asymmetry is calculated but even so, there is almost no mass asymmetry.

If the shell corrections are taken into account in evaluating nuclear masses, the agreement between theory and experiment is improved considerably. The mass is calculated as the sum of the (macroscopic) liquid-drop model mass expression at the (distorted) nuclear ground state and a microscopic correction, including shell plus pairing energies, which for doubly magic nuclei may be of the order of 10 MeV in magnitude. In fig. 9.7 theoretical and experimental results are compared. The maximum deviations are now reduced to approximately ± 2 MeV. In more recent fits of all measured masses for nuclei heavier than oxygen ($Z = 8$) (e.g. Möller *et al.*, 1992) a root-mean-square deviation somewhat smaller than 1 MeV is typically obtained.

In fig. 9.7, the large negative shell corrections around mass numbers $A = 130$–140 and $A = 200$–210 are due to the spherical closed shells, $Z = 50, 82$ and $N = 82, 126$ respectively. These nuclei are thus spherical. The regions of rather constant shell energy for $A = 150$–190 (rare earths) and $A = 230$–260 (actinides) are built from deformed nuclei.

The deformation can also be studied in more direct ways, for example by the scattering of α-particles. Information about the deformation can be deduced from the scattering cross sections as illustrated in fig. 9.8. The quadrupole and hexadecapole moments deduced from such experiments are compared with theoretical calculations within the modified oscillator model in fig. 9.9. The theoretical moments are calculated at the ground state minima of the potential energy surfaces. From the occupied wave functions, $|v\rangle$, at these deformations, the moments are extracted as

$$Q_{\lambda 0} = \sum_v \langle v \, |\mathcal{Q}_{\lambda 0}| \, v \rangle$$

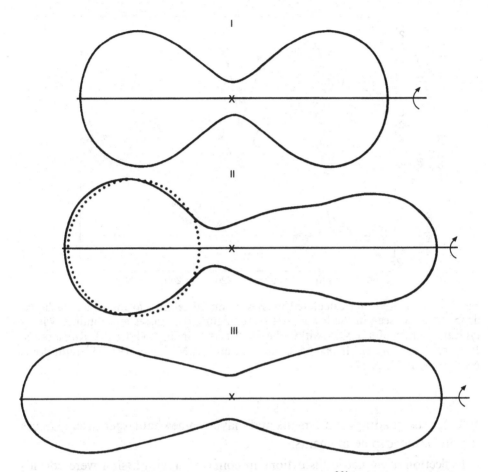

Fig. 9.6. Calculated shapes close to scission for the nucleus ^{264}Fm. In the calculations, a suitably defined 'necking coordinate' has been fixed and the three configurations correspond to minima with respect to other degrees of freedom. Case I resembles two touching ^{132}Sn nuclei and has a low shell energy caused by the $Z = 50$ and $N = 82$ magic gaps for spherical shape. Case III corresponds to a 'normal' fission while case II (with a half-volume sphere shown by dots) is a 'mixture' of I and III (from Pashkevich, 1988).

The quadrupole and hexadecapole operators $\mathcal{Q}_{20}$ and $\mathcal{Q}_{40}$ are defined as

$$\mathcal{Q}_{20} = \left(\frac{16\pi}{5}\right)^{1/2} r^2 Y_{20}(\theta, \phi)$$

$$\mathcal{Q}_{40} = r^4 Y_{40}(\theta, \phi)$$

The fits obtained in fig. 9.9 are rather typical with discrepancies smaller than

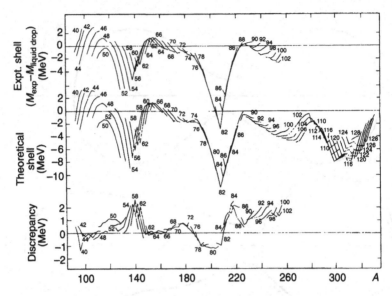

Fig. 9.7. Measured and calculated masses of nuclei relative to the spherical liquid drop value. As seen in the lower part of the figure, the masses are obtained with a typical accuracy of ± 1 MeV within the shell correction method (from I. Ragnarsson, *Proc. 6th Int. Conf. on Atomic Masses*, East Lansing, Michigan, 1979 (Plenum Press, New York, 1980) p. 87).

10% for the quadrupole moments and maybe somewhat larger discrepancies for the hexadecapole moments.

Reflection asymmetric distortions in connection with fission were considered above. For axial symmetry, such shapes are mainly described by a Y_{30}-term in the nuclear potential and, with $\lambda = 3$, this corresponds to octupole deformation. Also nuclear ground states might be octupole deformed although it is only recently that this issue has been considered in more detail (e.g. Leander *et al.*, 1982, Åberg, Flocard and Nazarewicz, 1990). The reason is that the calculated minima are generally very shallow or even do not show up in some standard calculations. For example, it seems important to use the alternative liquid-drop model (discussed in connection with fig. 4.6), which is softer towards higher multipoles. Furthermore, the single-particle potential should be realistic enough, for example of Woods–Saxon type. Potential-energy surface calculations all over the nuclear chart will then reveal reflection asymmetric minima in some specific regions. The deepest minima are obtained for heavy nuclei with $Z \approx 86$–90 and $N \approx 130$–140, but even in the 'best cases' the gain in energy due to reflection asymmetry is only around 1 MeV, see fig. 9.10. This should be compared with the energy

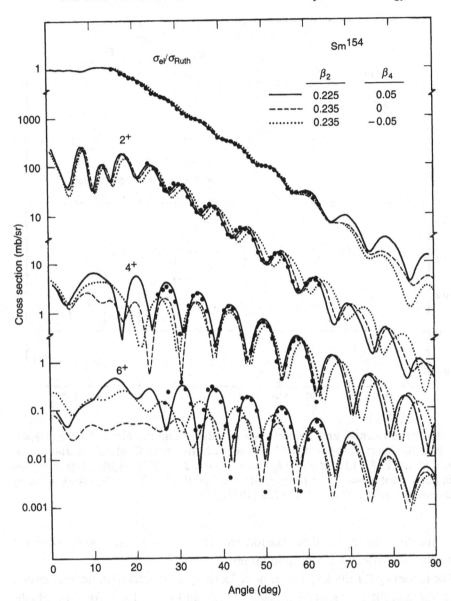

Fig. 9.8. The experimental scattering cross sections of 50 MeV alpha particles for Coulomb excitation of the 2^+, 4^+ and 6^+ excited states in ^{154}Sm. Furthermore, the elastic cross section relative to the Rutherford value is shown. The cross sections are plotted as functions of angle. A theoretical fit is made in terms of a charge shape as defined by β_2 and β_4. The best fit corresponds to $\beta_2 \approx 0.235$, $\beta_4 \approx 0.05$ (solid curves). (From N.K. Glendenning, *Proc. Int. School of Physics, 'Enrico Fermi'*, Course XL (Academic Press, New York, 1967) p.332.)

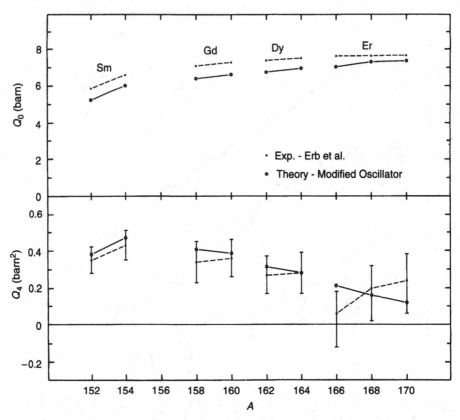

Fig. 9.9. Experimental and theoretically calculated quadrupole and hexadecapole moments for some rare-earth nuclei. The data are from Coulomb excitations as measured by K.A. Erb *et al., Phys. Rev. Lett.,* **29** (1972) 1010. The modified oscillator potential was used to calculate the equilibrium deformations as listed by I. Ragnarsson *et al., Nucl. Phys.* **A233** (1974) 329.

gain due to quadrupole deformation (relative to spherical shape), which is typically one order of magnitude larger.

The tendency for nuclei to become reflection asymmetric can be understood from the coupling of specific j-shells, (N, ℓ) and $(N - 1, \ell - 3)$, through the $Y_{3\mu}$ potential (or Y_{30} for axial shape) (Bohr and Mottelson, 1975). For example, for $Z \approx 86$–90, the $i_{13/2}$ orbitals couple with the $f_{7/2}$ orbitals and for $N \approx 130$–140, the $j_{15/2}$ orbitals couple with the $g_{9/2}$ orbitals, see figs. 8.3, and 8.5. These orbitals separate with increasing octupole deformation so that a region of lower level density is formed in an analogous way to that illustrated in fig. 9.5, where some other orbitals repel each other.

If the reflection asymmetric shape were really fully stabilised, it would strongly influence the spectrum, giving rise to a rotational band of alternating

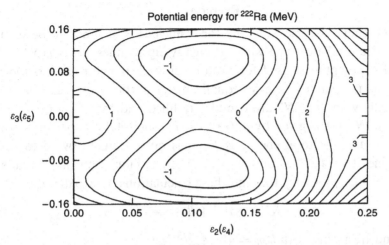

Fig. 9.10. Potential energy surface of ^{222}Ra calculated as a function of quadrupole ($\varepsilon_2(\varepsilon_4)$) and octupole ($\varepsilon_3(\varepsilon_5)$) deformation. A Woods–Saxon type single-particle potential has been used (more specifically, the folded-Yukawa potential). The energy gain due to quadrupole deformation is around 1.5 MeV and then an additional 1.2 MeV is gained because of octupole deformation (from Leander *et al.*, 1982).

parity, $0^+, 1^-, 2^+, 3^-, 4^+, \ldots$. Spectra of this kind have never been observed (at least not at very low spins), but in several nuclei, negative parity states are very low in energy. A very important consequence of the octupole deformation is that the centre of mass for the neutron distribution might be different from that of the proton distribution, i.e. the nucleus gets an intrinsic dipole moment with important consequences, in particular enhanced dipole radiation, see e.g. Leander *et al.* (1986). It seems clear that the concept of octupole deformation is very useful for our understanding of the spectroscopic properties of specific nuclei. This is so even though one could argue that it is probably difficult to find any nucleus that has acquired a shape of really permanent octupole deformation.

As indicated in chapter 5, it is also possible to do calculations within the more 'fundamental' self-consistent Hartree–Fock approach. These methods, although very time consuming, have also been applied to calculation of the fission barrier, where the two-peak character has been reproduced (Flocard *et al.*, 1974; Berger, Girod and Gogny, 1989). Also the absolute barrier heights are reproduced within a few MeV. The Strutinsky method can be formally derived from the Hartree–Fock equations (see Brack and Quentin (1981) for a review). From the calculational point of view, the Strutinsky prescription is much simpler and it has therefore made possible systematic calculations over large regions of the nuclear chart.

Exercises

9.1   The single-particle potential for a nucleus is assumed to be of harmonic oscillator type with the energy eigenvalues $e = \hbar\omega_0(N + 3/2)$. This potential is used for protons as well as neutrons and the nucleus is assumed to have equally many particles of each kind. The shells with $N \leq N^*$ are completely filled and the filling of the shell $N = N^* + 1$ is given by ρ $(0 \leq \rho \leq 1)$. The total number of particles is A and the total energy is E. It is now straightforward to determine $A = A(N^*, \rho)$ and $E = E(N^*, \rho)$. Elimination of N^* leads to $E = E(A, \rho)$. Carry through these calculations by expanding

$$N^* = \alpha A^{1/3} + \beta + \gamma A^{-1/3} + \dots$$

and from this, with $\hbar\omega_0 = 41 \cdot A^{-1/3}$ MeV,

$$E = \alpha' A + \beta' A^{2/3} + \gamma' A^{1/3} + \dots \quad \text{MeV}$$

Show that E can be split into one 'smooth part' $\tilde{E}$ and one shell correction part E_{shell} where

$$\tilde{E} = (41/8) \left[3(12)^{1/3} A + (18)^{1/3} A^{1/3} \right] \text{ MeV}$$

$$E_{\text{shell}} = (41/4) \left[12\rho(1 - \rho) - 1 \right] 12^{-1/3} A^{1/3} \text{ MeV}$$

Sketch E_{shell} and compare it with the experimental shell effects shown in figs. 3.9 and 9.7. Comments!

9.2   For a simple estimate of the single-particle effect behind the asymmetric deformations one might consider the matrix elements of the $\rho^3 Y_{30}(\theta_t, \varphi_t)$ operator (instead of $\rho^2 Y_{30}(\theta_t, \varphi_t)$, which is used in the modified oscillator). Apply the operator method of chapter 8 to find the selection rules for N' and n'_z in the matrix element

$$\left\langle N n_z \Lambda \left| \rho^3 Y_{30} \right| N' n'_z \Lambda \right\rangle$$

Calculate the distance between the corresponding orbitals as functions of ε for a pure oscillator potential. Evaluate the matrix element for those orbitals that come closest together for large ε. Compare with the single-particle diagrams of the modified oscillator, figs. 8.5 and 9.5.

9.3   The smearing function of the Strutinsky shell correction method is defined in such a way that long-range variations are preserved. Thus, with a correction function, $f_{\text{corr}}(u) = a_0 + a_2 u^2 + \dots a_m u^{2m}$, $(u = (e - e')/\gamma)$, a polynomial function $G(e)$ of order $2m + 1$ should remain unchanged by the smearing procedure. This makes it possible to derive $f_{corr}(u)$ by direct calculation.

(a) Use a polynomial $G = \alpha_0 + \alpha_2 e^2$ to determine a_0 and a_2. Sketch the corresponding smearing function.

(b) Show that with these coefficients, a_0 and a_2, a general polynomial of order 3 also remains unchanged.

(c) Determine a system of equations for a_i in the general case with G of order $2m$.

10

The barrier penetration problem – fission and alpha-decay

The deformation parameter ε (or β_2 discussed in chapter 4) is not enough to bring us to the point of scission. For example, $\varepsilon = 1.5$ corresponds to needle-like shapes. Indeed, independently of parametrisation, fission should be treated as a multi-deformation-parameter problem.

Let us, however, assume for the moment that the problem is a one-dimensional one and ε the relevant parameter. According to simple WKB theory the penetration probability for the penetration of a barrier is given by the expression (cf. appendix 10)

$$P \approx \exp\left(-\frac{2}{\hbar}\int_{\varepsilon'}^{\varepsilon''}[2B(V(\varepsilon) - E)]^{1/2}\,\mathrm{d}\varepsilon\right) \equiv \exp\left(-K\right)$$

where ε' and ε'' are the points of entry and exit of the fission barrier. For an initial excitation energy near that of the top of the barrier there exists an improved expression:

$$P = (1 + \exp K)^{-1}$$

The integral is the well-known action integral. Usually one is more familiar with the corresponding expression for the case that ε is replaced by the length coordinate x and B by the mass M of the penetrating particle. With our parameter choice, B takes the dimension of mass times (length)^2, or moment of inertia, as ε is dimensionless. The quantity $V(\varepsilon)$ represents the potential energy considered in the foregoing (cf. fig. 10.1). In the actual case, $B = B(\varepsilon)$ turns out to be strongly ε-dependent but, to make things simple, let us assume B to be constant, replacing $B(\varepsilon)$ by some kind of mean value. This mean value may not be too different from the value of $B(\varepsilon)$ at the saddle point, where the contribution to the integral is expected to be the largest. The quantity P is defined as the probability of penetration through the barrier for a given 'assault'. The 'assaults' correspond to the natural characteristic

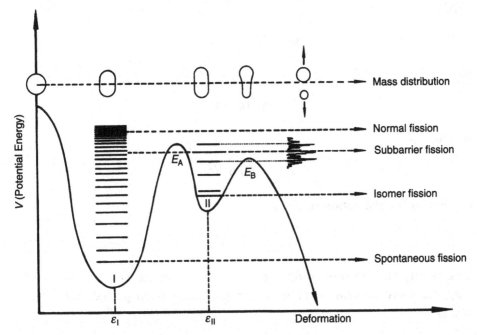

Fig. 10.1. Schematic illustration of different fission possibilities for a nucleus in the $A \approx 240$ region. The nucleus might undergo fission spontaneously from the ground state, it might be trapped in the second minimum before it goes to fission (isomer fission) or it might first get excited, for example by an impinging neutron (normal fission or sub-barrier fission). The corresponding nuclear shapes with preference for asymmetric deformations in the second barrier region are also exhibited.

zero-point vibrations of the nucleus. Thus the number of assaults, n, is usually equated to the frequency of vibration in the elongation coordinate (e.g. ε), so called β-vibration (cf. problem 10.2). One may therefore set $n = 10^{21}$ s^{-1}, corresponding to a vibrational frequency of $\hbar\omega_{\mathrm{vib}} \approx 1$ MeV. We have thus obtained an estimate of the life-time in units of seconds,

$$\tau \approx \frac{1}{n} \cdot \frac{1}{P} \approx 10^{-21} \exp K$$

which, converted to half-life, is

$$t_{1/2} = \ln 2 \cdot 10^{-21} \exp K$$

Consider first a simple parabolic barrier of height S over the energy minimum with the barrier top at $\varepsilon = \varepsilon_S$, situated half-way between ε' and ε'',

$$V(\varepsilon) = S - \frac{1}{2} C (\varepsilon - \varepsilon_S)^2$$

where thus

$$\varepsilon_S = (\varepsilon' + \varepsilon'') / 2$$

and

$$C = 8S / (\varepsilon' - \varepsilon'')^2$$

Introducing

$$\omega_f = (C/B)^{1/2}$$

one obtains for the action integral

$$K = 2\pi S / \hbar \omega_f$$

The quantity $\hbar\omega_f$ (of dimension energy) is usually called the 'transparency'. Using the more accurate expression for the penetration probability P one obtains

$$P = [1 + \exp(2\pi S / \hbar \omega_f)]^{-1}$$

For a purely parabolic shape the WKB approximation happens to give the exact result as shown by Hill and Wheeler (1953). In this formula S represents the barrier height, while the frequency ω_f contains both the barrier curvature C and the inertial parameter B (cf. the vibrational frequency for a particle in a parabolic potential well, e.g. problem 10.2).

From the Hill–Wheeler expression we obtain $P = 1/2$ for zero barrier. The zero barrier point accordingly is the point where, as a function of the excitation energy E, the penetrability has diminished to $1/2$ of that of the limit corresponding to infinite excitation energy. It is also the point of maximum change in penetrability with E.

Indeed, the Hill–Wheeler formula has long been used for the analysis of fission cross section data. From the dependence of the penetrability on excitation energy E, not only S but also an average $\hbar\omega_f$ appears experimentally accessible. Analyses of experimental data give S as dropping from 8 to 4 MeV when A goes from 230 to 250. Various empirical values of $\hbar\omega_f$ are available, centred around about 500 keV. Note, however, that this is a very simplified discussion because it is now well established that most nuclei in the $A = 230-250$ region have a two-peak barrier. Various calculations based on such a two-peak barrier have also been published (see Vandenbosch and Huizenga, 1973, for a review).

10.1 Application of the fission half-life formula to actinides

About 40 fission ground state half-lives are known for even–even transuranium elements. In addition a few even–even fission isomeric half-lives have been measured (most fission isomers being odd-A or odd–odd cases). This material provides an excellent set of data for testing our knowledge about the fission process. The different modes of the fission process are illustrated schematically in fig. 10.1.

Up to now very little has been said about the inertia parameter. For the multidimensional case the one-component function $B(\varepsilon)$ is replaced by a tensor function $B_{\alpha_i \alpha_j}(\alpha_1, \alpha_2, \dots)$ where the α_i denote the coordinates of deformation, e.g. ε, ε_4, ε_3, γ.... The action integral is then calculated along a trajectory L in the multidimensional deformation space

$$K(L) = \frac{2}{\hbar} \int_{s_1}^{s_2} [2(V(s) - E)B_s(s)]^{1/2} \, ds$$

where s specifies a point on the trajectory. The effective inertia function $B_s(s)$ along the trajectory is expressed as

$$B_s(s) \equiv B_{ss}(s) = \sum_{ij} B_{\alpha_i \alpha_j}(s) \frac{d\alpha_i}{ds} \frac{d\alpha_j}{ds}$$

The components of the inertia tensor $B_{\alpha_i \alpha_j}(s)$ are calculated from the nuclear wave function at the deformation s. We will not give any formulae but only point out that the problem of calculating $B_{\alpha_i \alpha_j}(s)$ is very similar to the calculation of the moment of inertia which will be discussed in appendix 14B.

It is now possible to calculate the action integral $K(L)$ along different trajectories L. So-called dynamical calculations are understood as a search for that trajectory $L_{\min}$ which minimises $K(L)$. Subsequently, $K(L_{\min})$ is inserted in the half-life formula given above, leading to a number, which can be compared with measured half-lives. Such calculations have been performed e.g. by Baran *et al.* (1981). They consider essentially the three deformation degrees of freedom ε, ε_4 and ε_3. Furthermore, they make sure that the fission half-lives are influenced by non-axial γ shapes only to a minor extent and therefore neglect this degree of freedom in their final results.

The theoretical fission half-lives of Baran *et al.* are compared with experimental ones in fig. 10.2. Considering the fact that no parameters have been fitted (in addition to the standard ones involved in the fission barrier calculations, see chapter 9), the agreement is really remarkable. The unusually long half-life for $N = 152$ isotones reflects the stability of the ground states mass associated with $N = 152$. This in turn is associated with the $N = 152$

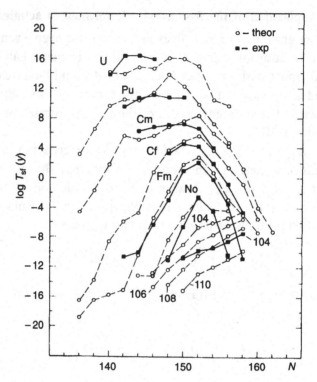

Fig. 10.2. Measured and calculated half-lives given in years for spontaneous fission of elements in the $A \approx 250$ region (from Baran *et al.*, 1981).

gap as observed for $\varepsilon = 0.2$–0.3 in fig. 8.5. The rather large discrepancy obtained for the heaviest Fm and No isotopes is probably associated with the fact that the alternative fission path discussed in connection with fig. 9.6 was not considered when fig. 10.2 was constructed, see e.g. Möller, Nix and Swiatecki (1989).

The predicted half-lives of some $Z = 106$–110 elements are also given in fig. 10.2. We will briefly discuss the synthesis of such elements in the last section of this chapter.

Some calculated and measured half-lives for fission from the isomeric minimum at $\varepsilon \approx 0.6$ are given in fig. 10.3. The calculated results are influenced by the relative depth of the second minimum compared with the second barrier, by the width of the second barrier and by the inertia parameter. In view of the uncertain single-particle orbitals at large deformations, the agreement between theory and experiment is at least as good as could be expected. The half-lives for ground state fission on the other hand are more sensitive to the properties at smaller deformations (especially the ground state

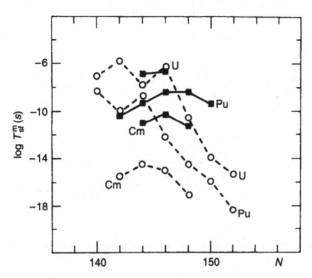

Fig. 10.3. Measured and calculated half-lives given in seconds for 'isomer fission' of some $_{92}$U, $_{94}$Pu and $_{96}$Cm isotopes (from A. Łukasiak *et al., Proc. 4th Int. Conf. on Nuclei far from Stability*, Helsingør, Denmark, 1981 (CERN 81-09, Geneva, 1981) p. 751).

mass). As the single-particle model has been designed mainly to describe the ground state properties, it seems rather natural that these half-lives are better reproduced than the isomeric ones.

10.2 Alpha-decay

One way to treat alpha-decay is to consider it as a barrier penetration problem essentially analogous to the fission process, see e.g. Rasmussen (1965). The decay probability is then roughly the product of the probability of alpha-particle formation and the probability of barrier penetration by the alpha-particle. It turns out that the first probability can be considered to be essentially the same in one nucleus and another. The barrier penetrability depends, however, in a smooth and systematic way on 1) the available energy 2) the charge of the daughter nucleus and to a minor extent 3) the mass or, equivalently, size of the daughter nucleus.

As shown in fig. 10.4 the potential far outside the nucleus is entirely determined by the Coulomb interaction,

$$V_C = \frac{ZZ'e^2}{4\pi\varepsilon_0 r}$$

where Z is the charge of the daughter nucleus and $Z' = 2$ is the charge of

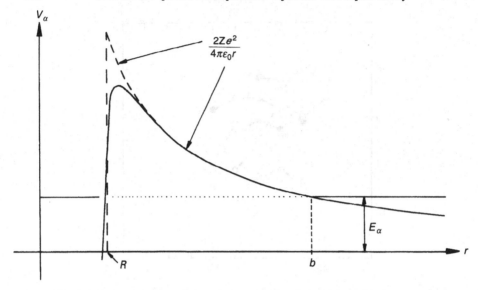

Fig. 10.4. The potential acting on the alpha-particle. The dashed curve shows the Coulomb barrier while the full curve exhibits a more realistic barrier where the effects of the nuclear forces are also accounted for.

the emerging alpha-particle. Inside the range of nuclear forces, extending from the density radius out to a distance of the order of $\hbar/m_\pi c$, the rise of the Coulomb potential is compensated by the attractive nuclear force field. The result is a barrier of the type shown in fig. 10.4. In the calculations we shall for simplicity assume a rise of the barrier given by V_C for diminishing r until an effective radius, R, where there is a sudden cut-off of the potential. We have thus

$$P = e^{-K}$$

with

$$K = \frac{2}{\hbar} \int_R^b \left[2M'_\alpha \left(\frac{2Ze^2}{4\pi\varepsilon_0 r} - E_\alpha \right) \right]^{1/2} dr$$

where E_α is the α-particle energy and b the classical turning point as shown in the figure,

$$b = \frac{2Ze^2}{4\pi\varepsilon_0 E_\alpha}$$

The reduced mass of the alpha-particle is denoted by M'_α ($\approx 0.98 M_\alpha$) and its energy by E_α. We have then to evaluate the integral

$$J = \int_R^b \left(\frac{1}{r} - \frac{1}{b}\right)^{1/2} dr$$

This is easily done by the substitution $r = b\cos^2 u$:

$$J = \sqrt{b} \left\{ \arccos\left(\frac{R}{b}\right)^{1/2} - \left[\frac{R}{b}\left(1 - \frac{R}{b}\right)\right]^{1/2} \right\}$$

For the approximation $R \ll b$ we find

$$J \approx \sqrt{b} \cdot \left[\frac{\pi}{2} - 2\left(\frac{B}{b}\right)^{1/2}\right]$$

which leads to the following expression for K

$$K = 3.92 Z E_\alpha^{-1/2} - 3.22 Z^{1/2} A^{1/6}$$

where we have inserted $R = r_0 A^{1/3}$ with $r_0 = 1.2$ fm. As we are mainly interested in heavy nuclei, we put $A = 2.5Z$ and thus get K as a function of E_α and Z only:

$$K = 3.92 Z E_\alpha^{-1/2} - 3.75 Z^{2/3}$$

In analogy with the fission decay case, the life-time for alpha-decay is calculated as

$$\tau \approx \frac{1}{x} \cdot \frac{1}{P}$$

where x is the number of assaults. From the motion of an alpha-particle in a square well potential of radius R, we estimate

$$x \approx 10^{22} \text{s}^{-1} \approx 10^{29} \text{ years}^{-1}$$

the inverted time of travel of the alpha-particle across the nucleus.

In this way, we arrive at a formula for the half-life

$$t_{1/2} = \ln 2 \cdot 10^{-29} \exp{(K)} \text{ years}$$

or

$$\log_{10}\left(t_{1/2}\right) = C_1 Z E_\alpha^{-1/2} - C_2$$

With $t_{1/2}$ expressed in years and E_α measured in MeV:

$$C_1 = 1.7 \text{ MeV}^{1/2}$$

$$C_2 = 29 + 1.6 Z^{2/3}$$

An analysis of alpha-decay data in terms of C_1 and C_2, as performed by Taagepera and Nurmia (1961) gives

$$C_1 = 1.61 \text{ MeV}^{1/2}$$

$$C_2 = 28.9 + 1.61Z^{2/3}$$

For nuclei in the $Z = 60$–100 region, this formula gives deviations that are generally smaller than one power of ten. In view of the very large variation of the experimental half-lives, about 10^{-15}–10^{15} years, this fit must be considered very good.

Up to now we have only considered the situation where the α-particle has orbital angular momentum $\ell = 0$ relative to the daughter nucleus. The case of $\ell \neq 0$ is, however, rather easily handled if a centrifugal barrier, $\ell(\ell + 1)\hbar^2/2M'_\alpha r^2$, is added to the Coulomb barrier in the barrier penetration calculation.

In the derivation above, we have implicitly assumed that in the nucleus there is always an alpha-particle ready to escape. This is of course not true and a correct treatment of the formation factor would increase our estimate of the half-life. A factor that will work in the opposite direction is deformation. The alpha-particle will then have a smaller barrier to penetrate at the poles of the prolate nucleus than at the equator and the overall effect will be a higher penetration probability. These two observations are consistent with the fact that the half-lives of the spherical nuclei around ^{208}Pb are generally underestimated by the Taagepera–Nurmia formula, while no systematic deviations are found for deformed nuclei.

10.3 The stability of superheavy nuclei

Let us now come back to the question of the existence of an island of superheavy elements as indicated schematically in fig. 10.5. From this figure, it becomes clear that the stability of such an island should be determined almost exclusively by shell effects, a feature shown in more detail in fig. 10.6. There, the calculated liquid-drop barrier of some heavy nuclei is compared with the total barrier that results if shell and pairing energies are also added. As was already discussed in chapter 4, the superheavy island around $Z = 114$ and $N = 184$ corresponds to a fissility parameter close to $x = 1$, i.e. the liquid-drop fission barrier has essentially vanished. Note, however, that the liquid-drop energy is almost constant in the region $\varepsilon \approx 0 - 0.4$. Therefore, with a deep shell energy minimum for spherical shape, and rather small

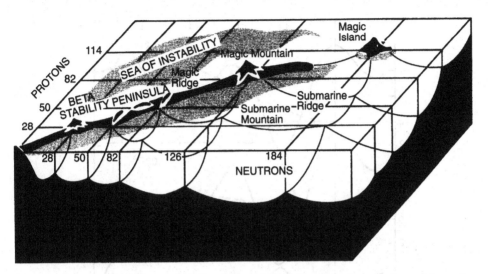

Fig. 10.5. An artist's view of the stability peninsula with the magic numbers for neutrons and protons marked on the x- and y-axes, respectively. The magic numbers give rise to ridges and mountains on the peninsula. Of special interest is the Magic Island, which is supposed to be created from combination of the extrapolated magic numbers, $Z = 114$ and $N = 184$ (partly from G.T. Seaborg and J.L. Bloom, *Scientific American*, April 1969, p. 57).

shell effects for deformed shapes, a substantial fission barrier reaching out to $\varepsilon = 0.6$–0.7 is created.

Calculated fission barrier heights in the heavy and superheavy region are shown in the form of a contour plot in fig. 10.7. This figure should mainly be taken as an illustrative example, keeping in mind that different extrapolations of the single-particle orbitals (see figs. 6.8, and 6.9) might lead to rather different results. The heaviest nuclei to have been synthesised so far are situated around $Z = 108$ and $N = 156$. Thus, it is by now well established that, in agreement with fig. 10.7, the barriers in this region are quite high.

Models that are able to describe the fission half-lives of heavy elements (e.g. fig. 10.2) can easily be extrapolated into the superheavy region. In the calculations presented in fig. 10.8, the fission half-life of $^{298}114$ is about 10^{15} years. The uncertainty of this number is illustrated by the fact that either a 1 MeV change in barrier height or a 5% change in barrier width or a 10% change in the mass parameter will lead to a change of about two orders in the fission half-lives. In view of this and also in view of other calculations, which have given rather different results, the fission life-times of fig. 10.8 must be considered uncertain within many orders of magnitude. For

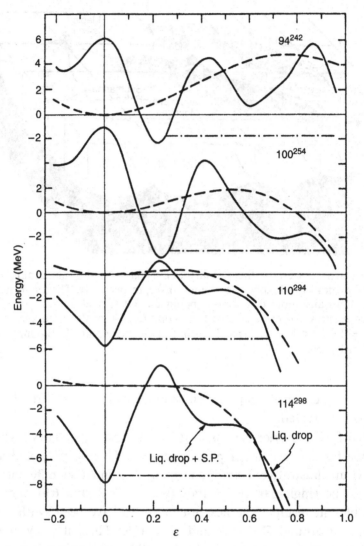

Fig. 10.6. Calculated barrier along the 'path of fission' for heavy and superheavy elements. Dashed lines mark the liquid-drop fission barrier. Solid lines are obtained after the inclusion of shell and pairing terms (from Nilsson *et al.*, 1969).

example, in one more recent study (Möller *et al.*, 1986), a longest half-life of 200 days is calculated in the superheavy region.

The stability of the superheavy nuclei is determined not only by fission but also by α- and β-decay. As the nucleus $^{298}114$ happens to fall almost on the line of β-stability (see fig. 3.7), many nuclei in this region should be β-stable (see fig. 10.8). Alpha-decay, on the other hand, is more of a problem and the

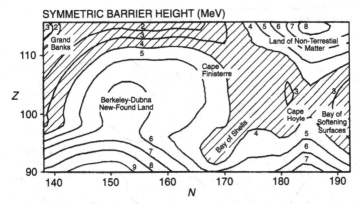

Fig. 10.7. Example of calculated fission barrier heights for nuclei in the heavy and superheavy region. The names given in the figure have been introduced to facilitate the reference to specific areas of the nucleid chart. The heaviest nuclei to have been synthesised so far are on the (Berkeley–Dubna) New-Found Land or even a little further to the 'north' where a lot of work has recently been done at the Darmstadt Heavy Ion Accelerator, see e.g. Armbruster (1985) (from R. Bengtsson, R. Boleu and S.E. Larsson, *Phys. Scripta* **10A** (1974) 142).

calculated total half-lives of fig. 10.8 are determined by α-decay for some nuclei and by fission for others. There is a very strong general trend that the α half-lives are short for proton-rich nuclei because α-decay will then lead to a daughter nucleus closer to β-stability (i.e. more strongly bound on average) while the situation is reversed on the neutron rich side (cf. problem 10.4). Considering both α-decay and fission, the largest half-life in fig. 10.8 is obtained for $^{294}110$, a result first obtained by Nilsson *et al.* (1969) and which is common for several different calculations (see e.g. Sobiczewski, 1974, 1978 and Nilsson, 1978).

Considering the fact that the total life-times in the superheavy region might be as long as 10^8–10^9 years it seems worthwhile to search for such elements on the earth or in cosmic rays impinging on the earth. Except for a long life-time, this would, however, also require that there exist some astrophysical process in which superheavy elements are formed. Indeed, most studies suggest that no such process exists (Boleu *et al.*, 1972; Howard and Nix, 1974) and no solid evidence for the existence of superheavy elements in nature has been reported although many different samples have been investigated (see e.g. Herrmann, 1980).

The remaining possibility is to synthesise superheavy elements in the laboratory (Herrmann, 1980). One feasible process seems to be fusion of two heavy ions. It is instructive first to consider how the heaviest elements

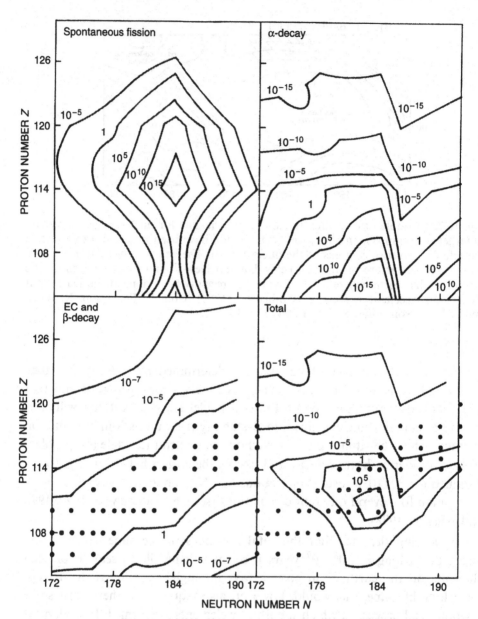

Fig. 10.8. Examples of calculated fission, α-decay and β-decay/electron-capture half-lives in the superheavy region. Nuclei that are β-stable are indicated by points. In the lower right hand figure, the total half-lives resulting from the three decay modes are indicated. The contour lines are marked by half-lives in years (from E.O. Fiset and J.R. Nix, *Nucl. Phys.* **A193** (1972) 647).

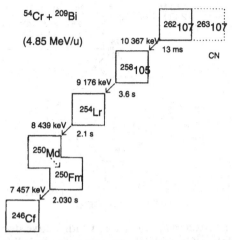

Fig. 10.9. Example of a decay chain observed in the irradiation of ^{209}Bi with 4.85 MeV/nucleon ^{54}Cr ions. First, one neutron is evaporated, then three consecutive α-particles followed by one positron and finally another α-particle. The end product is then the $^{246}_{98}$Cf$_{148}$ nucleus having a half-life of 36 hours (from G. Münzenberg *et al., Proc. 4th Int. Conf. on Nuclei far from Stability*, Helsingør, 1981 (CERN 81-09) p. 755).

known today are being synthesised. Thus, we show in fig. 10.9 how the fusion of ^{54}Cr and ^{209}Bi leads to the compound nucleus ^{263}107. After the emission of one neutron, some α-particles and one positron, one ends up with ^{246}Cf. In the time-scale of the experiment, this latter element is essentially stable, having a life-time of 36 hours. By measuring the different particles emitted and from previous knowledge of the decay chain from ^{258}105 it is possible to conclude that element ^{262}107 was really formed in the reaction. Furthermore, the energy of the α-particle gives the mass of ^{262}107 and it is of course also possible to measure the α-decay half-life.

One might then expect that it would be straightforward to form superheavy elements from fusion of two somewhat heavier elements. However, some different problems arise. The line of β-stability tends towards increasingly neutron-rich species with increasing mass. Therefore it is impossible to find two stable nuclei that can be combined to the desired neutron to proton ratio. If their neutron and proton numbers are simply added, one cannot achieve simultaneously that $Z = 114$ and $N = 186$. Some kind of compromise then has to be made. One reaction that has been suggested and tried (e.g. Oganessian *et al.*, 1978, Illige *et al.*, 1978, Armbruster *et al.*, 1985) is illustrated in fig. 10.10. With calculated life-times as input, the reaction chain that would result from fusion of ^{48}Ca with ^{248}Cm is shown. One notes

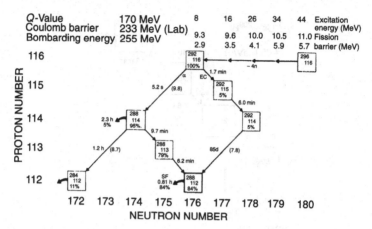

Fig. 10.10. Calculated neutron evaporation and radioactive decay chain after complete fusion of 255 MeV ^{48}Ca with ^{248}Cm using the half-lives of fig. 10.8. The estimated Q-value, i.e. the difference in total binding energy before and after the fusion, and the Coulomb barrier seen in the laboratory system is given. If the bombarding energy is transformed to the centre of mass system and compared with the Q-value, one easily calculates an excitation energy of 44 MeV for the ^{296}116 nucleus. This excitation energy should primarily be carried away by neutrons. Also listed are the fission barriers underlying the calculations of fig. 10.8 and of Randrup *et al.*, (1974). Note the large discrepancies between the two calculations where the latter one gives much lower fission barriers in this 'north-western' part of the super-heavy island (from G. Herrman, *Proc. 4th Int. Conf. on Nuclei far from Stability*, Helsingør, 1981 (CERN 81-09) p. 772).

that the compound nucleus is ^{296}116, i.e. four neutrons too little and two protons too much compared with the doubly magic ^{298}114. Additionally, most of the excitation energy inevitably created in the reaction is carried away by neutrons (cf. problem 11.4 below) leading further away from the centre of the island. The number of neutrons emitted is of course strongly connected with the bombarding energy and one of the intricate problems is to find the optimal bombarding energy. Fewer neutrons were emitted in the reaction of fig. 10.9, partly because of a comparatively low bombarding energy and partly because of strong binding for ^{209}Bi, being a neighbour of the doubly magic ^{208}Pb nucleus. Note also that owing to the Coulomb barrier (cf. problem 6.11) there is no prompt emision of protons.

Although the reaction ^{48}Ca + ^{248}Cm does not lead to the centre of the superheavy island, complete fusion would still produce quite long-lived species according to the estimate of fig. 10.8. No trace of the associated α-particle or fission products have, however, been observed. One has therefore been able to conclude only that the fusion cross section is extremely small, putting an upper limit on it. In view of the fact that the fusion of ^{54}Cr + ^{209}Bi (fig. 10.9) is just on the limit of being observable and one generally expects diminishing cross sections with increasing mass, this is not so surprising.

Another way to produce superheavy elements might be through collisions of very heavy elements leading to a transfer of a large number of nucleons, e.g.

$$^{238}_{92}U_{146} + ^{238}_{92}U_{146} \rightarrow {}^{298}114_{184} + ^{178}_{70}Yb_{108}$$

According to theoretical estimates, the cross section for such a reaction might be large enough to make observation of superheavy elements feasible. The uncertainties are, however, large and no evidence for production of superheavy elements in such reactions has been reported (see Herrmann, 1980, for references).

The hunt for superheavy elements has thus still not been successful but is going on with more and more sophisticated methods. It seems that theoretical calculations can only indicate that the chances for success are so large that it is worth going on but also that the uncertainties are large and we can neither expect nor exclude that any superheavy elements will be synthesised in the near future.

Exercises

10.1 Consider a one-dimensional model for fission where the nuclear shapes along the fission path are described by

$$R(\theta) = R_a \left[1 + a_2 P_2(\cos \theta) \right]$$

The corresponding fission barrier in the liquid-drop model was calculated in problem 4.2

$$E_f = \overset{0}{E_s} \left((1 - x)\frac{2}{5}a_2^2 - (1 + 2x)\frac{4}{105}a_2^3 + \ldots \right)$$

If we set $x = 0.76$, which is approximately valid for ^{238}U, we get

$$E_f = 0.096 \, \overset{0}{E_s} \cdot a_2^2 (1 - a_2)$$

We now consider the nucleus as consisting of two equal parts divided at $z = 0$ and use the distance between the two centres of mass as our coordinate, r. For large values of r, we will observe two fragments moving away from each other with the mass parameter B_r given by the reduced mass μ. Before the scission point, one expects a mass parameter that is bigger than μ and which increases with decreasing r.

We will now carry through a simple calculation to get some idea about the value of B_r for small values of r. To this end we set $B_r = k\mu$ where k is a constant. We furthermore neglect shell effects. The total length of the nucleus along the z-axis is $2R(\theta = 0) = 2R_a(1 + a_2)$. To get simple calculations, we put r equal to half of this value and neglect the difference between R_a and R_0, $r = R_0(1 + a_2)$; ($R_0 = r_0 A^{1/3}, r_0 = 1.2$ fm). Apply the WKB approximation to calculate the life-time of ^{238}U under these assumptions. The integral may be solved either numerically or analytically. Determine the value of k corresponding to the measured half-life of ^{238}U, $t_{1/2} = 10^{16}$ years. How is $t_{1/2}$ affected if B_r is changed by 10%? Comments?

10.2   Use the same model as in problem 10.1 but consider instead vibrations for small deformations. Let d_0 be the ground state value of r and use the potential energy shown in fig. 4.2 as a starting point to find an approximate expansion around the ground state minimum. Approximate the mass parameter by the reduced mass μ and by 10μ, respectively, to get an order of magnitude estimate of the energy of the lowest state for vibration in the r-coordinate, so-called β-vibration. Compare also with the number of assaults, n, that enters in the fission life-time formula.

10.3   Free particles of energy E and mass M approach a potential barrier of height V and width a. Show that the transmission coefficient T equals

$$T = \frac{(2k\kappa)^2}{(k^2 + \kappa^2)^2 \sinh^2(\kappa a) + (2k\kappa)^2}$$

The quantities k and κ are defined as $\hbar^2 k^2/(2M) = E$ and $\hbar^2\kappa^2/(2M) = V - E$.

10.4   Neglect the shell effects for heavy elements and use the simple binding energy formula of problem 3.7 in order to get a general idea of alpha half-lives on different sides of the line of β-stability. Consider

(a) the proton-rich nucleus $^{264}108$, which is situated in the region of the heaviest elements that have been synthesised,

(b) the nucleus $^{294}110$ close to β-stability and situated on the island of suggested superheavy nuclei,

(c) the neutron-rich nucleus $^{284}_{98}$Cf, which according to the calculations of fig. 10.7 has a rather high fission barrier.

Apply the Taagepera–Nurmia half-life formula. The binding energy of the α-particle is 28.296 MeV.

From fig. 3.9, one can estimate that shell effects alter the α-energies, E_α, by less than 1 MeV in deformed regions while corrections to E_α of up to 3 MeV might occur around closed shell nuclei. Calculate the corresponding correction to the α half-lives for the nucleus $^{294}110$.

Appendix 10A

Barrier penetration in one dimension with piece-wise constant barrier

We consider the one-dimensional barrier penetration problem of fig. 10A.1. Free particles of energy E and mass M enter from the left towards a constant barrier of height V extending from 0 to a. To the left of the barrier (region I) we have incoming and reflected waves with the k-number given by

$$\frac{\hbar^2 k^2}{2M} = E$$

or

$$\Psi^I = A\mathrm{e}^{ikx} + B\mathrm{e}^{-ikx}$$

In the barrier region (II) we have solutions of the type

$$\Psi^{II} = C\mathrm{e}^{-\kappa x} + D\mathrm{e}^{\kappa x}$$

with

$$\frac{\hbar^2 \kappa^2}{2M} = V - E \qquad (\text{where } V > E)$$

Finally in the region to the right of the barrier (III) we have only outgoing waves and thus the solution

$$\Psi^{III} = F\mathrm{e}^{ikx}$$

We will now be interested in the transmission coefficient, T, corresponding to the problem of how much the incoming amplitude A is reduced by the passage of the barrier

$$T = \left|\frac{F}{A}\right|^2$$

The solution to the problem is provided by the matching of solutions Ψ^I and Ψ^{II} at $x = 0$ and Ψ^{II} and Ψ^{III} at $x = a$. The result (the derivation is

178

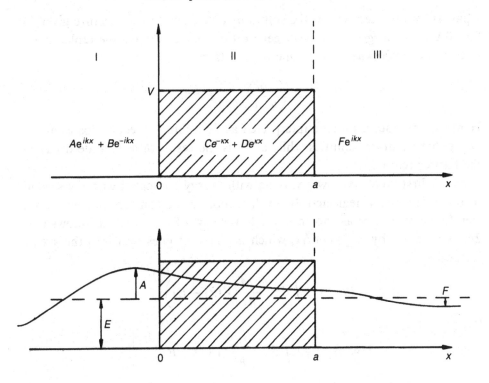

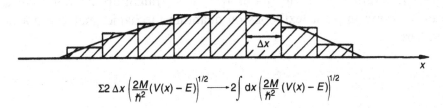

Fig. 10A.1. Schematic illustration of the quantum mechanical barrier penetration solution. The solution explicitly given in the upper figure is drawn schematically in the middle figure (apart from reflected waves). Note the exponential decrease of the amplitude in the barrier region. In the lower figure, it is shown that the essential features of the WKB solution are obtained if an arbitrary barrier is approximated by a step-function barrier.

requested in problem 10.3) is

$$T = \left| e^{-ika} \frac{2k\kappa}{2k\kappa \cosh(\kappa a) - i(k^2 - \kappa^2)\sinh(\kappa a)} \right|^2$$

Apart from reflected waves, the solution corresponds to the picture given in fig. 10A.1. For *large* κa, or *small* penetration, one can make the replacement $\sinh(\kappa a) \approx \cosh(\kappa a) \approx \frac{1}{2}e^{\kappa a}$. One then obtains

$$T = \left(\frac{4k\kappa}{k^2 + \kappa^2}\right)^2 e^{-2\kappa a}$$

Here the first factor may be interpreted as describing reflection losses at the two potential discontinuities while $e^{-2\kappa a}$ is the result of amplitude decay in the barrier region.

As the first factor is slowly varying with energy compared with the second factor, it is usually neglected. If we then replace the constant barrier with a step-function barrier as shown in the lower part of fig. 10A.1, it follows that $2\kappa a$ is replaced by $2\sum \kappa(x) \Delta x$, which in the limit goes over into the action integral

$$2\kappa a \rightarrow 2\int dx \left(\frac{2M}{\hbar^2}(V(x) - E)\right)^{1/2}$$

or

$$T \approx \exp\left[-2\int dx \left(\frac{2M}{\hbar^2}(V(x) - E)\right)^{1/2}\right]$$

as used in the main text.

We have thus derived the main ingredients of the WKB formula in a simple and intuitive way. The full derivation is straightforward but tedious. It can be found in most textbooks on quantum mechanics and will not be given here.

11

Rotational bands – the particle–rotor model

A general frame for the description of rotational states in nuclei was set in the beginning of the fifties by Bohr (1952) and by Bohr and Mottelson (1953). Rotation is a typical example of a collective degree of freedom in nuclei. A collective excitation is characterised by the coherent movement of a large number of nucleons. Thus, an elementary understanding of collective excitations is often achieved from macroscopic models. One example is nuclear fission, which could be considered as some kind of very large amplitude shape vibration. It is then also straightforward to introduce shape vibrations in general as a collective degree of freedom as illustrated in an elementary way in problem 10.2. In a more general context, shape vibrations can be described for example by the variation around the equilibrium value of the $\alpha_{\lambda\mu}$ parameters introduced in chapter 4. The most important and first non-trivial mode corresponds to $\lambda = 2$, quadrupole vibrations.

When describing nuclear quadrupole vibrations in the laboratory system, one has to introduce all the five $\alpha_{2\mu}$ shape parameters. These can, however, be transformed to a body-fixed system where two parameters describe deformations, namely in the ε_2 (or β_2) and the γ degrees of freedom (the γ parameter was introduced in chapter 8). The three additional parameters then describe the orientation of the body-fixed system, e.g. by the three Euler angles. These three parameters thus describe the rotational motion, which is treated in the present chapter and continued in chapter 12. Vibrations, on the other hand, will not be treated here but instead we refer to the literature, e.g. Rowe (1970) and Eisenberg and Greiner (1987).

When it comes to a quantum mechanical description, a further important observation is that one cannot define any collective rotation around a symmetry axis. This is seen from the fact that such a rotation would change only a trivial phase factor in the wave function (for example in the $e^{i\Lambda\varphi}$ part in the single-particle orbitals of a potential with cylindrical symmetry). Such an

181

unchanged wave function is in contrast to collective rotation. Instead, collective rotation is characterised by small angular momentum contributions from a large number of particles, i.e. the wave functions of these particles change slowly with increasing angular momentum.

What has been said above implies that only deformed nuclei can rotate collectively and, if the nucleus is axially symmetric, the only possible rotation axis is perpendicular to the symmetry axis. For collective rotation, it is then also possible to define *one* moment of inertia, $\mathscr{J}$, leading to the following Hamiltonian

$$H_{\rm rot} = \frac{\mathbf{R}^2}{2\mathscr{J}}$$

where $\mathbf{R}$ is the collective angular momentum. For pure collective rotation the total angular momentum (often referred to as the total spin) $\mathbf{I}$ is equal to $\mathbf{R}$. The spectrum then takes the form

$$E_I = \frac{\hbar^2}{2\mathscr{J}} I(I+1).$$

As only deformed nuclei exhibit rotational spectra, it should be possible to determine which nuclei are deformed from the occurrence of rotational bands. In practice, really pure rotational bands are never realised in nuclei but instead, rotations and vibrations are more or less mixed. Even so, with a not very strict definition of a rotational band, it becomes possible to define approximately which nuclei are deformed as exemplified in fig. 11.1.

The moment of inertia $\mathscr{J}$ can be extracted from measured rotational bands. The values for deformed nuclei in the rare earth region are exhibited in fig. 11.2 together with calculated values. The experimental values are generally 25–50% of the rigid body values and can be calculated with any accuracy only when pairing correlations are introduced (appendix 14B). A simpler way to get an estimate of $\mathscr{J}$ is within the two-fluid model (see e.g. Rowe, 1970) where it is assumed that only nucleons outside the largest possible central sphere give any contribution to $\mathscr{J}$ (problem 11.1).

It was discussed in chapter 6 how the valence particle outside a spherical core determines the ground state angular momentum. This is thus a typical single-particle effect and, similarly, several valence particles may partly or fully align their angular momentum vectors to build higher spin states. Also, in deformed nuclei, similar non-collective components may be present and in this chapter we will discuss the low-energy spectra of more or less well-deformed nuclei as a mixture of single-particle and collective components where the latter are treated macroscopically. With increasing spin it becomes necessary to consider the single-particle contribution from more particles

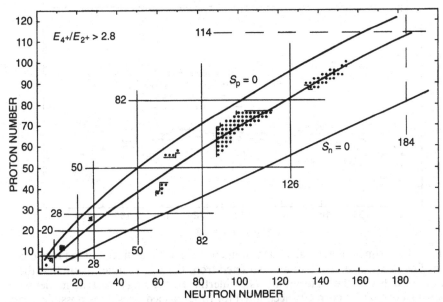

Fig. 11.1. Regions of deformed nuclei. The points represent even–even nuclei, whose excitation spectra exhibit an approximate $I(I+1)$ dependence, indicating rotational structure. An exact $I(I+1)$ dependence corresponds to $E(I=4):E(I=2)=3.33$. Practically all disturbances of the rotational motion will tend to decrease this value. The nuclei in the figure have been selected on the basis of the (rather arbitrary) criterion $E(I=4):E(I=2)>2.8$. The line of β stability and the estimated borders of instability with respect to proton and neutron emission are indicated (from Bohr and Mottelson, 1975, supplemented with data from M. Sakai, *Atomic Data and Nucl. Data Tables* **31** (1984) 399).

and at some point it seems more appropriate to consider also the collective component from a microscopic point of view. This will be discussed in chapter 12, both in the somewhat unrealistic but illustrative harmonic oscillator model and in more realistic models. In these calculations we will use the cranking model where collective and non-collective rotation are treated on the same footing. It then also becomes evident that one cannot really make a strict division between different ways to build angular momentum but that all kinds of intermediate situations occur.

11.1 Strong coupling – deformation alignment

For an odd nucleus, the specific features of the low-energy states are determined by the orbital of the odd nucleon. In chapter 8, it was found that, for such orbitals in a deformed axially symmetric potential, in addition to parity, only the projection of the angular momentum j on the symmetry axis, Ω, is

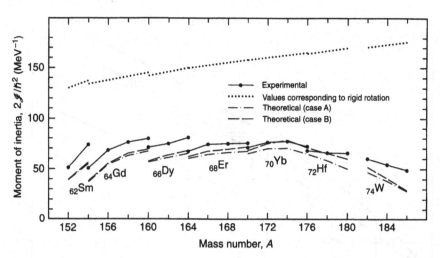

Fig. 11.2. Experimental and calculated moments of inertia of nuclei in the rare-earth region. The experimental values are extracted from the E_{2+} energies. Note that these values are far below the rigid moments of inertia. In the single-particle model with the pairing correlation correctly accounted for (dot–dashed lines), it is possible to get a fair agreement between theory and experiment. The two cases A and B correspond to somewhat different choices of the single-particle parameters (from Nilsson and Prior, 1961).

a preserved quantum number. As illustrated on the left in fig. 11.3, the total spin, **I**, is built as the sum of the spin of the odd particle, **j**, and the collective spin of the core, **R**. The core is built from all the paired nucleons. Thus, the collective energy for rotation of an axially symmetric nucleus around a perpendicular axis, the 3-axis being the symmetry axis, is calculated from

$$H_{\text{rot}} = \frac{\mathbf{R}^2}{2\mathscr{J}} = \frac{1}{2\mathscr{J}} \left[(I_1 - j_1)^2 + (I_2 - j_2)^2 \right]$$

$$= \frac{1}{2\mathscr{J}} \left[\mathbf{I}^2 - I_3^2 + \left(j_1^2 + j_2^2 \right) - (I_+ j_- + I_- j_+) \right]$$

The term $(I_+ j_- + I_- j_+)$ corresponds classically to the Coriolis and centrifugal forces. It gives a coupling between the motion of the particle in the deformed potential and the collective rotation. For small I it is justified to assume that this term is small and we need therefore consider only its diagonal contributions, i.e. the term $(I_+ j_- + I_- j_+)$ is treated in first order perturbation theory. This approximation, where it is assumed that the influence of the rotational motion on the intrinsic structure of the nucleus can be neglected, is generally referred to as the adiabatic approximation or the strong coupling limit. The selection rules for j_+ and j_- are $\Delta\Omega = \pm 1$. Each

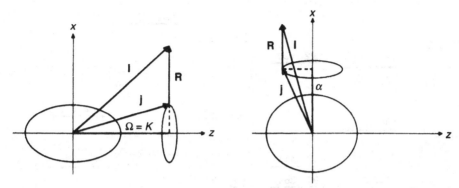

Fig. 11.3. Schematic illustration of the two extreme coupling schemes; deformation alignment (left figure) and rotation alignment (right figure) (from R.M. Lieder and H. Ryde, *Adv. in Nucl. Phys.*, eds. M. Baranger and E. Vogt (Plenum Publ. Corp., New York) vol. 10 (1978) p. 1).

orbital of the deformed potential is twice degenerate corresponding to the two possible signs of Ω. Thus, with the odd particle in one such orbital, it is only for $\Omega = \frac{1}{2}$ (or rather $\Omega = \pm\frac{1}{2}$) that the diagonal matrix elements of the $(I_+j_- + I_-j_+)$-term are different from zero (see below).

The projection of the total angular momentum on the nuclear symmetry axis is a preserved quantum number, which is given by K, see fig. 11.3. With no collective component along this axis, $\Omega = K$. The matrix elements of $(j_1^2 + j_2^2)$, the recoil term, depend only on the particle wave function, ϕ_ν. This means that they are constant for one rotational band. We will first consider situations where they furthermore are rather small so as a first approximation, we will neglect them.

The full Hamiltonian H has the form

$$H = H_{\text{sp}} + H_{\text{rot}}$$

where H_{sp} is the deformed single-particle Hamiltonian. Its eigenvalues are the single-particle energies e_ν

$$H_{\text{sp}}\phi_\nu = e_\nu\phi_\nu$$

as exhibited in figs. 8.3 and 8.5. The total energy is now obtained as

$$E_{IK} = |e_\nu - \lambda| + \frac{\hbar^2}{2\mathscr{J}}\left[I(I+1) - K^2\right], \quad K \neq \frac{1}{2}$$

where the single-particle energy is counted relative to the Fermi level energy λ (see fig. 11.4). As $I \geq K$, the spins $I = K, K+1, K+2, \ldots$ are observed. The application of this formula may be studied in fig. 11.4. In the lower part

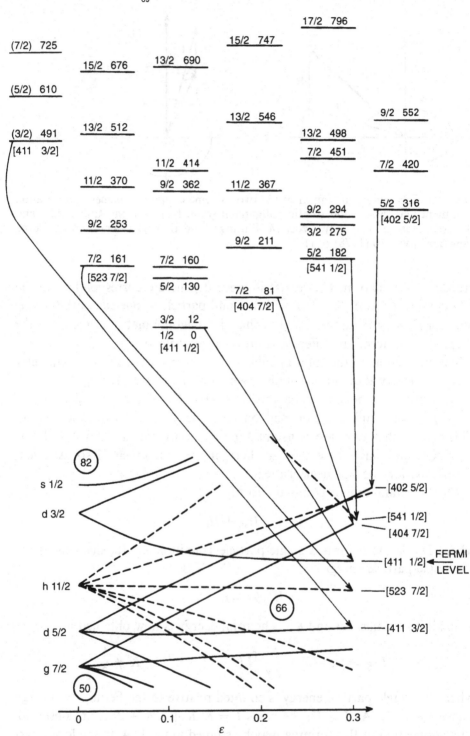

Fig. 11.4. For legend see opposite.

of this figure, the proton single-particle orbitals are exhibited as functions of the deformation coordinate, ε. For a given nucleus, the equilibrium value of ε can either be taken as an experimental quantity derivable from for example the measured quadrupole moment or it can be calculated with the methods described in chapter 9.

In the upper part of fig. 11.4, the measured low-energy spectrum of $^{165}_{69}\mathrm{Tm}_{96}$ is exhibited. For this nucleus, the equilibrium value of ε is approximately equal to 0.29. The even neutrons (96 of them) are assumed to be paired off two and two in orbitals 'Ω and $-\Omega$' to angular momentum zero. Similarly, the 68 protons are assumed to fill pairwise the 34 lowest orbitals. The 69th proton is then (for the ground state) placed in the 35th orbital, marked [411 1/2]. This is thus associated with $\Omega = \frac{1}{2}$. The ground state spin is also measured to be $\frac{1}{2}$ and a rotational band with $I = \frac{1}{2}, \frac{3}{2}, \frac{5}{2}, \ldots$ based on this orbital is identified.

At 81 keV of excitation energy there is another band starting with $I = \Omega = \frac{7}{2}$ (and having positive parity). This band is obtained by promoting the odd proton from [411 1/2] and up into [404 7/2]. The excitation energy, 81 keV, is associated with the energy difference in the single-particle diagram and described by $|e_v - \lambda|$ in the formula above. This energy is thus counted relative to the Fermi energy λ, where λ is given by the single-particle energy of the [411 1/2] orbital. A third band, having $K = \Omega = \frac{7}{2}$ and negative parity, is observed starting at 161 keV excitation energy. This band is realised by the promotion of one of the two protons from [523 7/2] to [411 1/2], in which latter orbital a pair state of compensating spins is formed. We may then call the $\frac{7}{2}^-$ state built on [523 7/2] a 'hole' state. From fig. 11.4, it is evident that such a hole state is associated with a positive excitation energy, thus justifying the absolute sign in the $|e_v - \lambda|$ term of the formula above.†

The other bands of $^{165}\mathrm{Tm}$ are now easily understood. They are obtained

† If pairing is also considered, the $|e_v - \lambda|$ term should be replaced by a $[(e_v - \lambda)^2 + \Delta^2]^{1/2}$ term, see chapter 14.

Fig. 11.4. (*opposite*) Calculated single-proton orbitals in the rare-earth region with the observed spectrum of $^{165}\mathrm{Tm}$ above. The usual spherical subshell notation is used for $\varepsilon = 0$. For $\varepsilon \neq 0$ the standard asymptotic notation is given for each orbital $[N n_3 \Lambda \Omega]$, N being the oscillator shell quantum number, n_3 the number of modes along the intrinsic 3-axis (the symmetry axis), Λ the value of the orbital angular momentum ℓ_3 along the 3-axis and Ω the value of the total angular momentum j_3 along the same axis. The Fermi level in the case of 69 protons is indicated. The experimental states are ordered in rotational bands and the orbital of the odd particle is indicated in each case.

simply by placing the odd proton in the orbitals indicated. The bands to the left are hole excitations, those to the right are particle excitations. A remaining problem is the $K = \frac{1}{2}$ bands, [411 1/2] and [541 1/2], which look somewhat peculiar. This is, however, what was already anticipated when it was found that the 'Coriolis term', $(I_+j_- + I_-j_+)$, gives diagonal contributions for such bands. To calculate these contributions we will first briefly discuss the wave functions.

The orientation of a body in space is described by the three Eulerian angles α, β and γ. Two angles are needed to describe the orientation of a body-fixed axis and one to describe rotations around that axis. This means that, for a rotationally symmetric nucleus, the latter angle becomes superfluous. We will not try to derive the wave function of the collective motion but simply state that it is described by a so called $\mathscr{D}$-function, $\mathscr{D}^I_{MK}(\alpha, \beta, \gamma)$. These functions are also used to describe transformations between differently oriented coordinate systems. The quantum number M is the projection of I on the laboratory z-axis. It is a trivial quantity to which we will pay no attention subsequently. With the intrinsic wave function of the odd particle given by ϕ_ν we get the total wave function as

$$\psi_{IKM} \propto \mathscr{D}^I_{MK}(\alpha, \beta, \gamma)\, \phi_\nu(\text{intr}) = \mathscr{D}^I_{MK} \sum_{Nj} a^\nu_{N\ell j\Omega}|N\ell j\Omega\rangle$$

where in the last step we have expanded the intrinsic wave function in an $|N\ell j\Omega\rangle$-basis

$$\phi_\nu = \sum_{Nj} a^\nu_{N\ell j\Omega}|N\ell j\Omega\rangle$$

In the present discussion, we confine ourselves to nuclei having axial symmetry with respect to the 3-axis and in addition reflection symmetry with respect to a plane perpendicular to the 3-axis (these restrictions may exclude some but not very many of the nuclei that are deformed in their ground states). The nuclear shape is then mainly described by ε and ε_4 while for example $\varepsilon_3 = 0$ (cf. chapter 9). With these symmetries, there is no way to distinguish operationally between a wave function ψ_{IMK} and one $R_1\psi_{IMK}$ that is rotated 180° with respect to the 1-axis, the nuclear body-fixed x-axis. We shall therefore be required to use a new redefined wave function that is invariant with respect to the R_1-operation; $(1 + R_1)\psi$ instead of ψ. It is easy to realise that the operation with R_1 changes K to $-K$ and in addition a phase factor is introduced. The derivation of this phase factor lies outside the scope of the present treatment so we give it without proof,

$$\psi_{IKM} \propto \sum_{Nj} a^v_{N\ell j\Omega} \left[\mathscr{D}^I_{MK}(\alpha,\beta,\gamma)|N\ell j\Omega\rangle + (-1)^{I-j}\mathscr{D}^I_{M-K}|N\ell j-\Omega\rangle \right]$$

It is now convenient to define a conjugate intrinsic state $\phi_{\bar{v}}$ where, however, different phase conventions are used in the literature. We will use the convention that makes the wave functions for two particles in a j-shell coupled to $I = 0$ (see chapter 14) particularly simple,

$$\phi_{\bar{v}} = \sum_{Nj}(-1)^{j-\Omega}a^v_{N\ell j\Omega}|N\ell j-\Omega\rangle$$

It now becomes possible to write the total wave function as

$$\psi_{IKM} \propto \left[\mathscr{D}^I_{MK}(\alpha,\beta,\gamma)\phi_v + (-1)^{I-K}\mathscr{D}^I_{M-K}(\alpha,\beta,\gamma)\phi_{\bar{v}} \right]$$

For $\Omega = \frac{1}{2}$, the first and the second terms of ψ_{IKM} couple through I_+j_- and I_-j_+. The matrix elements of $j_\pm$ are well known to be

$$\langle j\Omega|j_\pm|j\Omega\mp 1\rangle = [(j\pm\Omega)(j\mp\Omega+1)]^{1/2}$$

When the total spin I is projected, not on the laboratory axes, but in the rotating body-fixed axes, one can show that the '+' and '−' operators change character leading to the matrix elements

$$\langle IK|I_\pm|IK\pm 1\rangle = [(I\mp K)(I\pm K+1)]^{1/2}; \quad |IK\rangle \propto \mathscr{D}^I_{MK}$$

It is now straightforward to calculate the general expression for the energies of the rotational bands in the strong coupling approximation:

$$E_{IK} = E_K + \frac{\hbar^2}{2\mathscr{J}} \left[I(I+1) - K^2 + \delta_{K\frac{1}{2}}a(-1)^{I+\frac{1}{2}}\left(I+\frac{1}{2}\right) \right]$$

Here, a is the so-called decoupling parameter, which has a fixed value for each $\Omega = \frac{1}{2}$ orbital. It is calculated as

$$a = \langle \phi_v|j_+|\phi_{\bar{v}}\rangle = \langle \phi_{\bar{v}}|j_-|\phi_v\rangle = \sum_{Nj}(-1)^{j-\frac{1}{2}}\left(j+\frac{1}{2}\right)|a^v_{N\ell j\frac{1}{2}}|^2$$

where the last expression is independent of phase conventions. For an $\Omega = \frac{1}{2}$ band, the $\frac{3}{2}$ and $\frac{1}{2}$ states, the $\frac{7}{2}$ and $\frac{5}{2}$, etc. become degenerate for $a = -1$. Thus for the [411 1/2] band in fig. 11.4, a decoupling parameter $a \approx -0.8$ can be extracted. The [541 1/2] band is approximately described in the range $3 \leq a \leq 4$. However, in this case the adiabatic approximation with only one 'deformed orbital' considered and the Coriolis term, $\mathbf{I} \cdot \mathbf{j}$ treated in first order perturbation theory, is not very accurate. In the next section, we

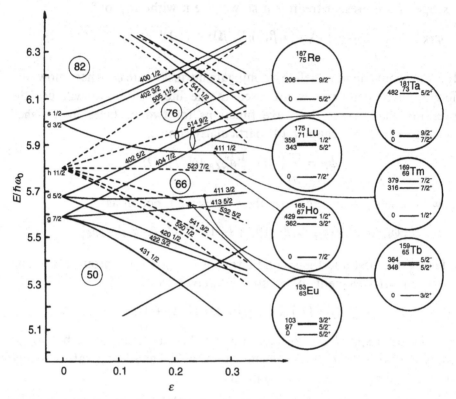

Fig. 11.5. A similar single-particle diagram as in fig. 11.4 with measured band heads of a number of odd-proton rare-earth nuclei to the right. In each case, a rotational band is built on the band heads in a similar way as for ^{165}Tm in fig. 11.4. The orbital of the ground state rotational band is indicated in each case (we are grateful to Sven Åberg who prepared this and the following figure).

will consider a different coupling scheme but, first, we will make some more comparisons between the present formalism and experimental spectra.

For a number of odd-Z nuclei with $Z = 63$–75, we show in fig. 11.5 the measured band heads together with the orbitals of the deformed shell model. On each band head, a rotational band is then built as exhibited for ^{165}Tm in fig. 11.4. The ground state ε-deformation for the nuclei in fig. 11.5 varies roughly as exhibited in the figure with ε being largest around $Z = 70$. In addition the equilibrium ε_4-value varies rather much, being negative for small Z and positive for large Z. This latter variation is not accounted for in fig. 11.5 where $\varepsilon_4 = 0$. In spite of this approximation, all the band head spins and corresponding energies come out more or less as expected from the level order in the deformed shell model.

Fig. 11.6 shows a similar comparison for odd-N nuclei with $N = 93$–107.

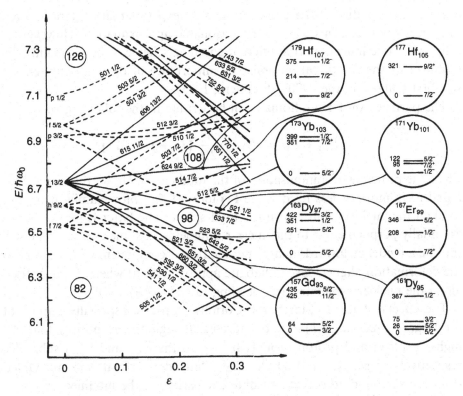

Fig. 11.6. Similar to fig. 11.5 but for neutrons instead.

Also in this case, the agreement between theory and experiment is very satisfying. In an early stage, the classification of such data as presented in figs. 11.4–11.6 played a very important role in the understanding of deformed nuclei, see e.g. Mottelson and Nilsson (1959a).

Coming back to even nuclei, we find that, for the ground state rotational band, the wave function that is properly invariant with respect to the R_1-operation takes the form

$$\psi_{I0M} \propto \mathscr{D}^I_{M0}(\alpha, \beta, \gamma) \left[1 + (-1)^I\right]$$

Thus, for odd I, the wave function disappears in agreement with the fact that only even spins, $I = 0, 2, 4, \ldots$ are observed.

11.2 Decoupled bands – rotation alignment

In the preceding section, we discussed situations where the rotational motion and the nuclear deformation are essentially uncoupled. The rotational frequency was not high enough to break the coupling scheme caused by the intrinsic deformation. The comparison with experimental spectra showed that, in many cases, this was a very reasonable approximation. However, in other cases one must expect a different coupling scheme, which is mainly determined by the rotational motion. The importance of this coupling scheme was first realised in the 1970's (Stephens, 1975).

The particle–rotor Hamiltonian can be written in the following way:

$$H = H_{sp} + H_{rot} = H_{sp} + \frac{\hbar^2}{2\mathscr{J}}\left(I^2 + j^2 - 2\mathbf{I}\cdot\mathbf{j}\right)$$

Thus, the principle of minimisation of the total energy shows that for a fixed I and for a more or less fixed j, the $\mathbf{I}\cdot\mathbf{j}$ term of the rotor Hamiltonian tries to align the intrinsic spin j with the total spin I. The latter is in most cases essentially perpendicular to the nuclear symmetry axis, the 3-axis. There will thus be a tendency towards a large perpendicular component of j contrary to the deformation aligned case (adiabatic approximation) where j is quantised along the nuclear symmetry axis (leading to $\langle j_1\rangle = \langle j_2\rangle = 0$).

The effects of the $\mathbf{I}\cdot\mathbf{j}$-term, the Coriolis term, are especially important for large j (and large I). In the rare-earth region, one notices that the high-j neutron and proton orbitals belong to the $i_{13/2}$ and $h_{11/2}$ subshells, respectively (figs. 11.5 and 11.6). These 'intruder' orbitals are uncoupled from the surrounding orbitals of different parity. In the modified oscillator potential, this is understood from the fact that they are pushed down in energy due to the ℓ^2- and $\ell\cdot$s-terms. The j quantum number of these subshells is therefore almost pure up to rather large ε-values. This is in contrast to most other orbitals which are built from a mixture of several different j-values.

We will now study in some detail the neutron $i_{13/2}$ orbitals and assume that j is pure, $j = 13/2$. A quadrupole deformation along the 3-axis, $\varepsilon Y_{20}(\hat{3})$, will then split the orbitals according to

$$e_\Omega = e_0 + \frac{1}{6}\varepsilon M\omega_0^2 \left\langle r^2\right\rangle \frac{3\Omega^2 - j(j+1)}{j(j+1)}$$

as was found in chapter 8 (problem 8.1). With $\left\langle r^2\right\rangle = \left(N+\frac{3}{2}\right)\hbar/M\omega_0$ where $N = 6$ and with $\hbar\omega_0 = 8$ MeV as a typical value for neutrons in the rare-earth region ($A = 160$) we get

$$e_\Omega = e_0 + 10\varepsilon\frac{3\Omega^2 - j(j+1)}{j(j+1)} \quad \text{MeV}$$

A rotational motion will try to break this coupling scheme and align the j-vector along the rotation axis instead. In the extreme case, the orbitals get quantised along the 1-axis. We will denote the projection of j along this axis by α, see right part of fig. 11.3. Then, such a state being denoted by ϕ_α can be expressed in the ϕ_Ω states (Stephens, Diamond and Nilsson, 1973):

$$\phi_\alpha = \sum_\Omega C_\Omega \phi_\Omega$$

We will not try to evaluate the transformation coefficients, which are given by $\mathscr{D}^j_{\alpha\Omega}(0,\frac{\pi}{2},0)$, with $j = 13/2$ (cf. problem 11.3). In the limiting case of full alignment along the 1-axis, $\alpha = 13/2$, one would expect that mainly the small Ω-values enter and the result is

$$\left|C_{\pm 1/2}\right|^2 = 0.21, \quad \left|C_{\pm 3/2}\right|^2 = 0.16, \quad \left|C_{\pm 5/2}\right|^2 = 0.09$$

$$\left|C_{\pm 7/2}\right|^2 = 0.03, \quad \left|C_{\pm 9/2}\right|^2 = 0.01, \quad \left|C_{\pm 11/2}\right|^2 \approx \left|C_{\pm 13/2}\right|^2 \approx 0$$

The Fermi level is now placed on the $\Omega = \frac{1}{2}$ orbital and the energies of some high spin states are calculated in the two extreme coupling schemes as functions of deformation. In the deformation aligned case ($K = \Omega = \frac{1}{2}$ band), it is convenient to write the Hamiltonian as

$$H = H_{\rm sp} + \frac{\hbar^2}{2\mathscr{J}}\left[I^2 + j^2 - 2I_3 j_3 - (I_+ j_- + I_- j_+)\right]$$

With the Fermi level on the $\Omega = \frac{1}{2}$ orbital, $H_{\rm sp}$ gives a zero energy contribution and we obtain

$$E = \frac{\hbar^2}{2\mathscr{J}}\left[I(I+1) + j(j+1) - 2\Omega K + a(-1)^{I+\frac{1}{2}}\left(I + \frac{1}{2}\right)\right]$$

where $j = 13/2$ and $\Omega = K = \frac{1}{2}$. In this expression, we have thus included the matrix element of $(j_1^2 + j_2^2)$, which was neglected in the preceding section. Here it is included to make the energy compatible with that calculated in the rotation aligned case below. For an orbital with pure j (and $\Omega = \frac{1}{2}$) the decoupling factor is trivially obtained as $a = (-1)^{j+\frac{1}{2}}(j + \frac{1}{2})$. Thus, in the case of $j = 13/2$, one obtains $a = -7$.

In the rotation aligned case, the energies for the $I = 13/2, 17/2, 21/2,$ ... states are calculated as

$$E = \sum_\Omega 2|C_\Omega|^2\left(e_\Omega - e_{\frac{1}{2}}\right) + \frac{\hbar^2}{2\mathscr{J}}[I(I+1) + j(j+1) - 2I\alpha]$$

where $j = \alpha = 13/2$ in the present case. The first term is the single-particle contribution, which gets larger the more the orbitals are spread apart, i.e. the larger the deformation becomes. This term will thus make the rotation aligned coupling scheme disadvantageous at large deformations. The dependence of the energy splitting, $e_\Omega - e_{\frac{1}{2}}$, on ε can be calculated from the first order expression given above or can be extracted from a figure like fig. 11.6.

We also must find a value for $\hbar^2/2\mathscr{J}$. An empirical relation (Grodzins, 1962) for the rotational 2^+ energies of even nuclei is $E_{2+} \approx 1225/(A^{7/3}\beta^2)$ MeV. With β transformed to ε we get $E_{2+} \simeq 1100/(A^{7/3}\varepsilon^2)$ MeV, i.e. for $A \simeq 160$:

$$\frac{\hbar^2}{2\mathscr{J}} = \frac{E_{2+}}{6} \approx \frac{1.3}{\varepsilon^2} \quad \text{keV}$$

The two energy expressions corresponding to the deformation aligned (strongly coupled) and rotation aligned coupling schemes are compared in fig. 11.7. This figure should mainly be taken to show the trends. With increasing spin I, increasing particle spin j and decreasing deformation, the rotation aligned coupling scheme becomes more favoured. This is especially the case when the Fermi level is in the region of low-Ω orbitals of a high-j shell.

In fig. 11.7, we only plot the so-called favoured states, $I = j, j+2, j+4, \ldots$. In the pure rotation aligned case, the total nuclear wave function is symmetric with respect to an R_1-rotation and, in a similar way as for an even nucleus, the wave function for the intermediate spin states, $j+1, j+3, \ldots$ disappears. Full alignment is hardly realised in any nucleus. Experimentally, one thus often observes also the $j+1, j+3, \ldots$ states but they come relatively higher in energy than the $j, j+2, \ldots$ states.

In the rotation aligned case, the spin projection on the rotation axis, α, equals j and the rotational energy can be written

$$E_{\text{rot}} = \frac{\hbar^2}{2\mathscr{J}} [I(I+1) + j(j+1) - 2I\alpha] = \frac{\hbar^2}{2\mathscr{J}} [(I-\alpha)(I-\alpha+1) + 2\alpha]$$

$$= \frac{\hbar^2}{2\mathscr{J}} R(R+1) + \text{constant}$$

where $R = I - \alpha$ describes the collective rotation. Thus, the energy spacings in a rotation aligned spectrum of an odd nucleus should be the same as in neighbouring even nuclei. This is nicely illustrated in fig. 11.8. With 57 protons in the La nuclei, the Fermi level is situated around the [550 1/2] orbital with $j \approx 11/2$ (fig. 11.5) and rotation aligned bands starting with $I = 11/2$ are formed.

The breaking of the deformation aligned coupling scheme is generally referred to as decoupling. The wave function of the particle is then distributed

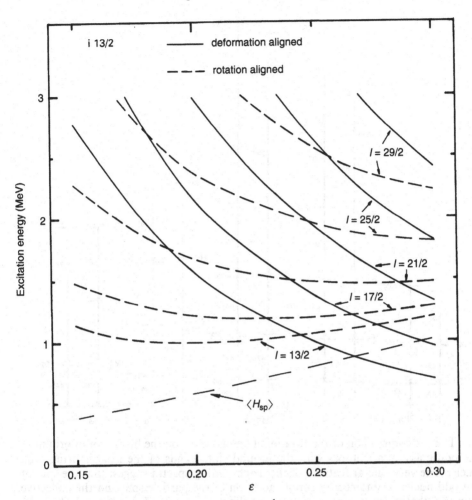

Fig. 11.7. With the Fermi level on the $\Omega = \frac{1}{2}$ state in an $i_{13/2}$ shell, the spectra of the favoured states in the two coupling schemes, deformation alignment and rotation alignment, are shown as functions of ε. The quantity $\langle H_{sp} \rangle$ is the energy required when the wave function of the odd particle is redistributed over the $i_{13/2}$ orbitals to get its spin vector aligned with the axis of rotation. One notes that small deformations and high spins tend to favour the rotation aligned scheme. If the particle–rotor Hamiltonian is fully diagonalised, a situation between the two simple models of the figure will result. However, many experimental spectra can be quite accurately described by one or the other of the two extremes.

over several 'deformed orbitals' and the spin of the particle is largely aligned along the collective rotation vector, **R**. In the idealised situation described in the present section, this alignment is complete. In the strong coupling scheme, the mixing of the $\Omega = \pm\frac{1}{2}$ orbitals correspond to a partial alignment. As the particle wave function is equally distributed over the $\Omega = \frac{1}{2}$ and the

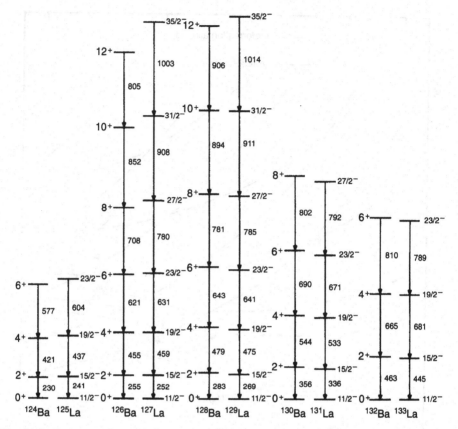

Fig. 11.8. A comparison of the rotational bands based on the $h_{11/2}$ proton orbital in the odd-mass $_{57}$La isotopes with the ground state bands of the neighbouring $_{56}$Ba nuclei. The very similar features of the neighbouring spectra suggest that the spin of the odd nuclei is obtained by simple addition of the particle spin and the collective spin, i.e. that these two spin vectors are aligned (from R.M. Lieder and H. Ryde, *Adv. Nucl. Phys.*, eds. M. Baranger and E. Vogt (Plenum Publ. Corp., New York) vol. 10 (1978) p. 1).

$\Omega = -\frac{1}{2}$ orbitals (the v and $\bar{v}$ orbitals), the alignment is easily calculated as

$$\langle j_1 \rangle = \left\langle \frac{1}{\sqrt{2}}(\phi_v + \phi_{\bar{v}}) | j_1 | \frac{1}{\sqrt{2}}(\phi_v + \phi_{\bar{v}}) \right\rangle$$

$$= \frac{1}{4}\left(\langle \phi_v | j_+ | \phi_{\bar{v}} \rangle + \langle \phi_{\bar{v}} | j_- | \phi_v \rangle\right) = \frac{1}{2}a$$

For an orbital $|j\Omega\rangle = |13/2\ 1/2\rangle$, it was found from the equation above that the decoupling factor $a = -7$, i.e. $|\langle j_1 \rangle| = 3.5$ in the strong coupling approximation. This should be compared with $\langle j_1 \rangle = \alpha = 6.5$ for full alignment in a $j = 13/2$ shell.

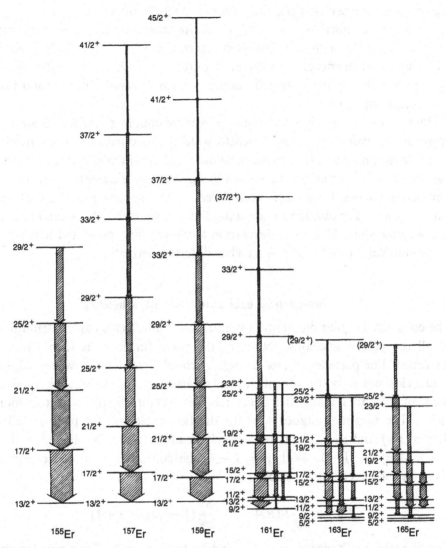

Fig. 11.9. Observed rotational bands based on the $i_{13/2}$ neutron orbitals in odd-mass $_{68}$Er isotopes. One observes a gradual change from a rotation aligned spectrum in ^{155}Er and ^{157}Er, to a deformation aligned spectrum for the lower spin states in ^{165}Er (from R.M. Lieder and H. Ryde, *Adv. Nucl. Phys.*, eds. M. Baranger and E. Vogt (Plenum Publ. Corp., New York) vol. 10 (1978) p. 1.).

The transition between the two coupling schemes is illustrated in fig. 11.9. The positive-parity spectra of the odd $_{68}$Er isotopes with $N = 89$–97 are shown. These isotopes change from being weakly deformed with the Fermi level around the $i_{13/2}$, $\Omega = \frac{1}{2}$ orbital for small N to larger deformations with the Fermi level higher up in the $i_{13/2}$ shell with increasing N. Consequently, the spectrum is essentially decoupled for $N = 89$–91, intermediate for $N = 93$ and strongly coupled for $N = 95, 97$. Note, however, that also for these latter isotopes the high spin favoured states, $I = 17/2, 21/2, 25/2, \ldots$ come relatively lower in energy than the unfavoured $15/2, 19/2, 23/2$ states. Thus, as expected, the rotation aligned coupling scheme becomes more important with increasing spin.

Here we have only discussed the two extreme coupling schemes. It should, however, be evident that it is straightforward to diagonalise the full particle–rotor Hamiltonian and thus to describe intermediate situations as for example the spectra of ^{161}Er and ^{163}Er shown in fig. 11.9. Furthermore, only axially symmetric shapes have been considered. For the generalisation of the particle–rotor Hamiltonian to non-axial shapes, we refer to Larsson, Leander and Ragnarsson (1978) for a derivation along the lines presented here or to Meyer-ter-Vehn (1975) for a somewhat different derivation.

11.3 Two-particle excitations and back-bending

The collective angular momentum vector, **R**, is built from small contributions of all the paired nucleons. None of the wave functions is then strongly disturbed. For particles in low-Ω high-j orbitals one must, however, expect tendencies, not only for odd nucleons but also for paired nucleons, to align their spin vectors along the collective spin vector (Stephens and Simon, 1972). The maximal aligned spin for the two nucleons in a pure j-shell is then $\alpha_1 = j$ and $\alpha_2 = j - 1$, respectively, leading to a total aligned spin of $\alpha = \alpha_1 + \alpha_2 = 2j - 1$. With $R = I - \alpha$, the collective rotational energy is given by (cf. preceding section):

$$E_{\mathrm{rot}} = \frac{\hbar^2}{2\mathscr{J}} R(R+1) = \frac{\hbar^2}{2\mathscr{J}} (I - \alpha)(I - \alpha + 1)$$

The alignment is however accompanied by the breaking of one pair leading to a configuration with 'two odd particles'. A rough estimate is therefore that the energy cost for breaking the pairs is approximately twice the odd–even mass difference, 2Δ (see chapter 14). Compared to this, the energy cost for redistributing the particle wave function over the different orbitals, as discussed in the preceding section, can be neglected. Furthermore, the pairing correlations will tend to decrease this energy.

In the present approximation, we thus get for the band with 'two aligned spins'

$$E \approx 2\Delta + \frac{\hbar^2}{2\mathscr{J}} (I - \alpha)(I - \alpha + 1); \quad I \geq \alpha$$

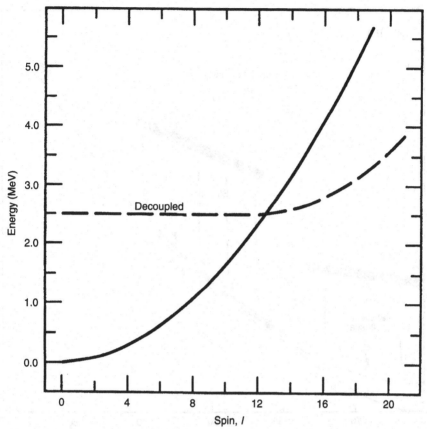

Fig. 11.10. Simple estimates are shown of the ground-band energy in an even–even $A \simeq 160$ nucleus (solid line) and the energy of a decoupled band based on two 'aligned' $i_{13/2}$ particles (dashed line) (from F.S. Stephens, *Proc. 4th Summer School on Nuclear Physics*, Rudziska, Poland, 1972, p. 190).

Total spin values smaller than the largest possible aligned spin can be obtained by a partial alignment with no collective rotation. Provided one pair is broken, this should lead to an energy $E \approx 2\Delta$. The resulting 'aligned' band is compared with the ground band in fig. 11.10.

The states having lowest possible energy for given spin are referred to as the yrast states. A typical yrast line for an $A \simeq 160$ nucleus is sketched in fig. 11.11. In this figure is also shown how the yrast levels can be studied. If two nuclei, e.g. $^{40}_{18}\text{Ar}_{22}$ and $^{124}_{52}\text{Te}_{72}$ collide in a non-central collision, a compound nucleus having a large excitation energy and a large angular momentum might be formed. By emission of e.g. four neutrons, which each carry away about 8 MeV of excitation energy (i.e. the neutron binding energy) a point some few MeV above the yrast line is reached.

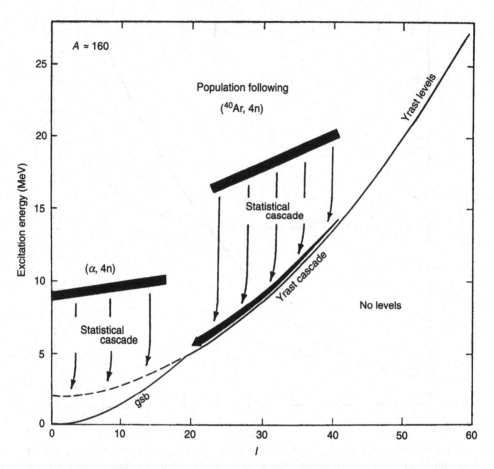

Fig. 11.11. Schematic illustration of which nuclear states are populated in reactions where either ^{40}Ar or an α-particle fuses with a target nucleus to form a compound system with $A \simeq 160$. Such reactions are best suited for the study of the yrast states, i.e. the lowest energy states for each spin I (partly from Stephens and Simon, 1972).

Some additional excitation energy might be carried away by a few so called statistical γ-rays and the yrast region is reached. The additional excitation energy is now carried by the rotational motion. For collective rotation, the compound nucleus will now de-excite mainly through E2 transitions, $(I + 2) \rightarrow I$, along the yrast line. A situation like in fig. 11.10 will then lead to E2 energies that are larger for spins $I = 8$–10 than for spins $I = 14$–16.

In fig. 11.11 is also illustrated that the real high spin states can only be reached in so called heavy-ion collisions where both the projectile and the target are heavy nuclei. If an α-particle is used as projectile, only lower spin states can be reached (see problem 11.4).

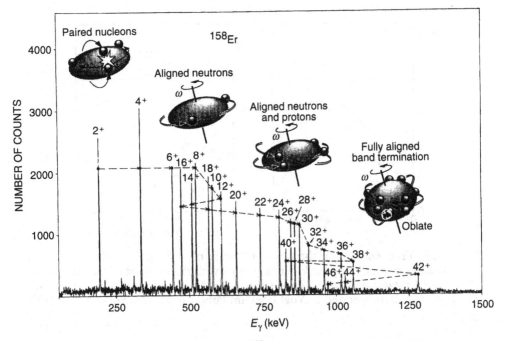

Fig. 11.12. Observed γ-ray energies of ^{158}Er formed in a reaction like the one illustrated in fig. 11.11. For $I \approx 14$ two $i_{13/2}$ neutrons become aligned resulting in a backbend while a second irregularity caused by the alignment of two $h_{11/2}$ protons is seen for $I \approx 32$. The features for $I \geq 38$ with the final band termination for $I = 46$ are discussed in chapter 12.

As seen in fig. 11.12, the experimentally observed E2 energies of ^{158}Er in the range $I = 12$–18 show the properties expected from the simple model of fig. 11.10. Of course, one must expect that the ground band and the decoupled band interact in the crossing region giving rise to a smoother transition between the two bands than shown in fig. 11.10. This is in agreement with the experimental spectrum of ^{158}Er. We must also remember that the model we have described is very much idealised. Still it seems to contain the main features of the physical effect.

The feature that the yrast E2 transition energies suddenly become smaller with increasing spin is generally referred to as back-bending (see fig. 11.12). When investigating such spectra, the yrast energies are often plotted in a somewhat different way. An effective moment of inertia as a function of the spin I can be obtained as

$$\frac{2\mathscr{J}}{\hbar^2} = \left(\frac{\mathrm{d}E}{\mathrm{d}I(I+1)}\right)^{-1} \simeq \left(\frac{E_I - E_{I-2}}{4I - 2}\right)^{-1}$$

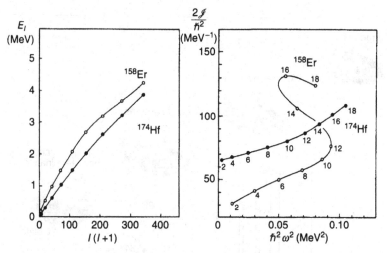

Fig. 11.13. Yrast energies in the $I = 0$–18 range of ^{158}Er and ^{174}Hf plotted versus $I(I+1)$ and corresponding back-bending plots with the moment of inertia $\mathscr{J}$ versus the squared rotational frequency, ω^2 (from R.M. Lieder and H. Ryde, *Adv. in Nucl. Phys.*, eds. M. Baranger and E. Vogt (Plenum Publ. Corp., New York) vol. 10 (1978) p. 1).

The canonical relation between the spin I and the rotational frequency ω is

$$\omega = \frac{\partial H}{\partial I}$$

Thus, in the quantum mechanical case it is natural to define

$$\hbar\omega = \frac{E_I - E_{I-2}}{[I(I+1)]^{1/2} - [(I-2)(I-1)]^{1/2}}$$

which formula is often simplified to

$$\hbar\omega = \frac{E_I - E_{I-2}}{2}$$

i.e., twice the rotational frequency is equal to the E2 transition energy.

A standard back-bending plot shows the moment of inertia $2\mathscr{J}/\hbar^2$ as a function of the squared rotational frequency. This is illustrated for ^{158}Er and ^{174}Hf in fig. 11.13. Note that while the yrast lines, E versus $I(I+1)$, look rather similar, the differences are blown up in the $\mathscr{J}$ versus ω^2 plot. Thus, ^{158}Er shows back-bending while ^{174}Hf does not.

The yrast states of ^{160}Yb and ^{164}Hf are shown in an alternative back-bending plot, I (or rather the component I_x) versus ω, in fig. 11.14. In the spin region $I = 10$–14, two $i_{13/2}$ neutrons get aligned for each nucleus. The second irregularity (up-bend) seen for ^{160}Yb at $I \approx 28$ and for ^{158}Er at

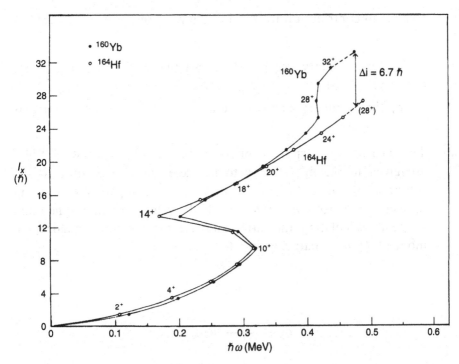

Fig. 11.14. Plot of the spin projection I_x ($\approx I$) versus $\hbar\omega$ for yrast transitions up to $I \approx 30$ in ^{160}Yb and ^{164}Hf (from L.L. Riedinger, *Phys. Scripta* **24** (1981) 312).

$I \approx 32$ (fig. 11.12) appears to be caused by alignment of two $h_{11/2}$ protons. Thus, the general features of these curves can be understood from the simple models we have discussed here. However, for any more detailed theoretical description, a proper treatment of, for example, the pairing correlations (chapter 14) becomes necessary.

Spectra of the type illustrated in figs. 11.12–11.14 have more often been described in terms of the cranking model, which is the subject of the coming chapter where, however, we concentrate on even higher spins. For more details on the description of intermediate spin states, we refer to Bengtsson and Frauendorf (1979) and Bohr and Mottelson (1977). Articles of review character have been written by for example Szymański (1983), de Voigt *et al.* (1983) and Bengtsson and Garrett (1984).

Exercises

11.1   Consider a nucleus of spheroidal shape. Calculate the moment of inertia, $\mathscr{J}$, for rigid rotation around a perpendicular axis. Compare with the value in a two-fluid model where a central sphere is assumed to give no contribution to $\mathscr{J}$. Find the ratio of the two moments of inertia if

    (a) the symmetry axis is 30% longer than the perpendicular axis ($\varepsilon = 0.25$)

    (b) the symmetry axis is twice as long as the perpendicular axis ($\varepsilon = 0.6$).

11.2   The measured low-spin members of the rotational band of ^{178}Hf are given in the figure. Try to fit these values according to the rotational formula, $E_I = (\hbar^2/2\mathscr{J})I(I+1)$. Compare the resulting moment of inertia with the rigid body value. Assume spheroidal shape in calculating the latter and use the measured quadrupole moment, $Q = 7.5$ barns, as input.

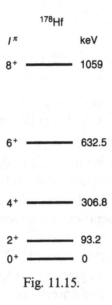

Fig. 11.15.

11.3   Find the expansion of an orbital with $j = 5/2$, $j_x = 5/2$ and $\ell = 2$ in terms of orbitals quantised along the z-axis.

11.4   Consider the situation illustrated in fig. 11.16. A nucleus with mass number A_1 is accelerated to react with another nucleus with mass number A_2. The impact parameter is equal to $(3/4)R_2$, see fig. 11.16. The aim of forming a rapidly rotating compound nucleus can only be attained if the energy of the projectile is neither too high (too high excitation energy) nor too low (the projectile will not overcome the Coulomb barrier of the target). Find the spin and the excitation energy of the compound nucleus if the projectile, its energy and the target are

(a) ^{4}He, 60 MeV and ^{160}Dy
(b) ^{40}Ar, 180 MeV and ^{124}Sn

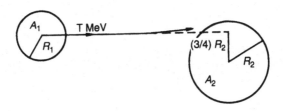

Fig. 11.16.

12

Fast nuclear rotation – the cranking model

At very high spin, one expects the Coriolis and centrifugal forces to disturb strongly the wave functions of many nucleons. As discussed in the introduction to the previous chapter, it then becomes desirable to treat all nucleons on the same footing.

In this chapter, we will introduce two models of this kind, namely the cranking model and the rotating liquid-drop model. The cranking model is first applied to the simple harmonic oscillator potential to illustrate some different concepts. Then the Nilsson–Strutinsky cranking approach corresponding to a combination of the two models is introduced. Within this approach, phenomena like band terminations and superdeformed high-spin states are discussed.

12.1 The cranking model

In the cranking model, the rotation is treated in the classical sense with the rotation vector coinciding with one of the main axes of the nucleus. It then turns out that, in this system, the nucleons can be described as independent particles moving in a rotating potential. In fact, the rotation degree of freedom enters in very much the same way as the deformation degrees of freedom, which were introduced in chapters 8 and 9. One important shortcoming of the cranking model is that the wave functions are not eigenstates of the angular momentum operator. Instead, the angular momentum is generally identified with the expectation value of its projection on the rotation axis.

The mathematical formulation of a rotating single-particle potential was first given by Inglis (1954). With the coordinates in the laboratory system given by x, y and z and those in the rotating system by x_1, x_2 and x_3, we get, for constant angular velocity, ω, around the x_1-axis

$$x_1 = x$$
$$x_2 = y \cos \omega t + z \sin \omega t$$
$$x_3 = -y \sin \omega t + z \cos \omega t$$

Apart from some phase-factor, the time-dependent wave functions in the two systems must satisfy

$$\psi^\omega (x_1, x_2, x_3, t) = \psi (x, y, z, t)$$

which leads to

$$\left(\frac{\partial \psi}{\partial t} \right)_{x,y,z} = \left(\frac{\partial \psi^\omega}{\partial t} \right)_{x_1,x_2,x_3} + \frac{\partial \psi^\omega}{\partial x_2} \frac{\partial x_2}{\partial t} + \frac{\partial \psi^\omega}{\partial x_3} \frac{\partial x_3}{\partial t}$$

With

$$\frac{\partial x_2}{\partial t} = \omega(-y \sin \omega t + z \cos \omega t) = \omega x_3$$

$$\frac{\partial x_3}{\partial t} = \omega(-y \cos \omega t - z \sin \omega t) = -\omega x_2$$

we now find

$$\frac{\partial \psi(x, y, z, t)}{\partial t} = \left(\frac{\partial}{\partial t} - i\omega \ell_1 \right) \psi^\omega (x_1, x_2, x_3, t)$$

where the angular momentum operator ℓ_1 is given by

$$\ell_1 = -i \left(x_2 \frac{\partial}{\partial x_3} - x_3 \frac{\partial}{\partial x_2} \right) = -i \left(y \frac{\partial}{\partial z} - z \frac{\partial}{\partial y} \right) = \ell_x$$

The equality between ℓ_1 and ℓ_x is easily proven by direct evaluation.

The relation for the time derivatives implies that the time-dependent Schrödinger equation for ψ

$$i\hbar \frac{\partial \psi(x, y, z, t)}{\partial t} = h\psi(x, y, z, t)$$

is transformed into

$$i\hbar \frac{\partial \psi^\omega}{\partial t} (x_1, x_2, x_3, t) = (h - \hbar \omega \ell_1) \psi^\omega (x_1, x_2, x_3, t)$$

for the wave function in the intrinsic system, ψ^ω. In these equations, the Hamiltonian is given by h to point out that it is a one-particle operator. This is in contrast to the total Hamiltonian, which is denoted by H (in the preceding chapters, no such distinction has been made and a capital H has been used also for the single-particle Hamiltonian).

The Schrödinger equation in the rotating system can now be solved in the standard way as an eigenvalue problem

$$(h - \hbar\omega j_1)\, \phi^\omega = e^\omega \phi^\omega$$

where the orbital angular momentum operator ℓ_1 has been generalised to cover also particles having an intrinsic spin and has thus been replaced by j_1 ($\mathbf{j} = \boldsymbol{\ell} + \mathbf{s}$). The Hamiltonian in the rotating system,

$$h^\omega = h - \hbar\omega j_1$$

is also referred to as the cranking one-particle Hamiltonian. The eigenvalues e_i^ω are referred to as the single-particle energies in the rotating system or more properly Routhians. This is so because the Hamiltonian in the rotating system does not overlap with the energy. The cranking one-particle Hamiltonian may be summed over all the independent particles of the system to obtain the total cranking Hamiltonian,

$$H^\omega = H - \hbar\omega I_1$$

Alternatively, the cranking Hamiltonian can be derived by direct use of the rotation operator $\mathscr{R} = \exp(-iI_x\omega t)$ (see e.g. de Voigt, Dudek and Szymanski, 1983) or from the Lagrangian (problem 12.1).

A simple way to obtain the cranking Hamiltonian is to minimise the energy

$$E = \langle \Psi | H | \Psi \rangle$$

under the constraint that the total spin

$$I = \langle \Psi | I_1 | \Psi \rangle$$

is fixed. The rotational frequency ω (or rather $\hbar\omega$) will then take the role of a Lagrangian multiplier, which, as seen from the derivation above, can be identified with the rotational frequency.

The energies of the particles are measured in the laboratory system and are calculated as

$$e_i = \langle \phi_i^\omega | h | \phi_i^\omega \rangle$$

where it should be observed that the time-independent wave functions, ϕ_i^ω, are not eigenvectors of the Hamiltonian, h. Similarly, the angular momentum is calculated as an expectation value

$$\langle j_x \rangle_i = \langle \phi_i^\omega | j_x | \phi_i^\omega \rangle = \langle \phi_i^\omega | j_1 | \phi_i^\omega \rangle$$

The total energy and the total spin are now given as sums over the occupied orbitals:

$$E_{sp} = \sum_{occ} e_i$$

$$I \approx I_x = \sum_{occ} \langle j_x \rangle_i$$

In a similar way as for deformed nuclei at spin zero (chapter 9), these summed quantities can on average be renormalised to liquid-drop behaviour. We will, however, first study the pure harmonic oscillator potential within the cranking model. For this potential, no renormalisation is necessary.

12.2 The rotating harmonic oscillator

Many of the effects observed or expected at high spin can be illustrated in the rotating harmonic oscillator potential. It is then very advantageous that the single-particle wave functions and corresponding energies are given by closed expressions (Valatin, 1956). In the present discussion of the harmonic oscillator, we mainly follow Cerkaski and Szymański (1979). We will thus introduce some approximations so that the minimal energy and corresponding shape for different spins I can also be determined analytically. With these approximations, the details of the solutions should not be given too much significance but the main trends are illustrated very nicely.

In the pure oscillator, the intrinsic spin is uncoupled from the spatial coordinates. We thus only consider the orbital angular momentum, which gives the cranking Hamiltonian

$$h^\omega = h_{osc} - \omega \ell_1$$

where

$$h_{osc} = -\frac{\hbar^2}{2m}\Delta + \frac{1}{2}m\left(\omega_1^2 x_1^2 + \omega_2^2 x_2^2 + \omega_3^2 x_3^2\right)$$

and

$$\ell_1 = (x_2 p_3 - x_3 p_2)$$

Boson creation and annihilation operators are now introduced in a similar way as in chapter 8:

$$x_i = -i\left(\frac{\hbar}{2m\omega_i}\right)^{1/2}(a_i^+ - a_i), \quad p_i = \left(\frac{\hbar m\omega_i}{2}\right)^{1/2}(a_i^+ + a_i)$$

for $i = 1, 3$ and

$$x_2 = \left(\frac{\hbar}{2m\omega_2}\right)^{1/2} (a_2^+ + a_2), \quad p_2 = i \left(\frac{\hbar m\omega_2}{2}\right)^{1/2} (a_2^+ - a_2)$$

With the special phases chosen (Bohr and Mottelson, 1975), one finds that, for example, the matrix elements of ℓ_1 are real. The Hamiltonian h^ω is obtained as

$$h^\omega = \sum_{i=1}^{3} \hbar\omega_i \left(a_i^+ a_i + \frac{1}{2}\right) - \omega \ell_1$$

where

$$\ell_1 = \frac{\omega_2 + \omega_3}{2(\omega_2 \omega_3)^{1/2}} (a_2^+ a_3 + a_3^+ a_2) - \frac{\omega_2 - \omega_3}{2(\omega_2 \omega_3)^{1/2}} (a_2^+ a_3^+ + a_2 a_3)$$

It is possible to make a transformation among the operators a_2^+, a_2, a_3^+ and a_3 (i.e. among the coordinates x_2, x_3 and the momenta p_2, p_3) to bring the Hamiltonian into the form of three uncoupled harmonic oscillators, but as mentioned above, here we will introduce some approximations to make the solution more transparent.

While the first term of the ℓ_1 operator couples orbitals within the same major oscillator shell, the second term couples orbitals of $(N, N \pm 2)$ shells (where, as usual, the N-shells are defined in the stretched basis, $N \equiv N_t$). For small and intermediate deformations, the energy spacing between such orbitals belonging to different N-shells is large. Furthermore, the coefficients of the two terms differ by a factor $(\omega_2 - \omega_3)/(\omega_2 + \omega_3)$, which is far below unity for reasonably small deformations. We thus conclude that the $\Delta N = 2$ couplings of the ℓ_1 operator are generally much less important than the $\Delta N = 0$ couplings. Consequently, we neglect the second term of the ℓ_1 operator.

The remaining part of h^ω is now diagonalised by a unitary transformation

$$a_2^+ = a_\alpha^+ \cos\phi + a_\beta^+ \sin\phi, \quad a_3^+ = -a_\alpha^+ \sin\phi + a_\beta^+ \cos\phi$$

which leads to

$$h^\omega = \hbar\omega_1 \left(a_1^+ a_1 + \frac{1}{2}\right) + \frac{\hbar\omega_2}{2} + \frac{\hbar\omega_3}{2}$$

$$+ \hbar a_\alpha^+ a_\alpha \left(\omega_2 \cos^2\phi + \omega_3 \sin^2\phi + \omega \frac{\omega_2 + \omega_3}{(\omega_2 \omega_3)^{1/2}} \cos\phi \sin\phi\right)$$

$$+ \hbar a_\beta^+ a_\beta \left(\omega_2 \sin^2\phi + \omega_3 \cos^2\phi - \omega \frac{\omega_2 + \omega_3}{(\omega_2 \omega_3)^{1/2}} \cos\phi \sin\phi\right)$$

$$+\hbar \left(a_\alpha^+ a_\beta + a_\beta^+ a_\alpha \right) \left((\omega_2 - \omega_3) \cos\phi \sin\phi - \omega \frac{\omega_2 + \omega_3}{2(\omega_2\omega_3)^{1/2}} \left(\cos^2\phi - \sin^2\phi \right) \right)$$

To get h^ω in the form of three uncoupled oscillators, we must require the mixed operators to disappear

$$(\omega_2 - \omega_3) \cos\phi \sin\phi - \omega \frac{\omega_2 + \omega_3}{2(\omega_2\omega_3)^{1/2}} \left(\cos^2\phi - \sin^2\phi \right) = 0$$

The angle ϕ is thus obtained as

$$p = \tan 2\phi = \frac{\omega}{\omega_2 - \omega_3} \frac{(\omega_2 + \omega_3)}{(\omega_2\omega_3)^{1/2}}$$

where the notation p ($= \tan 2\phi$) has been introduced as a measure of the rotational frequency, ω. We now get the Hamiltonian h^ω in the following form

$$h^\omega = \hbar\omega_1 \left(a_1^+ a_1 + \frac{1}{2} \right) + \hbar\omega_\alpha \left(a_\alpha^+ a_\alpha + \frac{1}{2} \right) + \hbar\omega_\beta \left(a_\beta^+ a_\beta + \frac{1}{2} \right)$$

with the frequencies of the normal modes given by

$$\omega_{\alpha,\beta} = \frac{1}{2} (\omega_2 + \omega_3) \pm \frac{1}{2} (\omega_2 - \omega_3) (1 + p^2)^{1/2}$$

The single-particle eigenvalues of h^ω are

$$e_\nu^\omega = \hbar\omega_1 \left(n_1 + \frac{1}{2} \right) + \hbar\omega_\alpha \left(n_\alpha + \frac{1}{2} \right) + \hbar\omega_\beta \left(n_\beta + \frac{1}{2} \right)$$

where n_1, n_α and n_β specify the number of quanta in the three normal-mode degrees of freedom.

A further quantity of interest is the expectation value of ℓ_1. The diagonal parts of this operator are easily obtained and thus, for an orbital characterised by the occupation numbers n_1, n_α and n_β

$$\langle \ell_1 \rangle = \langle n_1 n_\alpha n_\beta | \frac{p}{(1 + p^2)^{1/2}} \left(a_\beta^+ a_\beta - a_\alpha^+ a_\alpha \right) | n_1 n_\alpha n_\beta \rangle = \frac{p}{(1 + p^2)^{1/2}} (n_\beta - n_\alpha)$$

We will now consider total quantities of the A-particle system with the A particles filling the orbitals (generally those being lowest in energy) of the rotating harmonic oscillator. For this purpose we define the quantities

$$\Sigma_k = \sum_{\substack{\nu \\ \text{occ}}} \langle \nu | a_k^+ a_k + \frac{1}{2} | \nu \rangle = \sum_{\substack{\nu \\ \text{occ}}} \left(n_k + \frac{1}{2} \right)_\nu$$

The index k takes the values $k = 1, \alpha$ and β (or $k = 1, 2$ and 3 for $\omega = 0$)

and the summation runs over the occupied orbitals, $|v\rangle$. The total energy in the rotating system is given by

$$E^\omega = \sum_{\substack{v \\ \text{occ}}} \langle v|h^\omega|v \rangle = \hbar\omega_1\Sigma_1 + \hbar\omega_\alpha\Sigma_\alpha + \hbar\omega_\beta\Sigma_\beta$$

In the cranking model, the angular momentum is identified with the sum of the expectation values of ℓ_1

$$I = \sum_{\substack{v \\ \text{occ}}} \langle v|\ell_1|v \rangle = \frac{p}{(1+p^2)^{1/2}} (\Sigma_\beta - \Sigma_\alpha)$$

Thus, each configuration is associated with a maximum angular momentum, $I_{\max}$, with

$$I_{\max} = \Sigma_\beta - \Sigma_\alpha$$

The energy that will result from a measurement (in the laboratory system) is calculated as the sum of the expectation values of the static Hamiltonian h_{osc}:

$$E = \sum_{\substack{v \\ \text{occ}}} \langle v|h_{\text{osc}}|v \rangle = \sum_{v} \langle v|h^\omega + \omega\ell_1|v \rangle = E^\omega + \hbar\omega I$$

For a fixed configuration, i.e. fixed values of Σ_1, Σ_α and Σ_β, and for a fixed spin I, we now want to find the potential shape that minimises the energy E. For this purpose, the energy is rewritten in the form

$$E = \hbar\omega_1\Sigma_1 + \hbar\omega_2\tilde{\Sigma}_2 + \hbar\omega_3\tilde{\Sigma}_3$$

where

$$\tilde{\Sigma}_{2,3} = \frac{1}{2}(\Sigma_\alpha + \Sigma_\beta) \mp \frac{1}{2}\left(I_{\max}^2 - I^2\right)^{1/2}$$

This simple formula for the energy E is obtained only if a further approximation is made, namely

$$p = \frac{\omega_2 + \omega_3}{\omega_2 - \omega_3} \frac{\omega}{(\omega_2\omega_3)^{1/2}} \approx \frac{2\omega}{\omega_2 - \omega_3}$$

In the derivation, it is also useful to note that $p = I/(I_{\max}^2 - I^2)^{1/2}$ or $(1+p^2)^{1/2} = I_{\max}/(I_{\max}^2 - I^2)^{1/2}$.

For fixed values of $\Sigma_1, \tilde{\Sigma}_2$ and $\tilde{\Sigma}_3$, it is straightforward to find the minimum under the volume conservation constraint (cf. chapter 8):

$$\omega_1\omega_2\omega_3 = \overset{0}{\omega}_0^3$$

The following relation results:

$$\omega_1 \Sigma_1 = \omega_2 \tilde{\Sigma}_2 = \omega_3 \tilde{\Sigma}_3$$

The frequencies ω_1, ω_2 and ω_3 are obtained as

$$\omega_i = \overset{0}{\omega}_0 \left(\Sigma_1 \tilde{\Sigma}_2 \tilde{\Sigma}_3 \right)^{1/3} / \tilde{\Sigma}_i$$

(with $\tilde{\Sigma}_1 \equiv \Sigma_1$). These values of the frequencies correspond to a shape adjusted to minimise the energy E for any value of the spin $I(I \leq I_{\max})$ and for any configuration specified by Σ_1, Σ_α and Σ_β (or by Σ_1, Σ_2 and Σ_3 in the limit of $I = 0$). The minimised energy E can be written in the concise form

$$E = 3\hbar \overset{0}{\omega}_0 \left(\Sigma_1 \tilde{\Sigma}_2 \tilde{\Sigma}_3 \right)^{1/3} = 3\hbar \overset{0}{\omega}_0 \left[\Sigma_1 \left(\Sigma_\alpha \Sigma_\beta + \frac{1}{4} I^2 \right) \right]^{1/3}$$

For $I = I_{\max}$, i.e. for maximal spin within a configuration, we note that $\tilde{\Sigma}_2 = \tilde{\Sigma}_3$. Thus, the formula for the frequencies shows that $\omega_2 = \omega_3$, which corresponds to a potential being axially symmetric around the rotation axis. For such a shape, the rotation is not collective but instead built from individual nucleons having their spin vectors quantised along the rotation axis, so-called rotation around the symmetry axis.

If the formula for the frequencies is combined with the definitions of ε and γ (chapter 8), it is possible to calculate ε and γ as functions of spin I (see problem 12.3)

$$\varepsilon = \frac{3 \left(\Sigma_1^{-2} + \tilde{\Sigma}_2^{-2} + \tilde{\Sigma}_3^{-2} - \Sigma_1^{-1}\tilde{\Sigma}_2^{-1} - \tilde{\Sigma}_2^{-1}\tilde{\Sigma}_3^{-1} - \tilde{\Sigma}_3^{-1}\Sigma_1^{-1} \right)^{1/2}}{\Sigma_1^{-1} + \tilde{\Sigma}_2^{-1} + \tilde{\Sigma}_3^{-1}}$$

$$\tan \gamma = \frac{\sqrt{3} \left(\Sigma_1^{-1} - \tilde{\Sigma}_2^{-1} \right)}{\Sigma_1^{-1} + \tilde{\Sigma}_2^{-1} - 2\tilde{\Sigma}_3^{-1}}$$

We will finally consider the moment of inertia in the simple harmonic oscillator model. For collective rotation with a constant moment of inertia, $\mathscr{J}$, the energy is given by $E = (\hbar^2/2\mathscr{J})I^2$ (in the case of rotation around one axis, one should use I^2 rather than $I(I+1)$, which latter is the proper quantity for three-dimensional quantum-mechanical rotation). For a general function, $E = E(I)$, it seems natural to define moments of inertia from the derivatives (Bohr and Mottelson, 1981),

$$\frac{\hbar^2}{\mathscr{J}^{(1)}} = 2\frac{dE}{dI^2} = \frac{1}{I}\frac{dE}{dI} \approx \frac{E(I+1) - E(I-1)}{2I}$$

and

$$\frac{\hbar^2}{\mathscr{J}^{(2)}} = \frac{\mathrm{d}^2 E}{\mathrm{d}I^2} \approx \frac{E(I+2) - 2E(I) + E(I-2)}{4}$$

where we also indicate how to define these moments of inertia from measured transition energies, $E_\gamma = E(I+1) - E(I-1)$. Note that $\mathscr{J}^{(1)} = \mathscr{J}^{(2)} = \mathscr{J}$ when $E = (\hbar^2/2\mathscr{J})I^2$. The $\mathscr{J}^{(1)}$ moment of inertia is a direct measure of the transition energies while $\mathscr{J}^{(2)}$ is obtained from differences in transition energies. Except that $I(I+1)$ has been replaced by I^2 and the discrete differences are now centred around I, $\mathscr{J}^{(1)}$ is identical to the moment of inertia introduced in chapter 11 and shown in fig. 11.13.

For the harmonic oscillator in the present approximation, it is straightforward to calculate

$$\mathscr{J}^{(1)} = \frac{2\left(\Sigma_\alpha \Sigma_\beta + I^2/4\right)^{2/3}}{\Sigma^{1/3}} \frac{\hbar}{\overset{0}{\omega_0}}$$

$$\mathscr{J}^{(2)} = \mathscr{J}^{(1)} \frac{\Sigma_\alpha \Sigma_\beta + I^2/4}{\Sigma_\alpha \Sigma_\beta - I^2/12}$$

These values might be compared with the static rigid body moment of inertia

$$\mathscr{J}_{\mathrm{rig}} = M \sum_{\substack{\nu \\ \mathrm{occ}}} \langle \nu | y^2 + z^2 | \nu \rangle$$

If we follow the equilibrium shapes of the harmonic oscillator, we obtain (see problem 12.4)

$$\mathscr{J}_{\mathrm{rig}} = \frac{\Sigma_\alpha^2 + \Sigma_\beta^2 - I^2/2}{\left[\Sigma_1 \left(\Sigma_\alpha \Sigma_\beta + I^2/4\right)\right]^{1/3}} \frac{\hbar}{\overset{0}{\omega_0}}$$

It turns out that both $\mathscr{J}^{(1)}$ and $\mathscr{J}^{(2)}$ are generally below $\mathscr{J}_{\mathrm{rig}}$; at large deformations and small spins even considerably below. One could also note, however, that, for $I = I_{\max}$, $\mathscr{J}^{(1)} \equiv \mathscr{J}_{\mathrm{rig}}$.

The solution of the harmonic oscillator presented here is useful because it illustrates general features like shape changes and band terminations. On the other hand, the numerical values of moments of inertia or deformations are somewhat crude. Indeed, as mentioned above, it is possible to solve the full cranking single-particle Hamiltonian for the harmonic oscillator potential. With the resulting single-particle energies as input, it is then possible to calculate the total energy using the same procedure as above, i.e. the volume conservation condition is applied to find the deformations, ε and γ, which for each spin I minimise the total energy, E. In this case, however, no

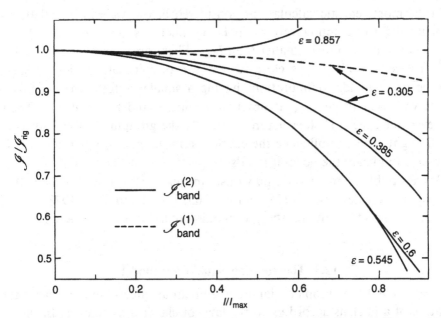

Fig. 12.1. The variation with spin of the moments of inertia $\mathscr{J}^{(1)}$ and $\mathscr{J}^{(2)}$ as obtained from the full solution of the cranked harmonic oscillator. The configurations, are labelled by their equilibrium deformations at spin zero, which are all axially symmetric. The moments of inertia and the spin are given in units of the rigid body moment of inertia at the spin zero deformation, $\mathscr{J}_{rig}$, and the maximum spin within the configuration, I_{max} (from I. Ragnarsson, 1987, *Phys. Lett.* **199B**, 317).

analytic expressions have been given for $E(I)$ but a numerical solution is straightforward. One difference (Troudet and Arvieu, 1979) compared with the approximate solution is that only those configurations that are not too deformed ($\Sigma_3/\Sigma_2 < 1.78$) for $I = 0$ really terminate, i.e. whose rotation becomes non-collective for $I = I_{max}$. The more deformed configurations do not terminate but, instead, they become more and more elongated for very high spins. With the exception of very deformed configurations, both the moments of inertia, $\mathscr{J}^{(1)}$ and $\mathscr{J}^{(2)}$, which start out equal to $\mathscr{J}_{rig}$ for $I = 0$, decrease with increasing spin, see fig. 12.1. In this figure, the different configurations, which are all axially symmetric for $I = 0$ ($\Sigma_x = \Sigma_y$), are labelled by their deformation ε at $I = 0$. The spin is given in units of I_{max} defined by $\Sigma_\beta - \Sigma_\alpha = \Sigma_3 - \Sigma_2$ independently of whether any real termination occurs or not, while the unit for the moments of inertia is $\mathscr{J}_{rig}(I = 0)$. With these units, the same figure can be applied to different mass regions. Note, however, that I_{max} is strongly dependent on deformation, e.g. I_{max} is much larger for an '$\varepsilon = 0.6$ configuration' than for an '$\varepsilon = 0.2$ configuration'.

In the more realistic calculations considered below, we will find that, for strongly deformed configurations in heavy nuclei (superdeformation), the spin will always be much smaller than I_{max}. Therefore, we might expect that $\mathcal{J}^{(2)}$ (as well as $\mathcal{J}^{(1)}$) stays close to $\mathcal{J}_{rig}$ for all spins of physical interest. On the other hand, configurations having a smaller deformation at $I = 0$ might very well reach I_{max} at observable spins (band termination). Then, however, the special shell structure caused by the grouping in the j-shells at $\varepsilon = 0$ might strongly influence the energies and the pure oscillator can only be used to indicate the general trends.

Besides publications quoted previously in this section, one could mention the papers by Zelevinskii (1975) and by Glas, Mosel and Zint (1978), where additional aspects of the rotating harmonic oscillator are considered.

12.3 The rotating liquid-drop model

We will now for a moment ignore the quantal effects and consider the rotation of a nucleus according to the laws of classical mechanics. In such a macroscopic model, the energy is given by

$$E_{macr}(E, N, \text{def}, I) = E(Z, N, \text{def}) + \frac{\hbar^2 I^2}{2\mathcal{J}(Z, N, \text{def})}$$

The energy $E(Z, N, \text{def})$ is taken as the static liquid-drop energy, which was treated in chapter 4. The variable 'def' denotes a number of deformation parameters, e.g. $\varepsilon, \gamma, \varepsilon_4 \ldots$. For stable nuclei the liquid drop energy has a minimum for spherical shape. This minimum is caused by the surface energy, which overcomes the deforming tendencies of the Coulomb energy.

In our discussion of the harmonic oscillator, we found that the dynamical moment of inertia was essentially equal to the rigid body value. In the case of independent nucleon motion, this is what is generally expected also for potentials other than the harmonic oscillator. The fact that the experimentally observed moment of inertia is smaller than $\mathcal{J}_{rig}$ for low I can be traced back to the pairing correlations. At higher spins, however, these correlations should disappear. As the rotating liquid-drop model is relevant only at relatively high spins, we will use the rigid body moment of inertia in connection with this model.

The rotational energy becomes smaller with increasing $\mathcal{J}$. Thus, with the rigid body value, configurations with the nucleons far away from the rotation axis are favoured. This means that the rotational energy tries to deform the nucleus and this tendency will become dominating for a large enough value of the angular momentum I.

For small values of I, the nucleus will behave in a similar way to the rotating earth and become flattened at the poles, i.e. oblate shape with rotation around the symmetry axis. With the rotation axis being the 1-axis and with the definitions of ε and γ given in chapter 8, this corresponds to $\gamma = 60°$ (there exists some confusion about the definition of the sign of the angle γ and consequently, oblate shape with rotation around the symmetry axis is sometimes referred to as $\gamma = -60°$). With increasing spin, the distortion of the nucleus will become larger and ε will increase (still at $\gamma = 60°$). The macroscopic energy of the nucleus ^{154}Sm, rotating with a spin of $I = 40$, is shown as a function of deformation in fig. 12.2.

Detailed calculations, as have been carried out by Cohen, Plasil and Swiatecki (1974), then show that, at sufficiently large angular momentum, the stability towards axial asymmetry is lost. For higher spins, most nuclei will for some intermediate spin values have a minimum for triaxial shape $(60° < \gamma < 0°)$ before, for even higher spins, the stability in the fission direction is lost. Thus, the nucleus divides into two fragments, which fly apart due to the centrifugal forces. The I-values where the transition to triaxial shape and where fission instability sets in, respectively, are shown in fig. 12.3. This figure also shows that, for heavy nuclei, these two I-values coincide. This means that as soon as the oblate regime becomes unstable the nucleus goes to fission.

12.4 An illustrative example of microscopic calculations of high spin states – ²⁰Ne

In the discussion of the modified oscillator potential (chapter 6), it was found that no ℓ^2-term was necessary for the light nuclei. This means that, apart from the $\ell \cdot$ s-coupling term, the single-particle potential of such nuclei is essentially a pure harmonic oscillator. Consequently, the harmonic oscillator has been used quite a lot for the description of light nuclei, especially the sd-shell nuclei. These are the nuclei with the valence protons and valence neutrons in the $d_{5/2}, s_{1/2}$ and $d_{3/2}$ shells (fig. 12.4), i.e. those having neutron and proton numbers in the range $N, Z = 8$–20.†

The rotating harmonic oscillator should thus be a very useful starting point for the study of high spin states in ²⁰Ne, which nucleus we will now discuss in some detail. Let us first point out that, for such a light nucleus, $I = 4$ or $I = 6$ are already very high spin states, i.e. they correspond to

---

† In the study of sd-shell nuclei, the symmetry group of the deformed harmonic oscillator, SU(3), has often been used for classification of different configurations. This same symmetry group has also played an important role in elementary particle physics, e.g. for the understanding of the protons and the neutrons as being built out of three quarks.

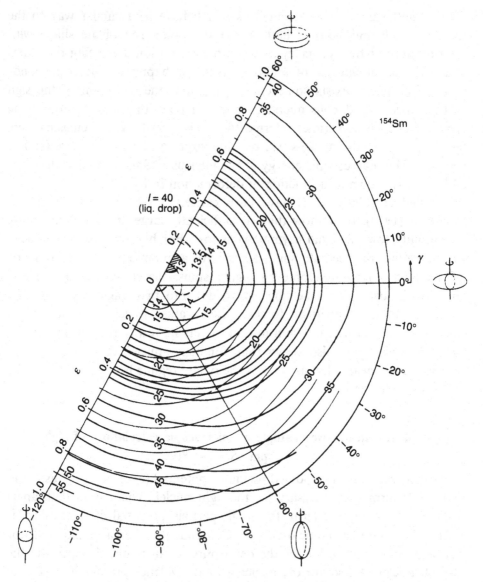

Fig. 12.2. Contour plot in the (ε, γ)-plane of the rotating liquid-drop energy calculated for the nucleus ^{154}Sm at $I = 40$. The rotation axis (defined as the 1-axis) is sketched for the different cases of axially symmetric shape (cf. fig. 8.6). The same nuclear shapes are formed in the three 60° sectors but the rotation axis coincides with the smaller ($\gamma = 0°$ to 60°), the intermediate ($\gamma = 0°$ to −60°) and the larger ($\gamma = -60°$ to −120°) principal axis, respectively. The numbers on the contour lines refer to MeV above the energy of a spherical liquid drop at $I = 0$ (from Andersson *et al.* 1976).

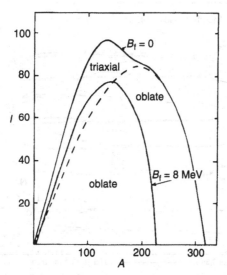

Fig. 12.3. Spin values in the rotating liquid drop model where the fission barrier disappears '$B_f = 0$' and is equal to 8 MeV, respectively. The dashed curve shows the spin value where the stability of the oblate regime ($\gamma = 60°$) is lost. Thus, between the dashed line and the ($B_f = 0$) line, the equilibrium shape is triaxial (from Cohen *et al.*, 1974).

high rotational frequencies (cf. problem 12.6). The calculated spin zero potential energy surface of ^{20}Ne has a minimum for prolate shape with $\varepsilon \approx 0.4$. Fig. 12.4 then shows that this corresponds to a situation where the $N = 1$ shell is completely filled and there are in addition two protons and two neutrons in the $N = 2$ orbital with asymptotic quantum numbers $|Nn_3\Lambda\Omega\rangle = |220\ 1/2\rangle$.

In the pure oscillator model, we described the nuclear configuration with

$$\Sigma_i = \sum_{\substack{v \\ occ}} (n_i + 1/2)_v ; \quad i = 1, 2, 3$$

As the 12 particles in the $N = 1$ shell have in total four quanta in each of the three Cartesian directions, while the four nucleons in the $N = 2$ shell have $n_3 = 2, n_1 = n_2 = 0$, it is easily found that the ground state of ^{20}Ne has $\Sigma_1 = \Sigma_2 = 14$ and $\Sigma_3 = 22$. The fact that $\Sigma_1 = \Sigma_2$ indicates that the configuration is axially symmetric and the only possible axis of collective rotation is thus perpendicular to the symmetry axis, e.g. the 1-axis. The rotation causes a mixing of the quanta in the 2- and 3-directions and the resulting normal modes were denoted by α and β in our discussion of the

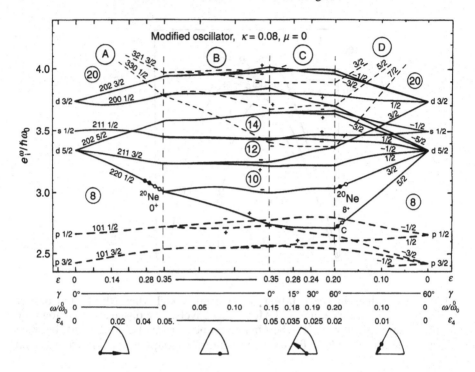

Fig. 12.4. Single-particle orbitals along a path in the $(\varepsilon, \varepsilon_4, \gamma, \omega)$ space as indicated schematically in the lower part of the figure. The path is chosen to illustrate how the orbitals can be followed when a prolate collective band goes to termination at oblate shape. The spherical origin of the orbitals at a typical low-spin deformation is traced in part A while in part B, rotation is switched on at constant deformation. At a frequency of $\omega / \overset{0}{\omega}_0 \simeq 0.15$ corresponding to $I \simeq 6$ in the ^{20}Ne ground band, the driving forces toward oblate shape become important. Thus, in part C the deformation is varied over the γ plane together with changes in the other parameters as they occur when a band approaches termination at $\gamma = 60°$. In part D, finally, the origin of the aligned oblate orbitals is traced, illustrating to which j shell they mainly belong and their aligned spin. The occupation of sd-shell orbitals in the ground state and in the terminating 8^+ state of ^{20}Ne is also indicated. It is interesting to note how the $Z = N = 10$ gap stays large all the way to the termination $(\varepsilon \simeq 0.20, \gamma = 60°)$ while this is not the case for the $N = Z = 12$ gap. Thus, we expect the aligned 8^+ state terminating the ground band in ^{20}Ne to be more favoured than the corresponding aligned 12^+ state in ^{24}Mg (revised from Sheline *et al.*, 1988).

rotating harmonic oscillator. The present configuration of ^{20}Ne should thus be denoted as $\Sigma_\alpha = 14, \Sigma_\beta = 22$ (and $\Sigma_1 = 14$).

In the harmonic oscillator approximation, it is now trivial to calculate the properties of the ground state configuration of ^{20}Ne from the explicit formulae given above. The maximum spin of the configuration is $I_{\max} = 8$

($I_{max} = \Sigma_\beta - \Sigma_\alpha$), a value that can also be deduced from more simple reasoning. With the $N = 1$ shell being completely filled, the particles in this shell do not contribute to the spin. Thus, eight protons and eight neutrons form an inert core. For particles in the $N = 2$ shell, the maximum ℓ-value is 2. The double degeneracy in each orbital makes it possible for the two protons and two neutrons to couple independently to $I = 4$. The four particles can thus give a total spin of $I = 8$. If the intrinsic spin of the particles is also considered, we get the situation illustrated in the upper right part of fig. 12.5. The two protons (or two neutrons) in the $N = 2$ orbitals (essentially $d_{5/2}$ orbitals) must have anti-aligned intrinsic spin vectors due to the Pauli principle. Thus, in this case also, the maximum spin is 8. The particles then have their spin vectors quantised along the rotation axis, which means that the nuclear state is axially symmetric around this axis. Furthermore, the four particles rotate mainly around the equator of the nucleus, giving rise to an oblate nuclear shape (fig. 12.5). This was also formally found for the pure oscillator. With the coupling between the different N-shells being neglected, the nuclear shape is always symmetric at the termination of a band.

The evolution of the proton or neutron single-particle orbitals for the ground band of ^{20}Ne is illustrated in fig. 12.4. To the far left in this figure, the splitting (and mixing) of the spherical subshells caused by prolate deformation is illustrated. At $\varepsilon \approx 0.35$, this leads to the orbitals appropriate for the ground state of ^{20}Ne where two protons and two neutrons fill the [220 1/2] orbital. The potential is now cranked around a perpendicular axis (the x-axis) leading to a splitting of the doubly degenerate orbitals and new eigenvalues $e_i^\omega(\omega)$. Without going into details, we should mention that apart from parity, one more symmetry (associated with rotation, 180°, around the cranking axis) survives so that the orbitals labelled by + and – (signature $\alpha = +1/2$ and $\alpha = -1/2$), respectively in fig. 12.4 remain uncoupled.

For cranking at a fixed deformation, the slope of the orbitals corresponds to the alignment, $\langle j_x \rangle = m$. This is seen from the relation

$$\langle j_x \rangle = -\frac{\partial e_i^\omega}{\partial \omega}$$

which is easily obtained from the cranking Hamiltonian. For prolate shape and small ω-values, the two branches of an $\Omega = 1/2$ orbital get an alignment of $\pm(1/2)a$ where a is the decoupling factor discussed in chapter 11 while in lowest order of ω, the $\Omega > 1/2$ orbitals show no alignment (no decoupling factor). Then with increasing rotational frequency ω, the coupling between the different orbitals means that all orbitals get a $\langle j_x \rangle$ different from zero. Note especially that the two orbitals emerging from [220 1/2] become

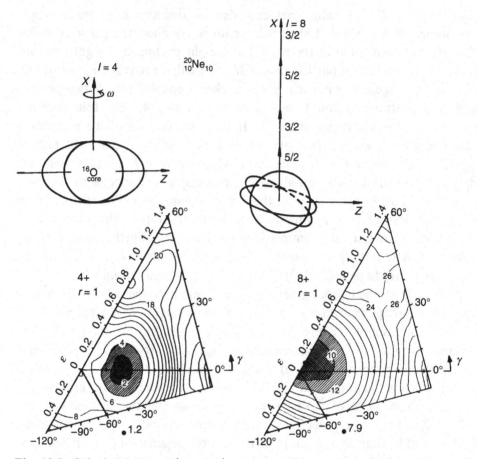

Fig. 12.5. Calculated $I^\pi = 4^+$ and 8^+ energy surfaces with inclusion of the shell energy for ^{20}Ne together with schematic illustrations of the configurations at the minima. The definition of ε and γ is the same as in fig. 12.2. The contour line separation is 2 MeV and the numbers on the lines refer to excitation energy above the spherical liquid drop at $I = 0$ (from Ragnarsson *et al.*, 1981).

strongly aligned at large ω. These are the orbitals occupied in the ground band of ^{20}Ne as discussed qualitatively above. Their strong alignment corresponds to a polarisation toward oblate shape. Thus, in the third section of fig. 12.4, the shape is followed through the γ-plane with slightly increasing rotational frequency ending up in an oblate nucleus 'rotating around its symmetry axis'.

The so-called rotation around the symmetry axis deserves some comments. Indeed, from the quantum mechanical point of view, no such rotation is possible as two states rotated by an angle φ relative to each other cannot be distinguished. Even so, the concept is useful when selecting favoured configurations of single-particle character. Indeed, rotation around a symmetry axis only corresponds to particle–hole excitations for a symmetric nucleus. One problem is then to select those excitations giving favoured energies for different spins (the yrast states). One convenient way to do this is from a diagram as illustrated on the left in fig. 12.6, where the single-particle energies e_i are plotted versus $m = \Omega, \Omega$ being the eigenvalue of j_x. The total spin is simply given by

$$I = \sum \left(m_i^{\text{part}} - m_i^{\text{hole}} \right)$$

and the excitation energy by

$$E^{\text{exc}} = \sum \left(e_i^{\text{part}} - e_i^{\text{hole}} \right)$$

where the sum runs over the particle–hole excitations. It is now easy to conclude that the most favoured states are selected from filling the orbitals below the straight line 'tilted Fermi surfaces' illustrated in the figure. In this way, states having lowest energy per spin unit will be selected, however only for some spin values. Intermediate spins will then be obtained from particle–hole excitations relative to the tilted Fermi surface.

It is straightforward to conclude that the same results are obtained from the cranking formalism. For 'rotation around a symmetry axis', $j_x = \Omega$ is a good quantum number and the eigenvalues of the cranking Hamiltonian are simply

$$e_i^{\omega} = e_i - \omega \Omega_i$$

corresponding to straight lines when drawn as functions of ω, see right panel of fig. 12.6. Then, if for different frequencies the lowest e_i^{ω} are selected, this is exactly analogous to choosing the points below the straight line Fermi surface (where ω corresponds to the slope of the line). This result is important because it now becomes possible to treat all kind of rotations, collective and particle–hole excitations, within the same formalism obtaining the yrast states from filling the lowest eigenvalues e_i^{ω} of the cranking Hamiltonian. Indeed, this is what we already used in our discussion of the harmonic oscillator. Note also that the slope of the tilted Fermi surface is proportional to ω, which in this case is only an auxiliary parameter. However, in analogy with the case of collective rotation, it appears natural to refer to it as rotational frequency.

Let us now go back to fig. 12.4 where the part to the right is analogous to the single-particle diagram in fig. 12.6. However, the deformation and the rotational frequency are varied simultaneously so that the spherical subshells are regained to the left in fig. 12.4. It is interesting to note that, by choosing

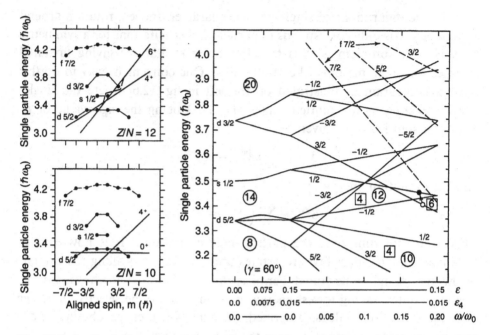

Fig. 12.6. In the left panels, the calculated single-particle energies in the sd-shell region are drawn versus their spin projection on the symmetry axis, m_i. The deformation is oblate at $\varepsilon = 0.15$. Sloping Fermi surfaces are then drawn to indicate how different spin states for particle numbers 10 and 12, respectively might be formed at this specific deformation. Whether these configurations are really formed in the spectra of ^{20}Ne and ^{24}Mg depends on whether they show up as minima in the energy surfaces or not, i.e. whether that specific deformation is favoured or not at the resulting spin value. The same information can be extracted from the diagram to the right where the spherical origins of the orbitals are first drawn and the 'rotating energies', $e_i - \omega m_i$, are then followed as functions of ω. The numbers enclosed in squares indicate total neutron or proton spin. Note that gaps develop for the same particle numbers and spins where straight line Fermi surfaces could be drawn in the e_i versus m_i diagrams. In both approaches is also illustrated (by a long-dashed line and by an arrow, respectively) how an $I = 5$ state for 12 particles might be formed as a particle–hole excitation from the $I = 6$ state. Note the large gap that develops to the far right for particle number 10 and the corresponding appearance of the e_i versus m_i diagram where there is lots of space for the $I = 4$ straight line Fermi surface. For particle number 12 on the other hand, no large gap develops and it becomes much more difficult to find space to draw straight line Fermi surfaces in the e_i versus m_i diagram.

the path in deformation/rotational frequency illustrated in fig. 12.4, it is possible to follow in a continuous way (no crossings between orbitals) the evolution of ^{20}Ne from its ground state to the aligned 8^+ state.

Using the methods discussed in the next subsection, it is possible to

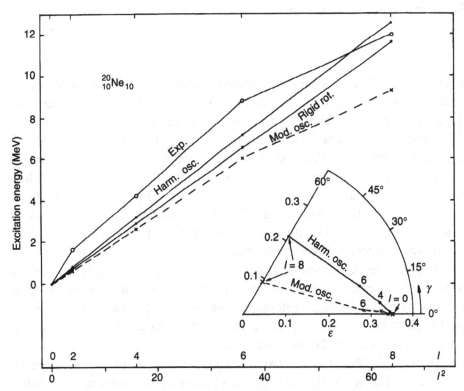

Fig. 12.7. Measured ground band energies of ^{20}Ne plotted versus I^2 compared with calculated energies in some different approximations. The curve marked 'Harm. osc.' shows the energies in the simplified solution of cranked harmonic oscillator while 'Mod. osc.' indicates the modified oscillator (fig. 12.5) model. For each spin, the energy is minimised as a function of deformation. The corresponding shapes are shown in the (ε, γ)-plane in the inset. The curve marked 'Rigid rot.' gives the energy $E = \left(\hbar^2 / 2 \mathscr{J}_{\mathrm{rig}} \right) I^2$ where $\mathscr{J}_{\mathrm{rig}}$ is kept constant, equal to the rigid body value at the ground state shape of the harmonic oscillator.

calculate potential energy surfaces also for $I \neq 0$. An example of such a calculation is given in fig. 12.5, where the $I = 4$ and $I = 8$ energy surfaces of ^{20}Ne are drawn. The ground state $I = 0$ shape of ^{20}Ne is calculated at $\varepsilon \simeq 0.35$, $\gamma = 0$. At $I = 4$, the disturbance caused by the rotation is rather small with an essentially unchanged deformation at the potential energy minimum. For $I = 8$, the rotational disturbances are much larger and, as anticipated above, the minimal energy shape is oblate with rotation around the symmetry axis ($\gamma = 60°$).

In fig. 12.7, the experimental yrast states are compared with the calculated energies within the harmonic oscillator and modified oscillator models.

Furthermore, we show the energy

$$E = \frac{\hbar^2}{2\mathscr{I}_{\mathrm{rig}}} I^2$$

where $\mathscr{I}_{\mathrm{rig}}$ is taken as a constant, namely the rigid moment of inertia of the harmonic oscillator at the $I = 0$ deformation (see problem 12.4). Except for the 2^+ energy, the calculated yrast line of the modified oscillator potential (Ragnarsson *et al.*, 1981) is in quite good agreement with experiment. It is especially satisfying that the relatively low energy of the $I = 8$ state is reproduced. This is in contrast to the harmonic oscillator calculations, which give a much too large E_{8^+} to E_{6^+} spacing. The equilibrium shapes in the two models are also shown in fig. 12.7. The main feature is that the shape remains essentially prolate up to the $I = 6$ state and that a large change in deformation occurs between $I = 6$ and $I = 8$.

If one goes to higher spins than $I = 8$ for ^{20}Ne, one expects an increase of the deformation. In the macroscopic description, this is understood as a result of the centrifugal forces. In the microscopic harmonic oscillator model, particles must be excited to higher shells to get spins above $I = 8$. This naturally leads to larger deformations.

The case of ^{20}Ne is particularly simple because of the axial symmetry ($\Sigma_1 = \Sigma_2$). For a triaxial configuration, there is the possibility to rotate around each of the three principal axis. Three different bands are thus formed but the stability and physical significance of the higher ones is unclear.

The ground state of ^{24}Mg with $\Sigma_1 = 16, \Sigma_2 = 20$ and $\Sigma_3 = 28$ is an example of a configuration that is triaxial in the harmonic oscillator approximation (problem 12.5). In the modified oscillator, however, the ground state comes out as essentially prolate (being soft towards γ-deformations) (Ragnarsson *et al.*, 1981; Sheline *et al.*, 1988). The shape evolution with increasing spin is then essentially the same as in the ground band of ^{20}Ne. Thus, figs. 12.4 and 12.6 can also be used for a qualitative understanding of ^{24}Mg. Note, however, the high level density at the terminating 12^+ state, indicating a competition between different configurations. This is in contrast to ^{20}Ne where the terminating 8^+ state is calculated to be energetically very favoured compared with other states of similar spin.

12.5 The shell correction method for $I \neq 0$

When the ground state potential energy has been calculated at some fixed deformation it should be possible to get the I-dependence simply by adding the rotational energy as extracted from the cranking model. Thus, for a prescribed spin I_0, the frequency ω_0 is determined so that

$$I_0 = \sum_{\substack{i \\ \text{occ}}} \langle j_x \rangle_i$$

Then the excitation energy is obtained as

$$E_{\text{exc}} = \sum_{\substack{i \\ \text{occ}}} e_i \big|_{\omega=\omega_0} - \sum_{\substack{i \\ \text{occ}}} e_i \big|_{\omega=0}$$

For example, in fig. 12.5, the Strutinsky shell correction method has been applied to calculate the $I = 0$ energy surface while the I-dependence at a fixed deformation has been calculated according to the above formulae. In practice, in each mesh point in deformation space, the cranking Hamiltonian is diagonalised for a number of ω-values. Subsequently, ω_0 and then E_{exc} are obtained from interpolation. In the energy surface of fig. 12.5, the energy has also been minimised with respect to ε_4 deformations.

Very often, however, simple summation to obtain E_{exc} might lead to undesired features. In general, this is caused by deficiencies in the single-particle potential so that the average behaviour of E_{exc} is unrealistic. For example, the (unphysical) ℓ^2-term in the modified oscillator potential corresponds to a velocity-dependence and leads to an average moment of inertia considerably larger than $\mathscr{J}_{\text{rig}}$. Similarly, in some parametrisations of the Woods–Saxon potential, the radius parameter is different from experimentally observed nuclear radii and, with $\mathscr{J} \propto r^2$, this might have rather drastic effects.

It is expected, however, that the *fluctuations* are more accurately described by the sums, cf. chapter 9. Therefore, it appears reasonable to retain only these fluctuations with the average behaviour governed by the rigid body moment of inertia. To this end we define (see e.g. Andersson *et al.*, 1976) a spin-dependent shell correction energy

$$E_{\text{sh}}(I_0) = \Sigma e_i \bigg|_{I=I_0} - \widetilde{\Sigma e_i} \bigg|_{\tilde{I}=I_0}$$

where the smoothed single-particle sums (indicated by '~') are calculated from a Strutinsky procedure essentially the same as that described in chapter 9. Subsequently, the total energy is calculated as the sum of the rotating liquid-drop energy and the shell energy,

$$E_{\text{tot}}(\bar{\varepsilon}, I) = E_{\text{L.D.}}(\bar{\varepsilon}, I = 0) + \frac{\hbar^2}{2\mathscr{J}_{\text{rig}}(\bar{\varepsilon})} I^2 + E_{\text{sh}}(\bar{\varepsilon}, I)$$

where $\bar{\varepsilon}$ is a shorthand notation for the deformation, $\bar{\varepsilon} = (\varepsilon, \gamma, \varepsilon_4, \dots)$.

In the definition of the shell energy, all quantities should be evaluated at the same spin I_0, i.e. the smoothed single-particle energy sum should be calculated at an ω-value giving a smoothed spin $\widetilde{\Sigma m_i} = I_0$. Thus, the ω-values in the discrete sum and in the smoothed sum are generally different and it becomes difficult to get any feeling for the variation of E_{sh} from an inspection of a single-particle diagram. However, it can be shown that the quantity

$$E_{\text{quasi-sh}}(\omega) = \Sigma e_i^\omega - \widetilde{\Sigma e_i^\omega}$$

with all quantities calculated at the same ω is numerically very similar to E_{sh}. An elementary discussion of this is given in Ragnarsson *et al.* (1978). The quantity $E_{\text{quasi-sh}}$ is defined exactly analogous to the static shell energy discussed in chapter 9. Thus, ω enters very much as a deformation parameter and we can take over all our experience from the static case; specifically that gaps in the single-particle spectrum give a favoured (negative) shell energy while a large level density leads to a positive shell energy, i.e. an unfavoured configuration.

12.6 Competition between collective and single-particle degrees of freedom in medium-heavy nuclei

We will now turn to heavier nuclei where, as seen in fig. 11.2, the moment of inertia extracted from the measured 2^+ to 0^+ energy spacing is less than 50% of the calculated rigid body value. We have already pointed out that the low value is due to the pairing correlations (the pairing correlations are less important in a light nucleus like ^{20}Ne). With increasing spin, the experimental moment of inertia becomes larger (fig. 11.13) and for the deformed rare-earth nuclei, it comes close to the rigid body value in the $I = 20$–30 region. This suggests that the pairing correlations are rather unimportant at these spins and the same conclusion is also reached from more fundamental theoretical considerations. The cranking model in the form in which we applied it to ^{20}Ne, with independent particles in a rotating potential, should then be applicable to heavy nuclei at high enough spins, let's say $I \geq 30$. For such high spins, the approximation of identifying the total spin with the projection on the rotation axis should also be quite accurate. The result from ^{20}Ne that the model seems to describe the spectrum quite reasonably all the way down to $I = 0$ or at least $I = 2$ is in some ways surprising. Indeed, the application of a rotating independent particle model to the $I = 0, 2, \ldots$ states of ^{20}Ne can hardly be justified theoretically.

Calculated potential energy surfaces for ^{160}Yb at different spin values

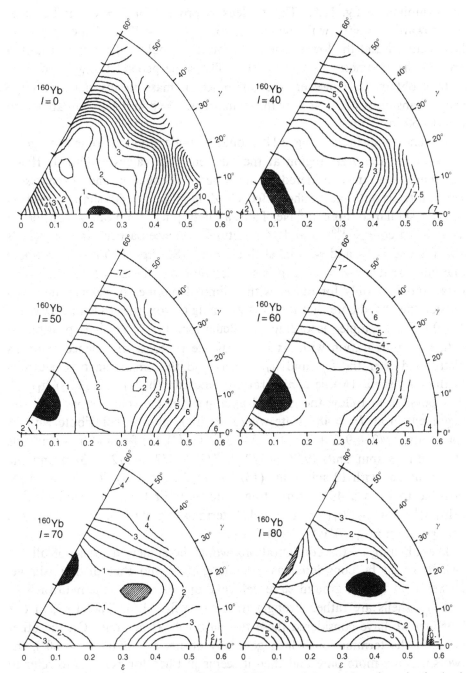

Fig. 12.8. Calculated $I = 0, 40, 50, 60, 70$ and 80 potential energy surfaces in the (ε, γ) plane with inclusion of the shell energy for $^{160}_{70}\text{Yb}_{90}$. The contour line separation is 1 MeV (from Andersson *et al.*, 1976).

are exhibited in fig. 12.8. The nucleus is prolate for $I = 0$ but becomes soft toward $\gamma = 60°$, with increasing spin. At $I = 50$, a shape transition has occurred and the lowest energy is found for oblate shape with rotation around the symmetry axis ($\gamma = 60°$). The same general trend as in ^{20}Ne is thus obtained. However, the difference in mass means that the large shape transitions occur at very different spins, $I = 8$ in ^{20}Ne compared with $I = 40$–50 in ^{160}Yb.

The energy surfaces of fig. 12.8 should be understood as some average of the states in the yrast region at the different spin values exhibited. If the different states are considered in more detail (Bengtsson and Ragnarsson, 1985), it turns out that the shell effects leading to the $\gamma = 60°$ shape transition are very similar to those in ^{20}Ne. Thus, at high spins, ^{160}Yb can be considered as a closed core of ^{146}Gd and 14 additional valence particles in the j-shells above the $Z = 64$ and $N = 82$ shell closures. (See figs. 11.5 and 11.6 where one notes that the $Z = 64$ gap is smaller than e.g. the $Z = 50$ and $Z = 82$ gaps, and it is only for nuclei with a limited number of valence nucleons outside the ^{146}Gd core that it shows any magic properties (Kleinheinz *et al.*, 1979).) For example, according to calculations for ^{160}Yb, it is possible to define a configuration with six $h_{11/2}$ valence protons, six valence neutrons distributed over the $f_{7/2}$ and $h_{9/2}$ shells and the remaining two neutrons in the $i_{13/2}$ shell. In this configuration denoted $\pi(h_{11/2})^6\nu(f_{7/2}h_{9/2})^6(i_{13/2})^2$, it is possible to follow the gradual alignment of the spin vectors until full alignment in an $I^\pi = 48^+$ state where the $h_{11/2}$ protons contribute with 18 spin units $(11/2 + 9/2 + 7/2 + 5/2 + 3/2 + 1/2)$, the $f_{7/2}$ and $h_{9/2}$ neutrons also with 18 spin units $(9/2 + 7/2 + 7/2 + 5/2 + 5/2 + 3/2)$ and the $i_{13/2}$ neutrons with 12 spin units $(13/2 + 11/2)$. Similarly, it is for example possible to form a 46^+ terminating state in ^{158}Er from the configuration $\pi(h_{11/2})^4\nu(f_{7/2}h_{9/2})^6(i_{13/2})^2$ and a 42^+ terminating state in ^{156}Er from the configuration $\pi(h_{11/2})^4\nu(f_{7/2}h_{9/2})^4(i_{13/2})^2$.

The labelling of the configurations would be somewhat easier if all the j-shells could be considered as essentially pure and thus distinguishable. However, at typical ground state deformations of $\varepsilon \geq 0.2$, the neutron $h_{9/2}$ and $f_{7/2}$ shells are rather strongly mixed (see fig. 11.6). Therefore, in our labelling we have considered these two subshells as one unit. On the other hand, at the termination, which generally occurs at $\varepsilon \simeq 0.1$ ($\gamma = 60°$), the two shells are more pure and then it seems justified for example to refer to the $\nu(f_{7/2}h_{9/2})^6$ structure as three $f_{7/2}$ neutrons coupled to $(15/2)\hbar$ and three $h_{9/2}$ neutrons coupled to $(21/2)\hbar$.

Starting from he configurations mentioned above, other terminating bands can be formed for example by redistributing the neutrons among the $(f_{7/2}h_{9/2})$

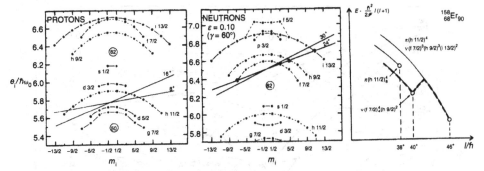

Fig. 12.9. Single-particle energies e_i for oblate shape, $\varepsilon = 0.10$, versus spin projection on the symmetry axis m_i together with a schematic illustration of the band structure in the closed core configuration of ^{158}Er. Sloping Fermi surfaces indicate the filling of the orbitals in the fully aligned $I_p = 16$ and $I_n = 30$ states and also in the cases of one proton and neutron, respectively, anti-aligned giving $I_p = 8$ and $I_n = 24$ states. It then becomes possible to form energetically favoured $46^+, 40^+, 38^+$ and 32^+ states, each associated with a terminating band as illustrated for the three higher spin values in the right part of the figure where the 40^+ state is assumed yrast and the 38^+ state is not. With no interaction, between the bands, an yrast line as shown by the thick dashed line results. Similarly, a sloping Fermi surface giving $I_n = 33^-$ can be drawn so that for example a favoured 49^- state might be formed (from Ragnarsson *et al.*, 1986).

and $i_{13/2}$ shells or by making holes in the proton core. Thus, in ^{158}Er we expect for example a $\pi(d_{5/2})^{-1}(h_{11/2})^5 \, \nu(f_{7/2}h_{9/2})^5(i_{13/2})^3$ configuration terminating in a 53^+ state and so on. It also seems possible to define band-terminating states where one spin vector points in the opposite direction relative to the total spin vector. As one example, consider an oblate $\pi(h_{11/2})^4$ $\nu(f_{7/2})^4(h_{9/2})^2(i_{13/2})^2$ configuration in ^{158}Er where one of the $f_{7/2}$ neutrons gives a spin contribution of $(-7/2)\hbar$, i.e. the four $f_{7/2}$ neutrons contribute only with $(7/2 + 5/2 + 3/2 - 7/2)\hbar = 4\hbar$, while the nucleons in the other subshell contribute with their maximal possible spin. The $f_{7/2}$ ($\Omega = m = \pm 7/2$) orbital is thus occupied by two particles, which can be understood from the fact that, at the oblate deformation in question, this orbital is far below the Fermi surface (see fig. 8.3 where the orbital is labelled [514 7/2] ($\varepsilon < 0$)).

As discussed already for the sd-shell nuclei, a more quantitative comparison between different aligned configurations is obtained from e_i versus m_i diagrams drawn at the appropriate deformation. This is illustrated for ^{158}Er in fig. 12.9 where some corresponding terminating bands are also shown schematically. Similarly, we show in fig. 12.10 how the neutron configuration of ^{158}Er, $(f_{7/2}h_{9/2})^6(i_{13/2})^2$, can be followed from the ground state to the terminating configuration with a neutron spin of $30\hbar$. Other terminat-

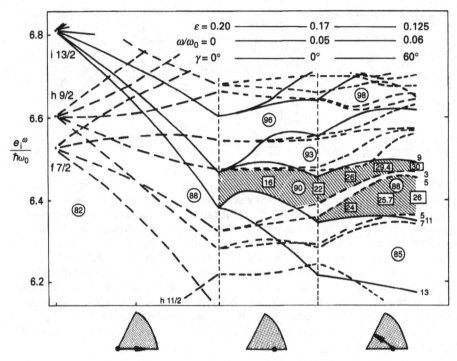

Fig. 12.10. Similar to fig. 12.4 but for neutrons around $N = 90$. Thus the left panel illustrates the spherical origin of the ground state orbitals while the other two panels show the orbitals in the $(\varepsilon, \varepsilon_4, \gamma, \omega)$ plane along an approximate path to termination. Particle numbers are given within rings and total neutron spin at specific deformations and rotational frequencies within rectangles. At the termination for oblate shape (right edge of the figure), twice the spin projection on the symmetry axis, $2m_i$, is indicated for some orbitals. It is interesting to note how the $N = 90$ gap survives all the way to the termination while a large $N = 88$ gap develops close to the termination.

ing neutron configurations are then obtained as particle–hole configurations with respect to this optimal configuration. Which configuration is considered optimal might, however, depend on the specific deformation and rotational frequency chosen.

The structure of ^{20}Ne is particularly simple because the ground state configuration is identical to the configuration terminating for $I^\pi = 8^+$. Fig. 12.10 shows that the situation is similar for the $N = 90$ neutron configuration. For the protons in the $A \approx 160$ nuclei on the other hand the terminating configurations come into the yrast region first for spin values in the $I = 30$–40 region while, at lower spins, other configurations are yrast. For example, from figs. 11.5 and 11.6, we may conclude that an approximate ground state configuration of ^{158}Er ($\varepsilon \approx 0.20$) is of the

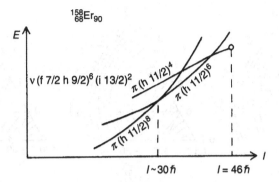

Fig. 12.11. Schematic illustration of the proton configuration yrast bands in ^{158}Er. As seen in fig. 11.5, in the ground state there are 8 protons in $h_{11/2}$ orbitals but with increasing spin, there is a tendency to close the $Z = 64$ core and thus to 'de-excite' protons from the $h_{11/2}$ shell above the gap to the $d_{5/2}$ and $g_{7/2}$ shells below the gap. In the configuration with all shells below the $Z = 64$ gap filled $(h_{11/2})^4$, an energetically favoured termination occurs for $I = 46$. In this aligned state, the spin contribution is $16\hbar$ from the protons and $30\hbar$ from the neutrons. As seen in fig. 12.10, the neutrons remain in the same configuration, $\nu(f_{7/2} h_{9/2})^6 (i_{13/2})^2$, for all spins up to the termination (from Ragnarsson *et al.*, 1986).

form $\pi(d_{5/2}g_{7/2})^{-4}(h_{11/2})^8\nu(f_{7/2}h_{9/2})^6(i_{13/2})^2$. Then with increasing spin, the deformation will develop towards decreasing ε-deformation and increasing γ-deformation leading to a band-crossing so that the configuration with two protons moved from $h_{11/2}$ to $(d_{5/2}g_{7/2})$ comes lower in energy before, at even higher spins, we reach the band-terminating configuration discussed above with a closed $Z = 64$ core. This situation is schematically illustrated in fig. 12.11. There we have omitted the low spin region, partly to indicate that at these spins the discussion above is only qualitatively but not quantitatively correct because of the neglect of pairing in the calculations. Note also that in each step two protons are exchanged between positive and negative parity orbitals in order to preserve the total parity.

Calculated potential energy surfaces close to termination in the $\pi(h_{11/2})^4$ $\nu(f_{7/2}h_{9/2})^6(i_{13/2})^2$ configuration of ^{158}Er are shown in fig. 12.12 while calculated energies for different configurations are compared in fig. 12.13. In fig. 12.14 the calculated evolution with shape of these different configurations is depicted. The bands denoted 1, 2 and 3 in figs. 12.13 and 12.14 are identical to the three bands shown schematically in fig. 12.11. The other bands then correspond to a redistribution of the valence nucleons over the open j-shells and/or making one or several holes first in the proton $Z = 64$ core and then also in the neutron $N = 82$ core.

With a fixed distribution of the particles over the j-shells, all configu-

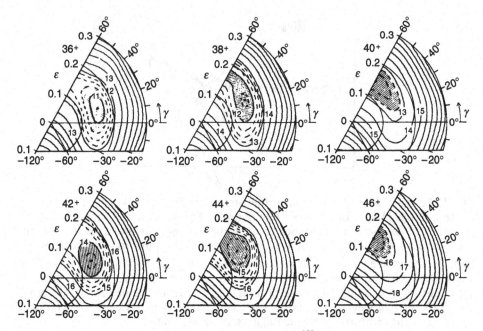

Fig. 12.12. Calculated potential energy surfaces for ^{158}Er in the configuration that terminates for $I = 46$. The configuration for the particles outside the $Z = 64$ and $N = 82$ cores is $\pi(h_{11/2})^4\nu(f_{7/2}h_{9/2})^6(i_{13/2})^2$ where the $f_{7/2}$ and $h_{9/2}$ shells are mixed so that no distinction can be made between neutrons in one or the other of these subshells. This means for example that the 'two configurations' terminating at $I = 40$ and $I = 46$ in fig. 12.9 are both considered and the minimum is found at oblate shape ($\gamma = 60°$) for these two spins (from Ragnarsson *et al.*, 1986).

rations will terminate sooner or later. However, it seems that it is only configurations with a few particles (or holes) distributed over rather many high-j subshells that are favoured in energy at the termination. Alternatively, this could be expressed by the fact that all valence particles should give an appreciable contribution to the total spin. Therefore, for example an $(h_{11/2})^6$ configuration does not really favour a termination because the two last protons give a spin contribution of only $(1/2 + 3/2)\hbar$ in the terminating state. For an $(h_{11/2})^8$ configuration the two last protons give a negative spin contribution and the (oblate) terminating state is even less favoured. One further observation is that particles rotating around the equator tend to make the nucleus oblate while holes of corresponding type will 'dig a hole' at the equator, thus trying to make the nucleus prolate (with 'rotation around the symmetry axis', i.e. $\gamma = -120°$). For example, coming back to the $(h_{11/2})^8$ configuration, it could as well be described as a $(h_{11/2})^{-4}$ configuration relative to a filled shell and combined with other high-j hole

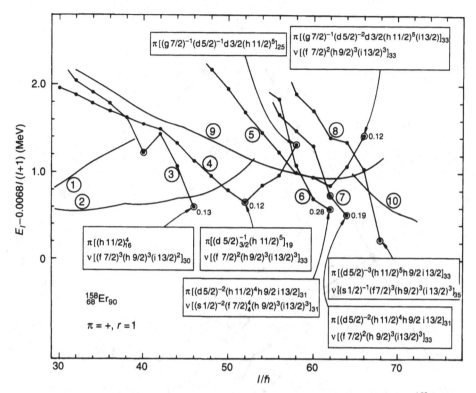

Fig. 12.13. Calculated energies of rotational bands in the yrast region of ^{158}Er. The bands labelled by 1, 2 and 3 are identical to the three bands of fig. 12.11. An open circle indicates an oblate aligned state ($\gamma = 60°$). For most such oblate states, the configuration relative to a ^{146}Gd core and the deformation ε is given (from Bengtsson and Ragnarsson, 1983).

configurations, a favoured prolate band-termination might be formed. Note, however, that high-j hole states and high-j particle states strive in opposite directions and therefore do not live very well together.

In figs. 12.13 and 12.14, we observe a mixture of bands which terminate and which essentially stay collective for all spins. The latter is the case not only for bands 1 and 2 but also for band 9, which is calculated as yrast or close-to-yrast in the $I = 50$–60 region. At these spins, one will also note that the centrifugal effects favouring larger deformations become more important and as suggested for example from fig 12.8 and discussed in the next section, some strongly deformed minima may come lowest in energy.

It seems that the calculated features discussed above are surprisingly well borne out in experiment. Thus, the observed (Simpson *et al.*, 1984, Tjøm *et al.*, 1985) γ-ray energies of ^{158}Er shown in fig. 11.12 are consistent with

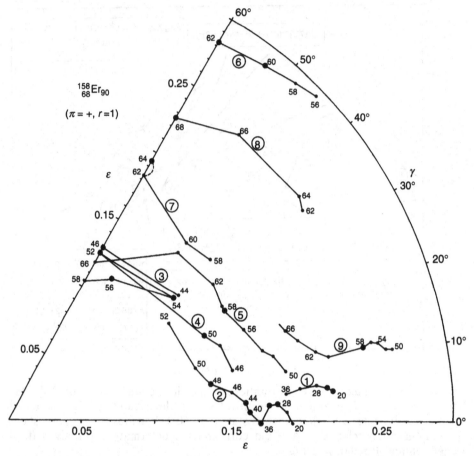

Fig. 12.14. Calculated shape trajectories for different rotational bands in the yrast region of ^{158}Er. The bands are labelled as in fig. 12.13 and the yrast states denoted by large filled circles. The configuration of some other bands can be extracted from the structure of the band-terminating states given in fig. 12.13 (from Bengtsson and Ragnarsson, 1983).

a crossing between bands 2 and 3 of fig. 12.13 at $I \approx 40$, which should be compared with $I \approx 44$ in the calculations. With a small adjustment of the single-particle parameters, we might even get full agreement. Evidence for band terminations has also been found in ^{158}Yb (Baktash *et al.*, 1985, Ragnarsson *et al.*, 1985) and ^{156}Er (Stephens *et al.*, 1985). In the latter nucleus, the positive parity ground band is collective up to $I \approx 30$ but around this spin, the spectrum changes character with somewhat irregular and comparatively small transition energies. Thus, when plotted as in fig. 12.13 relative to a rigid-rotation energy, the $I = 30$–42 structure comes down in energy with the 42^+ state very low-lying as predicted for a terminating state (Ragnarsson *et al.*, 1986). Indeed, $I^\pi = 42^+$ is identical to the maximum spin in the expected yrast configuration $\pi(h_{11/2})^4\nu(f_{7/2}h_{9/2})^4(i_{13/2})^2$. Note also the large gap associated with this neutron configuration to the far right in fig. 12.10.

Here, we have mainly discussed the nuclei with valence particles outside the ^{146}Gd core. However, starting from other closed shell nuclei similar analyses could be carried through and, as indicated above, also for holes in a closed core, an analogous formalism should be valid. The nuclei we have discussed here with $A \approx 160$ are, however, especially advantageous for several reasons. From the experimental side it is so because they are neutron deficient, which makes it easy to form high spin states from heavy ion reactions (fig. 11.11). Furthermore, nuclei with mass numbers $A = 100$–200 on the average can accommodate the highest spins as seen from fig. 12.3. From the theoretical side, the presence of several high-j shells just above the $Z = 64$ and $N = 82$ shell closures make the terminating bands especially favoured.

12.7 Shell effects at large deformation

In the preceding section, we discussed the case of a few valence nucleons outside closed shells leading to states of single-particle character at intermediate spin values. With more particles outside the core, the nucleus will stay collective to higher spins with only small shape changes. In any case, however, the centrifugal force will sooner or later become dominating as discussed within the liquid-drop model above and illustrated in figs. 12.2 and 12.3. Indeed, for nuclei with mass $A = 100$–150, the liquid-drop energy will be very soft over large regions of the deformation plane for spins $I \approx 50$–60. This means that the shell effects may play a very important role, creating minima at small but also at large and very large deformations. One example of this is seen for ^{160}Yb in fig. 12.8 where a minimum develops for $\varepsilon \approx 0.4$ and $\gamma = 20$–$30°$.

Because of the important role of the shell effects, it seems appropriate to consider their properties at large deformation in some detail. In general, one expects larger shell effects for axial symmetric shapes than for triaxial shapes. In the static harmonic oscillator approximation, this is understood from the fact that the quanta in the two perpendicular directions can be interchanged with no change in the single-particle energies. Consequently, large degeneracies occur as indicated in fig. 8.1. For example, with a two-fold spin degeneracy, the $n_z = 0$ orbitals with all quanta in the perpendicular

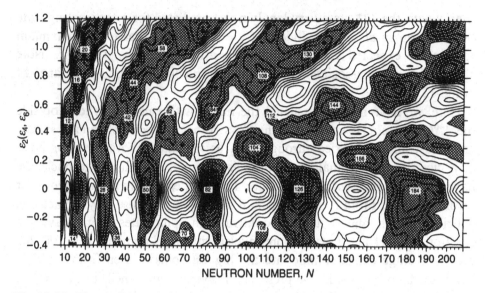

Fig. 12.15. The shell energy landscape for neutrons drawn as a function of neutron number and elongation. The shell energy is negative in shaded regions and the contour line separation is 1 MeV. Some neutron numbers in regions of favoured shell energy are indicated. The magic numbers for spherical shape are seen clearly. Another very prominent feature of the plot are the ridges that intersect the landscape, bending towards higher particle numbers with increasing prolate deformation. These ridges are caused by the numerous $n_z = 0$ orbitals, which are degenerate in the pure oscillator. In between the ridges, regions of negative shell energy are seen, suggesting very definite relationships between particle number and deformation for favoured configurations at large deformation (from I. Ragnarsson and R.K. Sheline, 1984, *Phys. Scripta* **29**, 385).

direction have a total degeneracy of $2(N+1)$ while those with one quantum in the polar direction ($n_z = 1$) have a total degeneracy of $2N$. In more realistic potentials like the modified oscillator, the $n_z = 0$ and $n_z = 1$ orbitals are not degenerate but most of them are still rather close together, see figs 8.2 and 8.5. In the latter figure, the regions of dense shading correspond to the $n_z = 0$ orbitals and those with less dense shading to $n_z = 1$ orbitals. Note that the orbitals emerging from the high-j intruder shells do not fit into this scheme as they are far below the other orbitals with the same n_z and $n_\perp$. An example seen in figs. 8.2 and 8.5 is the [505 11/2] orbital from the $N = 5$ $h_{11/2}$ subshell. The approximate degeneracy of the low-n_z orbitals is thus not applicable to the full oscillator spectrum as discussed by Ragnarsson *et al.* (1978). This has also been noted in the analysis of the Woods–Saxon potential by Dudek *et al.* (1987), who introduced the term pseudo-oscillator symmetry.

A more quantitative measure of the approximate degeneracies of the single-particle orbitals is obtained if the shell energy is plotted as a function of particle number and elongation, see fig. 12.15. Consider as an example the $n_z = 0, N = 5$ orbitals (fig. 8.5), which are found at the Fermi level for $N \approx 100$ at spherical shape, $N \approx 135$ at $\varepsilon = 0.3$ and $N \approx 185$ at $\varepsilon = 0.6$. The high level density along this path should then lead to a ridge of unfavourable positive shell energy, which is indeed the case. In a similar way, other ridges are created in fig. 12.15 from the $n_z = 0$, $N = 4$ orbitals, etc. while the approximate degeneracies of the $n_z = 1$ orbitals almost appear too small to give any discernible ridges in fig. 12.15.

In between the regions of high level density, there will by necessity be fewer single-particle orbitals leading to valleys in the shell energy landscape. This corresponds to a favoured shell energy, which is really what we are looking for. Thus, it is suggested from fig. 12.15 that the regions of low shell energy occur in a very regular and well-defined way at large prolate deformations. Furthermore, this general structure is the same for protons as for neutrons.

The feature of the harmonic oscillator spectrum that has been discussed most in the literature (e.g. Bohr and Mottelson, 1975 and references therein) is the large degeneracies that occur when all three frequencies are related through

$$\omega_x : \omega_y : \omega_z = a : b : c$$

where a, b, c are small integer numbers. Apart from spherical shape $(1 : 1 : 1)$ the most important ratios seen in fig. 8.1 correspond to $\omega_\perp : \omega_z = 2 : 1$ (cf. problem 5.1) and $\omega_\perp : \omega_z = 1 : 2$. In fig. 12.15, the prolate $2 : 1$ ratio shows up in the form of somewhat increased shell effects for $\varepsilon \approx 0.6$ compared with for other prolate shapes.

Note also that minima at $\varepsilon \approx 0.6$ (fig. 12.15) are seen at somewhat larger particle numbers than the shell gaps of the pure oscillator (fig. 8.1). For example, shell gaps at $N = 60$ and 80 correspond to a favoured shell energy of the modified oscillator at $\varepsilon \approx 0.6$ for $N \approx 60$–66 and $N \approx 80$–88. As illustrated in fig. 8.2 this is understood from the way the orbitals are influenced by the ℓ^2-term and to a smaller extent by the ε_4-term. It seems rather well established that the fission isomers discussed in chapter 9 show up because of a large negative shell energy for $\varepsilon \approx 0.6$ and $N = 140$–150 (fig. 12.15).

Fig. 12.15 is drawn for a static potential. The question is now what happens for a rotating potential. Indeed, the rotation will only influence the shell structure in a minor way (e.g. Ragnarsson *et al.*, 1980). The reason is

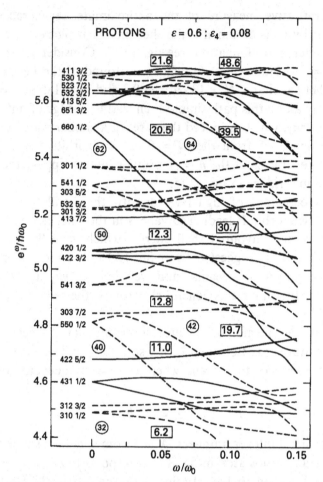

Fig. 12.16. The single-proton orbitals of the modified oscillator potential plotted as functions of rotational frequency. The deformation, $\varepsilon = 0.6$ corresponds to a 2 : 1 ratio of the symmetry axis and the perpendicular axis. The orbitals are labelled by the asymptotic quantum numbers at $\omega = 0$. At frequencies $\omega/\omega_0 = 0.05$ and 0.10, total spins for different particle numbers (shown in rings) are given within rectangles. Note that most orbitals are only weakly dependent on rotational frequency so that the regions of low and high level density are very little disturbed by the rotation (from Ragnarsson *et al.*, 1980).

that the rotation is strongly collective so that many orbitals contribute to the angular momentum but each orbital only marginally. Therefore, most orbitals get only a small $\langle j_x \rangle$ corresponding to a small slope in fig. 12.16 where the proton orbitals at $\varepsilon = 0.6$ are drawn. Orbitals having a small n_z are especially difficult to align. However, according to the discussion above, these are the orbitals that are essentially responsible for the regularity in

'SHELL ENERGY' : $\Sigma e_i^\omega - \langle \Sigma e_i^\omega \rangle$; $\varepsilon = 0.6$, $\varepsilon_4 = 0.08$; SCALE : 0.5 MeV

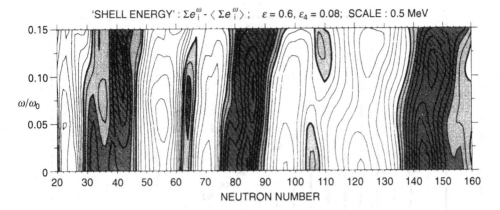

Fig. 12.17. The neutron shell energy drawn versus particle number and rotational frequency. The corresponding proton diagram is quite similar and as suggested from fig. 12.16, favoured particle numbers remain constant or are only slightly shifted at rapid rotation (from S. Åberg *et al.*, *Proc. Fourth IAEA Symp. on Phys. and Chem. of Fission*, Jülich, 1979 (IAEA, Vienna, 1980) vol. I, p. 303).

the shell effects. The fact that a few high-n_z (and high-N) orbitals get a stronger alignment does not alter the shell structure in any major way. This becomes evident from fig. 12.17 where the calculated neutron shell energy at $\varepsilon = 0.6$ is shown as a function of neutron number N and rotational frequency ω. It is evident that particle numbers of favoured and unfavoured shell energies are essentially the same independent of ω. This is apart from a general tendency that the valleys of favoured shell energy are slightly shifted towards higher particle numbers with increasing ω. This small shift is easily understood from the way in which the few orbitals showing strong alignment in fig. 12.16 cut through the large number of 'less aligned' orbitals. Note also that fig. 12.16 is drawn for protons while fig. 12.17 is drawn for neutrons. Even so the same particle numbers are favoured in both cases, e.g. $Z = 62-64$ in fig. 12.16 and $N = 62-64$ in fig. 12.17. This underlines once more that the main features of the shell effects are governed by the general symmetries and not by the detailed features of the single-particle potential. Therefore these main features are the same for protons as for neutrons.

12.8 Rotational bands at superdeformation

The discussion above suggests that it should be straightforward to obtain superdeformed states at high enough spins in theoretical calculations. This has also been the case with the predictions concentrating on the nuclei

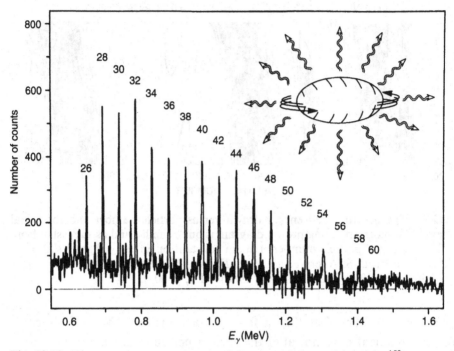

Fig. 12.18. The gamma-ray spectrum of the superdeformed band in ^{152}Dy as origi-nally identified in the 1986 Daresbury experiment (from Twin *et al.*, 1986).

around ^{152}Dy with $Z = 66$ and $N = 86$. The first observation of a superde-formed high-spin band in ^{152}Dy (Twin *et al.*, 1986) was still something of a surprise considering the many practical obstacles that had to be overcome. The experimental transition energies shown in fig. 12.18 are very regular, indicating large collectivity and no band-crossings. The 'magic' features at 2 : 1 deformation of ^{152}Dy suggest shell gaps and corresponding low shell energies at $Z = 66$ and $N = 86$ in qualitative agreement with figs. 12.16 and 12.17.

From the transition energies of fig. 12.18 it is straightforward to extract an average moment of inertia, which comes out very close to the rigid body value at 2 : 1 ($\varepsilon = 0.6$) deformation. Similarly, it has been possible to measure (Bentley *et al.*, 1987) average life-times of the high-spin states and thus the transition probabilities, which can be interpreted in the form of a quadrupole moment. Again, consistency with an approximate 2 : 1 deformation is obtained.

The full spectrum observed for ^{152}Dy is shown in fig. 12.19. As ^{152}Dy is close to the magic or semi-magic spherical nucleus ^{146}Gd, it is not unexpected that most of the states in the yrast region are of single-particle character

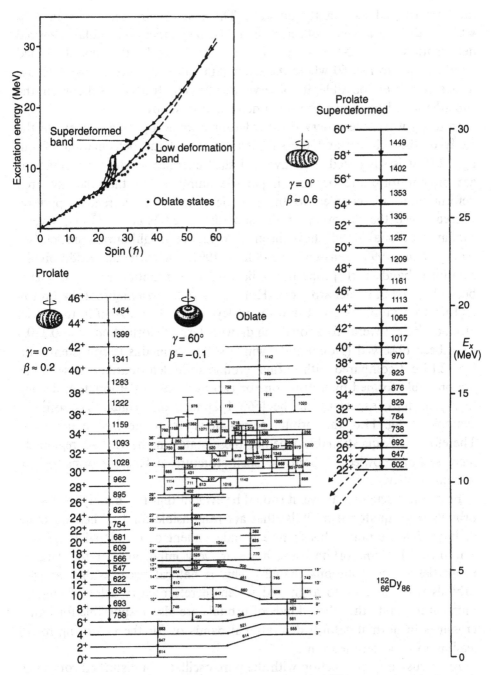

Fig. 12.19. The full spectrum observed for the nucleus ^{152}Dy showing the low-spin non-collective yrast states in the middle, a collective normal-deformed band to the left and the superdeformed band to the right. The inset in the upper left corner shows E versus I plotted in a schematic way for the different structures (from J.F. Sharpey-Schaffer, *Physics World*, Sept. 1990, p. 31).

(and interpreted as slightly oblate). These states are seen at spin $I \approx$ 40. Furthermore, one collective band corresponding to prolate normal deformation, $\varepsilon = 0.2$–0.3 is shown at $I = 46$. Finally, the superdeformed band extends to $I \approx 60$ where the exact spin values are not known because it has not been possible to observe the linking transitions between the superdeformed band and the small deformation states.

In many ways, the observed superdeformed band opens a new era in the study of the orbitals and the shell structure at large deformation. From fig. 12.15 it is suggested that favoured shell energies at large deformations are present for more or less all particle numbers but that the favoured deformations are coupled to the particle numbers in a regular pattern. Indeed, since the discovery of the superdeformed band in ^{152}Dy, several similar rotational bands have been observed in neighbouring nuclei (see Åberg et al., 1990; Janssens and Khoo, 1991, for reviews). Additionally, rotational bands with properties indicating a deformation of $\varepsilon \simeq 0.5$ have been observed in nuclei around $^{192}_{80}$Hg$_{112}$. It is very satisfying that the proton and neutron numbers do fit into the valleys of fig. 12.15. Also in the nuclei around ^{132}Ce, several rotational bands with large deformations $\varepsilon = 0.3$–0.4 have been observed (Nolan and Twin, 1988). Again this is consistent with fig. 12.15 but compared with normal ground state deformations these bands are probably formed by putting one or two particles in deformation driving high-j orbitals, especially in the [660 1/2] neutron orbital. As seen for example in fig. 11.6, this orbital comes down from the next major shell. Therefore, it is questionable whether these states in the Ce/Nd region do increase our knowledge about the shell structure at large deformations in any major way.

From the discussion above, it should be evident that it is mainly the low-n_z orbitals (the equatorial orbitals) that are responsible for the shell structure at large deformation. This feature is independent of rotational frequency. The high-j large-n_z orbitals are, however, very important for the detailed properties of the rotational bands. This is natural because they are the orbitals that are easy to align and which therefore carry a lot of angular momentum. Note that these kinds of orbitals are also responsible for band-crossings in normal-deformed rotational bands or for the polarising forces leading to band terminations.

As discussed in connection with the pure oscillator, a disturbed rotational band can be analysed in the form of different moments of inertia. As the exact spins for the superdeformed bands are not known, it is not possible to extract $\mathscr{J}^{(1)}$ with any certainty. On the other hand, $\mathscr{J}^{(2)}$, which is a measure of the relative change in transition energy, can be calculated from the relation

$$\mathscr{I}^{(2)}/\hbar^2 = \frac{1}{\mathrm{d}^2 E/\mathrm{d}I^2} \approx \frac{4}{[E_\gamma(I+1) - E_\gamma(I-1)]}$$

where $E_\gamma(I) = E(I+1) - E(I-1)$ is the transition energy for the transition, $(I+1) \to (I-1)$. When comparing with theory, we note that

$$\mathscr{I}^{(2)}/\hbar^2 = \left(\frac{\mathrm{d}^2 E}{\mathrm{d}I^2}\right)^{-1} = \frac{\mathrm{d}I}{\mathrm{d}\omega} = \sum_{\mathrm{occ}} \frac{\mathrm{d}\langle j_x \rangle}{\mathrm{d}\omega}$$

where the last equality is valid in the cranking approximation. Furthermore, we have used the canonical relation,

$$\omega = \frac{\mathrm{d}E}{\mathrm{d}I}$$

to obtain the rotational frequency from the spectrum. Thus, in the cranking approximation, $\mathscr{I}^{(2)}$ is a sum from single-particle terms measuring how the alignment $\langle j_x \rangle$ changes with rotational frequency. It is then instructive to consider how the alignment occurs in a high-j shell, see fig. 12.20 drawn for a $h_{11/2}$ shell at $\varepsilon = 0.25$. The general features of this figure remain for other high-j shells and also for larger deformation. At large deformations, however, the different orbitals get aligned at a higher frequency. As $\mathscr{I}^{(2)}$ measures the increase in alignment, it is evident that the lowest high-j orbital contributes at a very small frequency (fig. 12.20), the next at a somewhat higher frequency and so on.

Realistic calculations illustrating how $\mathscr{I}^{(2)}$ is built for the superdeformed bands in the $A \approx 150$ region are shown in fig. 12.21. Note that, for many particles in a shell, $\mathscr{I}^{(2)}$ is essentially constant similar to what was found in the pure oscillator where no orbital gets a strong alignment. Also, the first, second and third orbitals in a j-shell contribute as anticipated above. The fourth orbital becomes anti-aligned at low frequencies but it gives a positive contribution to $\mathscr{I}^{(2)}$ at higher frequencies. This means that, at superdeformation, high-j shells with four particles or more will contribute with an approximately constant value to the $\mathscr{I}^{(2)}$ moment of inertia.

From the analysis of the observed $\mathscr{I}^{(2)}$ moment of inertia for the superdeformed bands in $^{146}_{64}\mathrm{Gd}_{82}$–$^{153}_{66}\mathrm{Dy}_{87}$, it has been possible to extract probable configurations for most of these bands (Bengtsson *et al.*, 1988, Nazarewicz *et al.*, 1989). The configurations are specified by the number of particles in high-j (or rather high-N) shells, which according to the discussion above give characteristic contributions to $\mathscr{I}^{(2)}$. A large number of superdeformed bands have also been identified in the Hg/Pb region. These nuclei show a very rich structure as indicated from the plot of calculated bands in fig. 12.22.

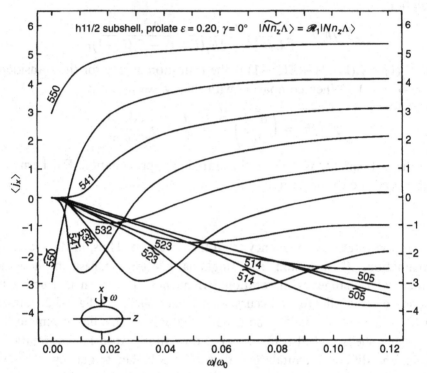

Fig. 12.20. Single-particle angular momentum component along the rotation axis for the $h_{11/2}$ orbitals at prolate shape, $\varepsilon = 0.2$. The orbitals are labelled by their asymptotic quantum numbers. The two-fold degeneracy is broken (cf. fig. 12.16) and the two sets of noninteracting orbitals (having different signature), are drawn with and without a 'tilde'. As discussed also in chapter 11, it is the polar orbitals at the bottom of the shell which become strongly aligned. The figure would be very similar for another high-j shell like $i_{13/2}$ or $j_{15/2}$ and it would also be qualitatively unchanged at a larger deformation, however with the alignment becoming more gradual (from Andersson et al., 1976).

Recently, superdeformed rotational bands with identical or almost identical transition energies have been identified in several neighbour nuclei both in the $A = 150$ and in the $A = 190$ regions. This has inspired a lot of experimental as well as theoretical investigations so we will discuss it in some more detail.

12.9 Identical bands at superdeformation

The first identification of almost identical bands in neighbouring nuclei at superdeformation came very much as a surprise. Indeed, as the spins are not known, what is really known with certainty from experiment is that

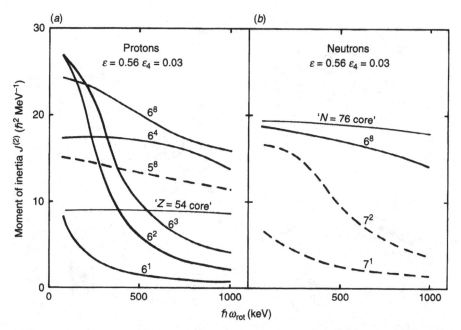

Fig. 12.21. At a deformation, typical for superdeformation in the $A = 150$ region, the contribution to the $\mathscr{J}^{(2)}$ moment of inertia from a) protons and b) neutrons in high-j shells is plotted versus the rotational frequency. The contribution from n particles in an N-shell is denoted as N^n. The lower orbitals in the $N = 6$ shell belong to $i_{13/2}$ and those in $N = 7$ to $j_{15/2}$ but as the j-shells are appreciably mixed at the large deformation, the labelling by N-shells is preferred. Furthermore, the contributions from a $(Z, N) = (54, 76)$ core are shown. Note that this core contribution is essentially constant while large fluctuations are seen in the high-j part (from Bengtsson *et al.*, 1988).

the $\mathscr{J}^{(2)}$ moments of inertia are identical, i.e. when the spins are plotted versus transition energies (fig. 12.23), the slopes are identical. Near-identical superdeformed rotational bands have been found both in the Dy/Gd region and in the Hg/Pb region (see Janssens and Khoo, 1991, for a review). Furthermore, it has been noticed recently (Baktash *et al.*, 1992) that normal deformed rotational bands also have $\mathscr{J}^{(2)}$ values, which are surprisingly similar in many cases.

The surprisingly large number of identical bands might suggest that some 'new symmetry' is involved but no such symmetry is known at present. Indeed, we are far from a more general understanding of the identical bands. Even so, it seems appropriate to discuss some of the specific features and how they can be described in simple models. We will concentrate on the Dy/Gd region where the number of identical bands is rather small (at

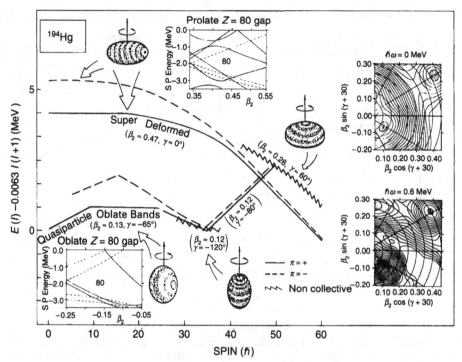

Fig. 12.22. Calculated excitation energy versus spin for ^{194}Hg illustrating a rich spectrum of collective and non-collective rotations, which evolve and co-exist near the yrast line. One insert shows the shell gap at oblate shape, which stabilises an oblate ground state configuration that terminates at prolate shape ($\gamma = -120°$) for spins just above $I = 30$. Another insert shows the $Z = 80$ gap at large prolate deformation, which is mainly responsible for the superdeformed configuration. Energy surfaces to the right indicate the corresponding minima at no rotation and at a high rotational frequency (M.A. Riley *et al.*, *Nucl. Phys.*, 1990, **A512**, 178).

present) but those cases that have been observed are very distinct, extending over a large range of frequencies. This is illustrated in fig. 12.23 where the transition energies of ^{152}Dy are compared with those of superdeformed bands in neighbouring nuclei. The figure is drawn assuming specific values for the spins. If all the bands were shifted up or down by the same value in spin, this would change nothing in our conclusions. However, the relative spin assignments are crucial and although very reasonable, we must remember that they have not been measured.

The excitation energy versus spin, $E(I)$, for the superdeformed bands can (at least locally) be approximated by the parabola

$$E(I) \approx E_0 + AI(I+1)$$

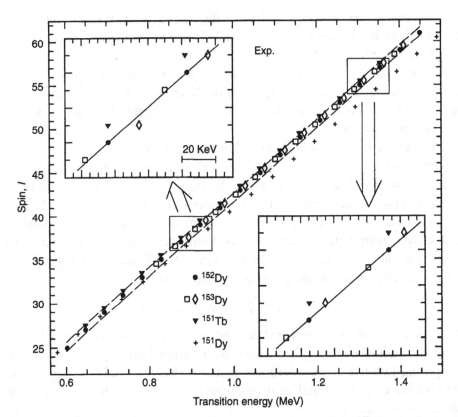

Fig. 12.23. Observed transition energies at superdeformation for ^{152}Dy, two bands in ^{153}Dy and the second band in ^{151}Tb. The bands are drawn as spin I versus transition energy E_γ. The functional dependence is very close to linear as indicated by straight dashed lines. In the insets, different parts are magnified with straight lines through the ^{152}Dy points. One finds that, within about 1–2 keV, the ^{152}Dy and ^{153}Dy data points (except one transition in ^{153}Dy) follow the same curve while the ^{151}Tb and ^{152}Dy data points have identical E_γ values. In the main figure, the ^{151}Dy data points are also included to show 'large' differences between superdeformed bands.

to a high accuracy. This means that the relation between the spin I and the quadrupole transition energies, E_γ, is approximately linear, namely

$$E_\gamma(I) = E(I+1) - E(I-1) \approx A(4I+2)$$

as comes out in fig. 12.23. Consequently, the energy difference between two consecutive transitions is roughly constant

$$\Delta E_\gamma(I) = E_\gamma(I+1) - E_\gamma(I-1) \approx 8A$$

Consider now the case that the excitation energy, $E(I) - E(I = 0)$, is described by the same function for a rotational band in an even and an odd nucleus. Then, when plotted as in fig. 12.23, all transitions will lie on the same line but because the spins are half-integer in the odd nucleus and integer in the even, the points of the two bands will be displaced relative to each other. Thus, with the even nucleus band as reference, the points in the odd nucleus will be displaced upwards or downwards by $\Delta E_\gamma/4$ depending on whether the spins in the odd nucleus are $0.5\hbar$ larger or smaller than the spins in the even nucleus.

It turns out that two bands in ^{153}Dy are identical to the yrast superdeformed band in ^{152}Dy in the way described here, i.e. that both the upward and the downward shift are realised. In fact, such bands come out from the most simple realisation of strong coupling discussed in chapter 11. In a simplified cranking model, assuming constant deformation and no pairing, such bands result if the orbital of the odd particle shows no alignment. This is easily seen if, starting from the cranking Hamiltonian, the expectation value with respect to the single-particle state $|i\rangle$ is taken:

$$e_i = e_i^\omega + \hbar\omega\langle j_x\rangle_i$$

With $\langle j_x\rangle_i = 0$, $e_i = e_i^\omega = $ constant, the total energy E_{sp} is changed by a constant while the total spin I remains unchanged (cf. fig. 12.24). When comparing two nuclei, an additional factor is that $\hbar\omega_p$ and $\hbar\omega_n$ depend on the number of protons and neutrons (chapter 8). This is, however, a small correction, see below.

For an orbital with $\langle j_x\rangle = 0$ independently of rotational frequency ω, the two branches with different signature will be degenerate but depending on in which of these orbitals the odd particle is put, the spin values realised are increased (signature $\alpha = 1/2$) or decreased (signature $\alpha = -1/2$) by $0.5\hbar$. Examples of orbitals that approximately fulfil the requirement of $\langle j_x\rangle = 0$ in fig. 12.16 are [303 7/2] in the $Z = 40$ region or [413 5/2], which is the 67th orbital at $\omega = 0$. For neutron particle states above '$N = 86$', the [402 5/2] orbital (cf. fig. 8.3) is of similar nature and could thus be responsible for the two bands in ^{153}Dy.

Even in more complete cranking calculations including shape polarisation, e.g. using the harmonic oscillator or the modified oscillator potential (see e.g. Ragnarsson, 1990; Szymański, 1990), it turns out to be rather easy to get out near-identical bands in calculations. Let us illustrate this by considering the rotating oscillator model in some detail. Thus, fig. 12.24 presents a comparison between calculated rotational bands in a 'core nucleus' (^{152}Dy) and 'core plus valence particle nucleus' (^{153}Dy), where the valence particle is placed in different orbitals. The differences between the calculated bands are illustrated as the difference between the I versus E_γ curves (of the type drawn

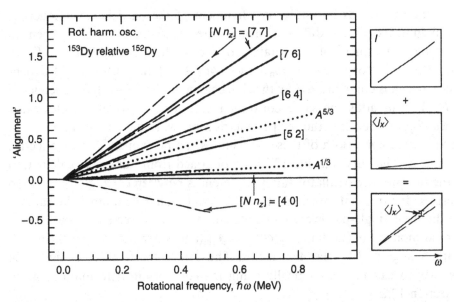

Fig. 12.24. Changes in transition energies ($E_\gamma = 2\hbar\omega$) caused by the addition of one particle to a core nucleus as calculated in the rotating harmonic oscillator. The core is defined from the asymptotic orbitals, which are filled in the superdeformed band of ^{152}Dy according to current theories. Its deformation is $\varepsilon = 0.57$. In the figures to the right is indicated schematically that, if a particle with alignment $\langle j_x \rangle$ is added to a core at constant deformation, this alignment (effective) is obtained from the differences in the I versus ω curves for the two bands. Thus, $\langle j_x \rangle$ as calculated from the simple expression given in section 12.2 is drawn by dashed lines for the orbitals considered (assumed to have $n_x = n_y$). Then we also give the calculated differences in transition energies (expressed as an effective alignment) between the core and the 'core plus particle' in a complete solution of the rotating oscillator. The important factors, in addition to the single-particle alignment, are the individual shape minimisation and the addition of an $A^{1/3}$ factor (see dotted line) because of the $A^{1/3}$-dependence of $\hbar\omega_n$. Note that the shape minimisation tends to 'decrease the scattering' between the different orbitals and that the average effect of all orbitals comes close to an $A^{5/3}$-dependence.

in fig. 12.23), i.e. as a relative alignment between the different bands. It is immediately obvious that, considered as functions of rotational frequency ω, all differences are essentially linear. Furthermore, with the valence particle in an equatorial orbital ($n_z = 0$), essentially identical bands are created while the bands in the two nuclei become increasingly different with increasing n_z of the valence orbitals.

In our simplified treatment of rotating oscillator (section 12.2), $\langle j_x \rangle \approx$ constant $\cdot\ \omega$. This relation and the relation $I = \mathscr{J} \cdot \omega$ for the core are illustrated to the right in fig. 12.24 and it is then also shown how they add. Thus, in this approximation assuming constant deformation, the difference between the rotational bands is simply the alignment of the particle. Consequently, the alignments for the different orbitals are drawn in fig. 12.24. Then, in a complete treatment of the rotating oscillator there are additional effects. First, the oscillator constant $\hbar\omega_n$ varies with $A^{1/3}$ (and a $(N-Z)/A$ dependence, which is compensated by the sign change in the $(N-Z)/A$ dependence of $\hbar\omega_p$). This adds the same constant factor for all orbitals as illustrated in the figure. Second, the shape polarisation will change the moment of inertia of the core. Both of these corrections can be expressed as an effective alignment. The curves in fig. 12.24 are obtained from numerical calculation in the full rotating oscillator taking differences between calculated transition energies. It turns out, however, that the result would be almost the same if the different single-particle alignments curves were corrected by the change in rigid moment of inertia, $\mathscr{J}_{\mathrm{rig}}(\mathrm{core}) - \mathscr{J}_{\mathrm{rig}}(\mathrm{core+particle})$ due to the $\omega = 0$ shape change and by the $A^{1/3}$ factor. Thus, the main features of fig. 12.24 are easy to calculate analytically using the equilibrium deformations given in section 12.2.

One could also note from fig. 12.24 that, if a mean value of all orbitals is taken observing that there are more equatorial orbitals than polar orbitals, the result will be an approximate $A^{5/3}$-dependence as expected for a rigid moment of inertia. In more realistic nuclear potentials, the properties of the equatorial orbitals are about the same as in the pure oscillator. As these are the most common type of orbitals, one would expect a large number of identical bands at superdeformation from this point of view. On the other hand, one might question whether the approximation of pure single-particle motion in a mean field is realistic or not.

Coming back to fig. 12.23, if we accept the 'explanation' given above for the bands of ^{153}Dy that are identical to the band in ^{152}Dy, the other identity between the ^{152}Dy band and one band in ^{151}Tb is straightforward to explain although it appears even more strange at first sight. The fact is that, in this case, a rotational band in an even nucleus and a band in an odd nucleus have transition energies that are indeed *identical*, i.e. not displaced due to the different quantisation of the spin values. Within the scheme we have described, this is understood as caused by an orbital that shows an alignment already at no rotation, i.e. an $\Omega = 1/2$ orbital with a decoupling factor $a \neq 0$ (cf. section 11.2). For an equatorial orbital of this kind, the frequency-dependence of the alignment is essentially independent of the initial alignment. Thus, we can use the same arguments as above if we only add a constant factor corresponding to the $\omega = 0$ alignment. The identical transition energies in the bands in ^{152}Dy and ^{151}Tb then require an initial alignment of $\langle j_x \rangle = -0.5$ (corresponding to $|a| = 1$) for the active orbital (the minus sign results because the particle is taken away, i.e. a hole excitation, when going from ^{152}Dy to ^{151}Tb). Indeed, in the single-particle diagram, the orbital [301 1/2] has essentially the desired properties even though the decoupling factor comes out somewhat smaller than $a = 1$ in most calculations using different potentials.

It was the observation (Byrski *et al.*, 1990) of the identical bands in ^{152}Dy and ^{151}Tb (and a similar identity between the yrast band in ^{151}Tb and one excited band in ^{150}Gd) that really focused attention on such bands (see e.g. Stephens *et al.*, 1990) and started a lot of theoretical investigations. This was so even though the identical bands in ^{152}Dy and ^{153}Dy were already known, as were also two identical superdeformed bands in the $A = 190$ region, the latter, however, over a range of fewer transitions. In the best cases, the identity of the transition energies is good within 1–2 keV extending over about 15 transitions with energies about 600–1600 keV. These numbers might, however, give the impression that the identity is even more strange than it really is. We must remember that all the superdeformed bands are very regular and follow essentially the same curve in an I versus E_γ diagram. This is illustrated in fig. 12.23 where we also give the transition energies for ^{151}Dy. The superdeformed band in ^{151}Dy is understood as being formed when one $N = 7$ neutron is removed from ^{152}Dy. Considering that these bands are found in neighbour nuclei, they are unusually different. This indicates that, if a more inert orbital is either empty or filled in two bands in neighbouring nuclei, the bands by necessity have to be rather similar. Even so, the extreme identity observed appears very strange. Furthermore, in view of the accuracy obtained in nuclear calculations in general, it is indeed surprising that the simple theories discussed here seem to describe the experimental situation so well. One would expect that different correlations not accounted for, especially pairing, would make the very detailed comparison between theory and experiment impossible. The models introduced here seem to be useful mainly in the Dy/Gd region. The superdeformed bands in the Hg/Pb region could not really be described within this scheme assuming pure single-particle degrees of freedom with no pairing. Different ideas on how the identical bands in Hg/Pb nuclei could be understood have been published e.g. by Stephens *et al.* (1990) and Azaiez *et al.* (1991). There is however no established understanding of these bands, see e.g. Baktash *et al.* (1993).

If the mechanisms for creating identical bands discussed here are qualitatively correct, it should also be possible to describe differences between bands that are not identical. One might say that to invent a theory which gives identical bands, or even bands that differ by some smooth quantity, is not

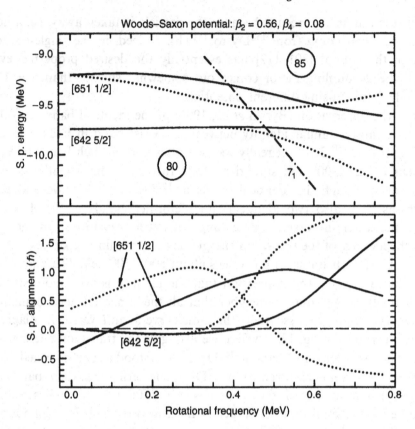

Fig. 12.25. The orbitals calculated in the Woods–Saxon potential at superdeformation in the $N = 80$–85 region are drawn versus rotational frequency ω in the upper figure while the alignments of the orbitals labelled by asymptotic quantum numbers [651 1/2] and [642 5/2] are shown in the lower figure.

so difficult. A better test of some theory is its ability to describe non-smooth differences between the observed quantities. Then, the superdeformed bands observed in $^{146-149}$Gd are especially favourable because some of these bands show a band-crossing while others do not.

When searching for orbitals at 2 : 1 deformations and neutron numbers $N = 82$–85, which could give rise to an observable crossing, the only reasonable candidates seem to be the $N = 6$ orbitals [642 5/2] and [651 1/2]. As should be evident from fig. 8.3, drawn for somewhat smaller deformations, these orbitals come close together for $\varepsilon = 0.5$–0.55. Other crossings occur between orbitals from different N-shells and appear to interact much less than observed in experimental bands. The details of this single-particle crossing are illustrated in fig. 12.25, as calculated in the Woods–Saxon model. The single-particle orbitals are drawn in the upper figure where a very small energy interval is considered so that only the two signature branches of these two orbitals together with the lowest $N = 7$ orbital are seen (cf. fig. 12.16, which shows a much larger energy interval but where the crossings are drawn somewhat schematically. Consider e.g. the crossing between the [532 5/2] and [541 1/2] orbitals around $Z = 60$ in fig. 12.16 which in many ways is similar to the crossing in fig. 12.25). In the lower part of fig. 12.25, the alignments $\langle j_x \rangle$ of the orbitals are drawn. These alignments are proportional to the slopes in the upper figure.

In a way analogous to the identical bands, we now consider (Haas *et al.*, 1993) the differences between the transition energies of two bands in neighbouring nuclei with one orbital either filled or empty. For the orbitals of fig. 12.25, the calculated differences are drawn in the lower part of fig. 12.26. It is evident that this figure has the same structure as the alignments in fig. 12.25 and it is straightforward to see which orbital is either empty or filled when comparing two bands. The differences when comparing the two figures arise mainly from the fact that the Woods–Saxon potential has been used in one figure and the modified oscillator in the other. Furthermore, in fig. 12.26, we compare rotational bands which have been minimised in deformation independently while in fig. 12.25, the single-particle alignment $\langle j_x \rangle$ is shown. The comparison shows that, in the present formalism corresponding to single-particle motion in a mean field, it is the alignment of the specific orbitals which is the important factor and that e.g. deformation changes between different bands will only lead to minor corrections.

In the upper panel of fig. 12.26, the differences in transition energies between the observed bands are drawn. The large similarity between experiment and theory in fig. 12.26 seems to be very strong evidence that our interpretation of which orbitals are active is really correct. In drawing the experimental figure, one has to make specific assumptions about the relative spins but now it seems possible to turn the argument around, claiming that the good agreement between theory and experiment means that we have determined these relative spins. This would mean that, if it becomes possible to measure the spin values in one superdeformed band, we might extract the spins also for the bands in neighbouring nuclei. Indeed, in a recent paper (Atac *et al.*, 1993), it has been claimed that the spins in the superdeformed band of ^{143}Eu have been measured. At present, very few superdeformed bands are known in neighbouring nuclei of ^{143}Eu so it does not seem possible to carry out a similar analysis around ^{143}Eu as for $^{146-149}$Gd.

The cases we have chosen in fig. 12.26 are not really typical but more

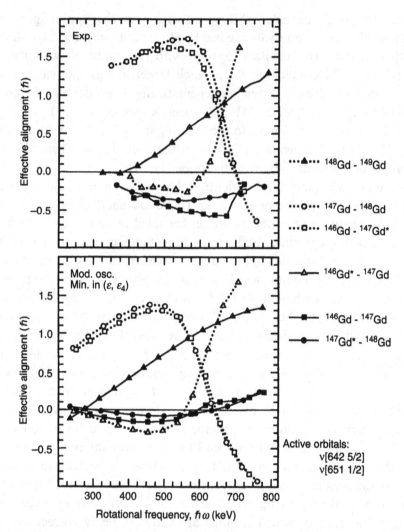

Fig. 12.26. The upper figure shows the relative transition energies for superdeformed bands in $^{146-149}$Gd where an asterisk indicates the 'second' band in that nucleus. The differences are plotted as (effective) alignments extracted as indicated in the lower right panel of fig. 12.24 ($\omega = E_\gamma/2$). Effective alignments extracted in the same way from rotational bands calculated in the modified oscillator are shown in the lower figure. A comparison with fig. 12.25 shows that it is the orbital that is labelled by [651 1/2] at $\omega = 0$ and has $\langle j_x \rangle > 0$ (the signature $\alpha = -1/2$ branch), which is being filled in the calculations when going from ^{147}Gd to ^{148}Gd or from ^{146}Gd to ^{147}Gd* etc. A comparison between the upper and lower figure strongly suggests a one to one correspondence between those orbitals used in the calculations and those active in the observed bands.

specific cases where theory and experiment seem consistent almost to the fine details. Even so, the comparison suggests that the superdeformed bands are really the best chance to see pure single-particle effects in nuclei. In the future, we could hope that a large number of rotational bands will be observed at different deformations (where favoured proton and neutron shell effects are 'in phase'). Thus, it might be possible to map the single-particle orbitals all the way from the ground state to two separated fragments. In this sense, nuclei are a very special laboratory in our study of quantum many-body physics.

Some more details of the cranking calculations have been reviewed in a recent paper (Bengtsson *et al.*, 1991) where also computer codes on a floppy disc are provided. The theoretical model behind the calculated energies shown in fig. 12.22 is somewhat different from the simple cranking model presented here. Specifically, the pairing interaction (chapter 14) is included. This should make the calculations more realistic in most cases but it has the disadvantage that they become less transparent and it becomes difficult to plot energy surfaces as functions of the physical quantity I or to follow the evolution of fixed configurations. Thus, the energy surfaces shown in fig. 12.22 are a mixture of different configurations (similar to fig. 12.8 for ^{160}Yb) and as they are drawn for a fixed rotational frequency, the spin I might be different at different deformations. Attempts to overcome these deficiencies have been made recently (Bengtsson, 1989). One could also note that, even within the simplified model described here, all the structures of fig. 12.22 come out (see calculations on ^{187}Au by Bengtsson and Ragnarsson, 1985) although in a somewhat qualitative way for low spins.

The recent discoveries of the superdeformed bands have thus made it possible to test theoretical predictions of the single-particle structure and the shell effects at large deformation. Fig. 12.15 suggests that favoured shell effects are present at large deformations for essentially all particle numbers. Thus, we would expect that, correlated with the particle number, rotational bands will be identified for essentially all deformations up to very elongated shapes, e.g. at 3 : 1 axis ratio. In this way it should be possible to scan large parts of fig. 12.15 and test how well the predictions are realised. One can then also start to ask more detailed questions about how large deformations and/or rotation disturb the nuclear quantal system.

Exercises

12.1   A particle of mass m is subject to the laws of classical mechanics. The motion of the particle can be described either in a laboratory system, x, y and z or in another system x_1, x_2 and x_3, which rotates with a constant angular frequency, ω, around the x-axis (equal to the x_1-axis). The potential energy only depends on the coordinates in the rotating system, $V = V(x_1, x_2, x_3)$. Find the Hamiltonian in the rotating system and compare with the cranking Hamiltonian.

12.2   The nuclear single-particle potential is built from the mutual inter-action of the individual nucleons. Therefore, it is important that the shape built from the nuclear density distribution is similar to the potential shape. In a non-rotating harmonic oscillator, this is easily verified. Thus, one finds that, if the energy of an arbitrary configuration is minimised with respect to deformation, the ratio of expectation values, $\langle x_1^2 \rangle : \langle x_2^2 \rangle : \langle x_3^2 \rangle$ is the same for the density distribution as for the potential. Show this!

12.3   In the discussion of the rotating oscillator model, the self-consistent frequencies were derived as $\omega_i = \overset{0}{\omega}_0 \left(\Sigma_1 \tilde{\Sigma}_2 \tilde{\Sigma}_3 \right)^{1/3} / \tilde{\Sigma}_i$. Use this expression together with the definitions of the quadrupole deformation parameters, ε and γ, to calculate how these latter parameters vary with spin, I.

12.4   Derive the static moment of inertia

$$\mathscr{I}_{\text{stat}} = M \sum_{v_{\text{occ}}} \left\langle v \left| y^2 + z^2 \right| v \right\rangle$$

for independent particles in a rotating harmonic oscillator potential.

12.5   In the harmonic oscillator model, the ground state energy is given as

$$E = 3\hbar \overset{0}{\omega}_0 \left(\Sigma_x \Sigma_y \Sigma_z \right)^{1/3}$$

as derived in the main text. Use this formula to determine the distribution of quanta in the ground state of ^{24}Mg. Then, apply the simplified formulae for the cranked harmonic oscillator to calculate $E(I)$, $\varepsilon(I)$ and $\gamma(I)$ for the three bands that result from rotation around the three principal axes.

12.6   Calculate the rotational frequency for rigid rotation of a spherical ^{20}Ne nucleus at angular momentum $I = 8$. Which angular momentum will result for a spherical $A = 160$ nucleus that compared with ^{20}Ne has

(a) the same rotational frequency;

(b) the same velocity at the nuclear surface?

13

The nucleon–nucleon two-body interaction

We have already discussed the nucleon one-body potential as the coherent external field exerted on one nucleon due to the presence of all the others. We shall now go on to analyse the basic characteristics of the underlying *two-body interaction* (the strong interaction).

Knowledge of this interaction rests, in part, on the observation of the properties of very simple nuclear systems. Historically, interest centred very much on the deuteron (consisting of a bound state of a neutron and a proton). Also, simple systems such as the trinucleon systems lend themselves to a direct test of the internucleon interaction.

In part the knowledge – and this is now the overwhelmingly important source – derives from a study of the properties of the scattering of protons against protons and protons against neutrons.

To reproduce the remarkably constant quantity of 8 MeV binding per nucleon encountered all over the periodic table, one has to postulate that nuclear forces have a very short range. We thus conclude as a general gross feature, a nucleon–nucleon interaction (or two-body potential) of the character exhibited in fig. 13.1 in terms of the inter-nucleon distance r.

The range, as roughly defined by fig. 13.1, we associate with the distance b marked in the figure. It is found empirically to be of order $b = 1.4$ fm. This is exactly the Compton wave length of the pi-meson (pion) or $b = \hbar/m_\pi c$. This in turn is indicative of the fact that at this relatively long range (i.e. in the outer regions of the interaction potential) the important agent, transmitting the interaction between the nucleons, is the pi-meson.

As shown in the figure, $V(r)$ is maximally attractive inside 1 fm while for very short distances the nucleon–nucleon interaction becomes repulsive. This is needed to account for the general finding of a constant nuclear density. This feature of the nuclear system has its counterpart in the constant density of solids. Here the interaction forces (largely Van der Waals forces) are

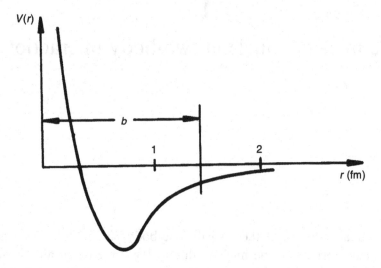

Fig. 13.1. Schematic illustration of the radial dependence of the nucleon–nucleon interaction.

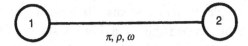

Fig. 13.2. Meson exchange between two nucleons.

attractive generally but at short distances become repulsive leading to a potential minimum characteristic of the interaction distance in the crystal.

13.1 Meson exchange

The assumption usually made is that in the nucleus the nucleon–nucleon interaction is of two-body character corresponding to the exchange of mesons (π, ρ or ω) as indicated in fig. 13.2. We thus assume that only two nucleons at a time are involved in the exchange of mesons. In a Feynman picture, with a time axis pointing upwards, we draw it symbolically like in fig. 13.3. The diagram, with the time arrow pointing upwards, may depict a proton (P) transforming into a neutron (N) and emitting a π^+-meson. The latter is subsequently absorbed, the neutron thereby becoming a proton.

We have already remarked that there is an exponential fall-off of the interaction with distance. For long distances this occurs approximately as $(1/r)\exp{(-r/\alpha_s)}$ (cf. appendix 13A). The range α_s may be related to a mass

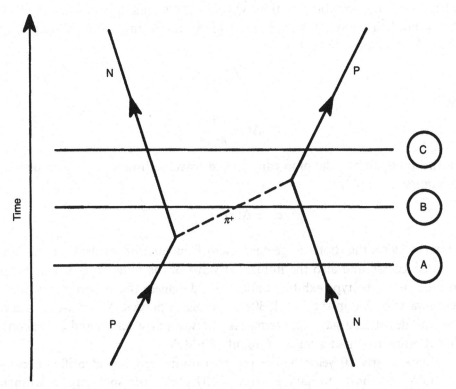

Fig. 13.3. Feynman diagram illustrating the exchange of a π^+-meson between two nucleons.

m of the transmitted agent and the fundamental constants $\hbar$ and c as

$$\alpha_s = \frac{\hbar}{mc}$$

Thus the mass of the agent, in our case mostly the π-meson, is reflected in the range of the interaction. One can understand how this comes about by the following simple argument.

Compare the energy at the initial state, marked 'A' in fig. 13.3,

$$E_A = 2M_N c^2 + E_N(\text{kin}) + E_P(\text{kin})$$

with the energy of the intermediate state marked 'B'

$$E_B = 2M_N c^2 + m_\pi c^2 + E_\pi(\text{kin}) + E_N(\text{kin}) + E_N(\text{kin})$$

Neglecting the kinetic-energy terms, which are small possibly apart from $E_\pi(\text{kin})$, we find

$$E_B \approx E_A + m_\pi c^2$$

The energy conservation is thus violated by a quantity of the order of $\Delta E = m_\pi c^2$. This violation is permissible during a time Δt consistent with the indeterminacy relation

$$\Delta t \cdot \Delta E \simeq \hbar$$

or

$$\Delta t \simeq \frac{\hbar}{m_\pi c^2}$$

In this time interval the pion can at most travel a distance α_s, which defines the range

$$\alpha_s \simeq \Delta t \cdot c = \frac{\hbar}{m_\pi c}$$

In the 1930s the short-range character of the nucleon–nucleon interaction became known and also the numerical value of the range. From theoretical arguments of the type exhibited above, the Japanese theoretical physicist H. Yukawa was able in the late 1930s to predict a particle, which he named a *'meson'* (denoting a particle *intermediate* between an electron and a nucleon). In his estimate it had a mass of about 100 MeV.

It took nearly 10 years before the *pion* (pi-meson) was identified experimentally and shown to have a mass of 137 MeV, corresponding to a range $\alpha \simeq 1.4$ fm. (In the meantime another particle, the *muon*, was discovered. This was first considered to be the Yukawa particle but was soon found to have very different properties from those predicted. Actually it does not interact by the strong interaction at all and is also not a meson but a lepton, a close relative of the electron.)

Obviously there is some asymmetry between the NN and PP systems and the NP system if we consider one-meson exchange only. Thus if the meson is charged (as π^+ or π^-), it is only in the case of a neutron and a proton that this first-order exchange can take place. For a neutron pair or a proton pair to lowest order only a π^0 meson can be exchanged. This first-order diagram is thus highly charge- or (as we shall later say) isospin-dependent. It is only in terms of second-order diagrams, involving the exchange of two mesons (for instance allowing the intermediate creation of a Δ-hyperon) that the approximate symmetry, encountered in nature between the NN, the PP and the NP systems, can be restored, see fig. 13.4. In this figure we have thus assumed that in the intermediate state the nucleon may go over into one of the spin 3/2 members of the nucleon 'family', the Δ-particle. The symmetry between NN and PP on the one hand and NP on the other found in nature leads us to conclude that it is the two-pion-exchange processes that

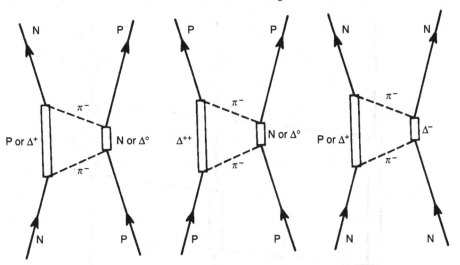

Fig. 13.4. Feynman diagrams illustrating two-pion exchange.

are mainly responsible for the nuclear binding. This is supported by detailed analysis.

Let us also say a few words about the (unimportance of) three-body forces, which one sometimes depicts as in fig. 13.5. It turns out that (fortunately for theoretical physicists who have to compute for a living) these effects are not very important and in nuclei heavier than ^{3}He or ^{3}H they are generally neglected.

Some more details about the meson theory are given in appendix 13A, while in the continuation of this chapter, we will instead treat the nucleon–nucleon interaction on a more phenomenological basis. We will discuss some invariance principles, which make it possible to draw some general conclusions about the analytic structure of the nucleon–nucleon interaction, its dependence on spin, charge and momentum. We are then left with some free constants, determining the radial dependence, which are obtained by fitting experimental scattering data to theory. An important background for our discussion is the symmetry of a two-particle wave function, which we will discuss in some detail and in this context introduce the concept of isospin. First we will, however, end this subsection with a brief discussion about the connection between the meson theory and modern theories of elementary particle physics.

The proton and the neutron belong to a family of particles referred to as baryons. The baryons are described as being built out of three quarks, qqq, which are held together by a gluon field. Similarly, the mesons are

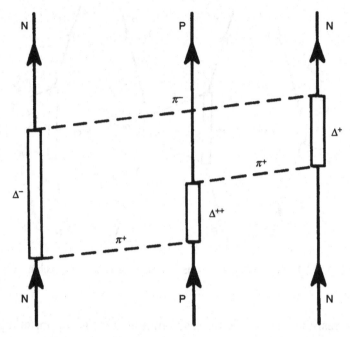

Fig. 13.5. Three-body interaction involving the exchange of π-mesons between two neutrons and a proton with intermediate creation of Δ-particles.

built from a quark–antiquark pair, q$\bar{\text{q}}$. It appears that quarks can only exist in these two combinations, qqq and q$\bar{\text{q}}$, and no free quark has been observed. One can consider the quarks as being held together by springs, as symbolised in fig. 13.6. The force appears to rise linearly with distance r, more or less indefinitely. If some energy is contributed for the purpose of quark separation, this energy may be available for the creation of a quark–antiquark pair. This pair is then broken and the released antiquark is hypothesised to join the separating quark to form a meson. The meson is thus ejected from the nucleon while the created quark joins the two original quarks to recreate an undisturbed baryon. One can talk about a quark 'bag' preventing the quarks escaping from the proton or neutron.

   With the introduction of the quark and gluon picture a problem arises. Why is it still useful to speak of neutrons and protons inside the nucleus? The quark bag picture also provides an 'explanation' for this finding. The bag walls thus rise fast enough to prevent the nucleons fusing. Furthermore, the usefulness of meson exchange as the main vehicle for nucleon–nucleon interaction goes well with the fact that no gluon and no solitary quark can get out of the bag. Only in the form of a q$\bar{\text{q}}$ pair is quark exchange between baryons possible.

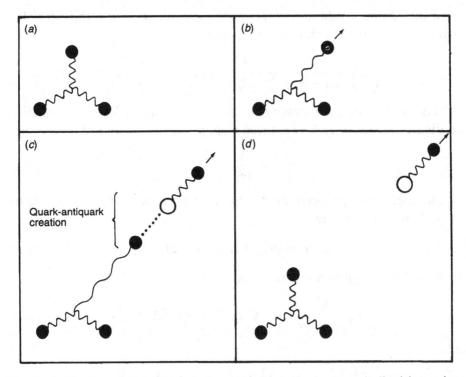

Fig. 13.6. The interaction between three quarks (a baryon) symbolised by springs. Energy contributed to this system might be used for creation of a quark–antiquark pair where the antiquark together with one of the original quarks forms an emitted meson.

13.2 Relative coordinates and symmetries for the two-particle problem

Consider generally two particles of mass m_1 and m_2 described by coordinates $\mathbf{r}_1$ and $\mathbf{r}_2$, velocities $\dot{\mathbf{r}}_1$ and $\dot{\mathbf{r}}_2$ (and spins σ_1 and σ_2). Obviously the problem will be one where the relative position $\mathbf{r} = \mathbf{r}_1 - \mathbf{r}_2$ and *relative* velocities $\dot{\mathbf{r}} = \dot{\mathbf{r}}_1 - \dot{\mathbf{r}}_2$ enter, apart from maybe σ_1 and σ_2. We have therefore to consider the elementary transformation to centre-of-mass coordinates (cf. problem 13.1).

With an interaction that only depends on the relative distance, $r = |\mathbf{r}_1 - \mathbf{r}_2|$, the Schrödinger equation takes the form

$$\left(-\frac{\hbar^2}{2m_1}\Delta^{(1)} - \frac{\hbar^2}{2m_2}\Delta^{(2)} + V\left(|\mathbf{r}_1 - \mathbf{r}_2|\right) - E \right) \Psi\left(\mathbf{r}_1, \mathbf{r}_2\right) = 0$$

The centre-of-mass coordinate,

$$\mathbf{R} = \frac{1}{m_1 + m_2}\left(m_1\mathbf{r}_1 + m_2\mathbf{r}_2\right)$$

is now introduced in addition to the relative coordinate, **r**. It is then straight-forward to express the Schrödinger equation as

$$\left(-\frac{\hbar^2}{2}\frac{1}{(m_1 + m_2)}\Delta_R - \frac{\hbar^2}{2\mu}\Delta_r + V(r) - E\right)\Psi = 0$$

where Δ_R and Δ_r are the Laplace operators in the centre-of-mass and relative coordinates, respectively. The reduced mass μ is related to m_1 and m_2 through

$$\frac{1}{\mu} = \frac{1}{m_1} + \frac{1}{m_2}$$

For the case of two particles having the same mass, $m_1 = m_2 = m$, one finds $\mu = m/2$. With the ansatz

$$\Psi = \psi(X, Y, Z)\phi(x, y, z)$$

the Schrödinger equation separates into

$$-\frac{\hbar^2}{2(m_1 + m_2)}\Delta\psi(X, Y, Z) = E_R\psi(X, Y, Z)$$

and

$$\left(-\frac{\hbar^2}{2\mu}\Delta + V(r)\right)\phi(x, y, z) = E_r\phi(x, y, z)$$

where $E = E_r + E_R$. We have thus reduced the two-body problem to a one-body problem with all the complication entering through the potential, which so far has been assumed to depend only on the relative distance r.

We have in addition a problem of symmetry. Obviously the neutron (N) and proton (P) are different particles and therefore, the NP system is apparently not restricted by the Pauli principle applicable to fermions of the same kind. For the NN and PP system on the other hand the Pauli principle must be observed in the construction of the wave functions. Thus, the wave function, built out of ordinary space and spin space

$$\psi = \phi(r)\chi(\sigma_1, \sigma_2)$$

must be antisymmetric, i.e. the wave function must change sign under interchange of the two protons (or two neutrons). The symmetry of the spatial wave function is determined by the eigenvalue of the angular momentum operator, which in the relative coordinates takes the form

$$\mathbf{L} = \mathbf{r} \times \mathbf{p} = \mu \mathbf{r} \times \dot{\mathbf{r}}$$

Even values of L then correspond to even wave functions, which are thus

Table 13.1. *Possible states defined by internal spin S, orbital angular momentum L, total angular momentum J and parity π applicable to the NP (neutron–proton) and NN and PP systems, respectively. In the last column, the corresponding isospin is given. Only those states having L ≤ 3 are indicated.*

	S	L	J^P	Symmetry	Notation	Isospin T
NP only	1	0	1^+	symmetric	$^3S_1$	
	1	2	$1^+, 2^+, 3^+$	in	$^3D_{1,2,3}$	0
	0	1	1^-	spin + position	$^1P_1$	
	0	3	3^-		$^1F_3$	
NN PP and NP	1	1	$0^-, 1^-, 2^-$	antisymmetric	$^3P_{0,1,2}$	
	1	3	$2^-, 3^-, 4^-$	in	$^3F_{2,3,4}$	1
	0	0	0^+		$^1S_0$	
	0	2	2^+	spin + position	$^1D_2$	

symmetric in the spatial coordinates while odd values of L give antisymmetric wave functions.

The total spin is either $S = 1$ (triplet) or $S = 0$ (singlet), whose wave functions take the form (problem 13.2)

$$\chi_m^{S=1} = \begin{cases} \alpha(1)\alpha(2) & , m = 1 \\ \beta(1)\beta(2) & , m = -1 \\ (1/\sqrt{2})\,[\alpha(1)\beta(2) + \beta(1)\alpha(2)] & , m = 0 \end{cases}$$

$$\chi_0^{S=0} = \frac{1}{\sqrt{2}}\,[\alpha(1)\beta(2) - \beta(1)\alpha(2)]$$

It is evident that the triplet wave function is symmetric in the spin variables while the singlet wave function is antisymmetric. Thus, for identical particles, even L must be combined with $S = 0$ and odd L with $S = 1$. These wave functions, and also those applicable to only the neutron–proton system are listed in table 13.1. Also the possible values of the total spin, $\mathbf{J} = \mathbf{L} + \mathbf{S}$, and the parity are given in this table. The standard notation is used, $^{2S+1}(L)_J$, where the $L = 0, 1, 2 \ldots$ states are denoted by $S, P, D, \ldots$.

In table 13.1 we have already, anticipating the discussion in a later section of this chapter, introduced the isospin classification denoting the states symmetric in space *and* spin as $T = 0$ states while the others are classified as $T = 1$.

13.3 The deuteron and low-energy nucleon–nucleon scattering data

In the 1940s and early 1950s information about the nucleon–nucleon interaction came largely from studying the simplest non-trivial nucleus, the deuteron, denoted d or ^{2}H, consisting of a neutron and a proton. For the deuteron the most important properties, known since the 1930s are the following

binding energy	$E_B = 2.25$ MeV
spin, parity	$J^\pi = 1^+$
isospin	$T = 0$
magnetic moment	$\mu = 0.8574$ n.m. $= \mu_p + \mu_n - 0.0222$ n.m.
quadrupole moment	$Q = 2.82 \times 10^{-3}$ barn

As the NP system is stable, while the NN (and PP) are not, we must look for the deuteron ground state among states available only to NP (and not to NN and PP). It thus appears obvious (table 13.1) that the deuteron ground state should be either $^3S_1$ or $^3D_1$ (or maybe a combination of these). For the potentials we have so far studied, the oscillator or the infinite-square well, the $L = 0$ state has always come out as the lowest state. For the deuteron $^3S_1$ should therefore be a reasonable first guess. The assumption that $L = 0$ is also in agreement with the fact that the measured magnetic moment is very close to the sum of the proton and neutron magnetic moments. From the fact that $^1S_0$ *does not exist bound* while $^3S_1$ is bound we have here to draw the *important conclusion* that the *nucleon–nucleon* interaction is necessarily *spin-dependent*. We could choose to distinguish between a *singlet-spin* and a *triplet-spin* potential.

Much more information about the nucleon–nucleon interaction has been obtained from the scattering of proton and neutron projectiles against protons and neutrons. The scattering cross section (the number of scattered particles per steradian angle† and per incoming flux of particles) depends critically on the relative angular momentum of the projectile–target system. For very high angular momentum the particles never come sufficiently close for the nuclear interaction to be felt. In fact for energies up to 10 MeV the main contribution comes from particles moving in relative S-states i.e. states having $L = 0$. One can then analyse the data in terms of two parameters for the total-spin 0 (singlet) and the total spin 1 (triplet) cases. These often used parameters are the so-called 'scattering length' and 'effective range'. We shall not discuss these more general terms here but consider the same information

---

† The angle referred to is the angle between ingoing and outgoing particle.

converted into two other more easily conceptualised parameters, namely the depth V and width b of approximate square well potentials.

From an analysis of scattering data, one finds that the triplet potential is much deeper but has a somewhat smaller range than the singlet potential. It should also be remarked that it is of course only when the effect of the Coulomb repulsion is disentangled from the PP scattering effects that the near symmetry among the NN, PP and NP cases becomes apparent.

It is easy to derive an approximate value of the depth of the triplet potential from the weak deuteron binding energy, $E_B = 2.25$ MeV. For a state having angular momentum $L = 0$, the Schrödinger equation inside a square well of depth V_t and range b_t takes the form

$$\left[-\frac{\hbar^2}{2\mu}\Delta + V_t \right] \phi_t(\mathbf{r}) = E_t\phi_t(\mathbf{r})$$

valid for the space part of the triplet case wave function

$$\psi = \chi_0^1(\sigma_1\sigma_2) \cdot \phi_t = \frac{1}{\sqrt{2}}(\alpha\beta + \beta\alpha) \cdot \phi_t(\mathbf{r})$$

We may estimate the product $V_t b_t^2$ by just noting the fact that inside the square well box we have a sine wave solution

$$\phi_{\text{in}} \sim \sin Kr \qquad (r < b_t)$$

where

$$\frac{\hbar^2 K^2}{2\mu} = V_t - E_B$$

Now in order to have a bound state we should have a solution with an exponential fall-off for $r > b_t$, i.e. outside the square well

$$\phi_{\text{out}} \sim \exp(-\kappa r)$$

where

$$\frac{\hbar^2\kappa^2}{2\mu} = E_B$$

In passing we may note that the numerical value, $\kappa^{-1} = 4.3$ fm, shows that the neutron and the proton are quite far away from each other a large fraction of the time.

The usual boundary condition has the familiar form

$$K \cot Kb_t = \kappa$$

As κ is much smaller than K, we can find an approximate solution

$$Kb_t \simeq \frac{\pi}{2}$$

which means that we neglect E_B compared with V_t. We have thus

$$V_t b_t^2 \simeq \frac{\pi^2}{4}\frac{\hbar^2}{2\mu} \simeq \frac{\pi^2\hbar^2}{4M}$$

The right hand side can be estimated to be about 100 MeV fm². Based on a range of 1.43 fm, the Compton wavelength of the pi-meson, the cited number would imply a potential depth, $V_t \approx 50$ MeV. From proton scattering data alone the product in question comes out to be 146 MeV fm². Although there is an agreement as to the order of magnitude, the sizeable discrepancy reflects the fact that the potential model is too rough and oversimplified.†

It is very easy to construct a spin-state-dependent potential within the available formalism. It is readily apparent that the operator $\boldsymbol{\sigma}_1 \cdot \boldsymbol{\sigma}_2$, where $\boldsymbol{\sigma}_1$ and $\boldsymbol{\sigma}_2$ are the spin operators of particles 1 and 2 respectively, is an eigenoperator of the triplet and singlet wave functions such that

$$\boldsymbol{\sigma}_1 \cdot \boldsymbol{\sigma}_2 \chi_m^1 = 1\chi_m^1$$

and

$$\boldsymbol{\sigma}_1 \cdot \boldsymbol{\sigma}_2 \chi_0^0 = -3\chi_0^0$$

where χ_m^1 and χ_0^0 are the triplet spin and singlet spin wave functions. We may now utilise this convenient operator for our aim to construct a spin-dependent two-nucleon potential

$$V = V_0(r) + V_\sigma(r)\boldsymbol{\sigma}_1 \cdot \boldsymbol{\sigma}_2$$

For the total triplet and singlet eigenfunctions $\chi_m^1 \phi_t(\mathbf{r})$ and $\chi_0^0 \phi_s(\mathbf{r})$, one can separate off the spin parts and arrive at the two equations that determine the spatial wave functions $\phi_t(\mathbf{r})$ and $\phi_s(\mathbf{r})$:

$$\left(-\frac{\hbar^2}{2\mu}\Delta + V_0(r) + V_\sigma(r)\right)\phi_t(\mathbf{r}) = E_t\phi_t(\mathbf{r})$$

$$\left(-\frac{\hbar^2}{2\mu}\Delta + V_0(r) - 3V_\sigma(r)\right)\phi_s(\mathbf{r}) = E_s\phi_s(\mathbf{r})$$

---

† In this connection one could also mention that the deuteron problem was the subject of much study in the early 1940s. A potential that represented a large improvement and fulfills the requirements of a diffuse surface of the proper depth is the Hulthén potential (see e.g. L. Hulthén and M. Sagawara, *Encycl. of Physics*, ed. S. Flügge (Springer Verlag, 1957) vol. XXXIX, p. 1)

$$V(r) = -V_0\frac{e^{-\mu r}}{1 - e^{-\mu r}}$$

to which an exact solution exists, the Hulthén radial wave function

$$\frac{u(r)}{r} = [2\kappa(\kappa + \mu)(2\kappa + \mu)]^{1/2} \cdot \frac{1}{\mu r}e^{-\kappa r}\left(1 - e^{-\mu r}\right)$$

where κ is defined above.

In this way we obtain an effective $V_t(r) \neq V_s(r)$, as required by experiments.

Before we go on to discuss more the character of the nucleon–nucleon force, it is convenient to introduce the isospin formalism to treat the NN, PP and NP interaction on an equal footing.

13.4 The isospin formalism

We have already mentioned the symmetry encountered within the nucleon family both as far as the mass values of the neutron and proton are concerned and as to the interaction. It is therefore natural to speak of the neutron and the proton as being just two 'charge states' of the nucleon particle.

Let us consider the proton as the 'isospin-up' state and the neutron as the 'isospin-down' state, or in other words:

$$\text{the proton:} \quad \begin{pmatrix} 1 \\ 0 \end{pmatrix}$$

$$\text{the neutron:} \quad \begin{pmatrix} 0 \\ 1 \end{pmatrix}$$

The wave function of a proton with position $\mathbf{r}_1$ and spin σ_1 and a neutron with position $\mathbf{r}_2$ and spin σ_2 can be written

$$\psi\left(\mathbf{r}_1 = \mathbf{r}_\mathrm{p}, \sigma_1 = \sigma_\mathrm{p}, (\tau_1)_3 = 1, \mathbf{r}_2 = \mathbf{r}_\mathrm{n}, \sigma_2 = \sigma_\mathrm{n}, (\tau_2)_3 = -1\right)$$

The isospin component τ_3 is thus by definition $+1$ for the proton and -1 for the neutron.†

The Pauli principle says that the exchange $1 \rightarrow 2$ for two protons or two neutrons respectively gives rise only to a phase -1 for the entire wave function containing the position and spin coordinates. We formulate *the generalised Pauli principle* by postulating that the exchange $1 \rightarrow 2$ shall always give a change of sign for the total wave function containing *position+spin+isospin*.

There is an almost complete analogy between the spin case and the isospin case. This means that we can take over what we know from the spin problem. Thus there are only four independent matrices, which we may choose as

$$\begin{pmatrix} 1 & 0 \\ 0 & 1 \end{pmatrix}, \begin{pmatrix} 0 & 1 \\ 1 & 0 \end{pmatrix}, \begin{pmatrix} 0 & -i \\ 1 & 0 \end{pmatrix} \quad \text{and} \quad \begin{pmatrix} 1 & 0 \\ 0 & -1 \end{pmatrix}$$

The first is the unit matrix and the latter three we shall call τ_1, τ_2, τ_3. These matrices look identical with σ_1, σ_2 and σ_3 but refer to a different space.

† This is the standard definition in particle physics while in nuclear physics the opposite definition with $\tau_3 = 1$ for neutrons and $\tau_3 = -1$ for protons is often used.

Now introduce a matrix $\mathbf{t} = \frac{1}{2}\boldsymbol{\tau}$. For the proton we have $t = \frac{1}{2}$ and $t_3 = \frac{1}{2}$ (t_3 = component of isospin along the 3-axis) and for the neutron: $t = \frac{1}{2}$, $t_3 = -\frac{1}{2}$. The charge is related as $q = t_3 + \frac{1}{2}$. Thereby the 3-axis in isobaric space and the physical meaning of t_3 are defined. For several nucleons we may form a total $\mathbf{T}$ as

$$\mathbf{T} = \sum_p \mathbf{t}_p$$

$$T_3 = (\mathbf{t}_1)_3 + (\mathbf{t}_2)_3 + \ldots = (Z - N) \cdot \frac{1}{2}$$

We now study a wave function containing spacial coordinates, spin and isospin. If a state is symmetric with respect to a change $\mathbf{r}_1\sigma_1 \to \mathbf{r}_2\sigma_2$, it follows from the generalised Pauli principle that it is *antisymmetric* with respect to a change $(\tau_1)_3 \leftrightarrow (\tau_2)_3$ and vice versa:

$$\psi(12) = \psi_{\text{symm}}\,(\mathbf{r}_1\sigma_1\mathbf{r}_2\sigma_2)\,\xi(T = 0)$$

where ξ refers to isospin. This is an isosinglet state with (cf. the spin case)

$$\xi(T = 0) = \frac{1}{\sqrt{2}}\left[\begin{pmatrix}1\\0\end{pmatrix}_1\begin{pmatrix}0\\1\end{pmatrix}_2 - \begin{pmatrix}0\\1\end{pmatrix}_2\begin{pmatrix}1\\0\end{pmatrix}_2\right]$$

This state can thus only be realised by a NP system (and not by NN or PP), which is consistent with the fact that the combined space and spin state is symmetric.

Let us now assume that we have a state, antisymmetric in $\mathbf{r}$ and σ, which as a consequence is symmetric in τ;

$$\psi(12) = \psi_{\text{antisymm}}\,(\mathbf{r}_1\sigma_1\mathbf{r}_2\sigma_2)\cdot\xi\begin{pmatrix}&+1\\T = 1,\; T_3 =&0\\&-1\end{pmatrix}$$

This is an isotriplet state

$$\xi(T = 1, T_3 = 1) = \begin{pmatrix}1\\0\end{pmatrix}_1\begin{pmatrix}1\\0\end{pmatrix}_2 \;;\; \text{PP system}$$

$$\xi(T = 1, T_3 = 0) = \frac{1}{\sqrt{2}}\left[\begin{pmatrix}1\\0\end{pmatrix}_1\begin{pmatrix}0\\1\end{pmatrix}_2 + \begin{pmatrix}0\\1\end{pmatrix}_1\begin{pmatrix}1\\0\end{pmatrix}_2\right] \;;\; \text{NP system}$$

$$\xi(T = 1, T_3 = -1) = \begin{pmatrix}0\\1\end{pmatrix}_1\begin{pmatrix}0\\1\end{pmatrix}_2 \;;\; \text{NN system}$$

The charge-*independence* principle now implies that the interaction is *independent* of T_3. The replacement of the eigenvalue of T_3 by another number thus should change nothing in the interaction. While thus T_3 has a direct physical meaning (charge state), on the other hand T_1 and T_2 have no apparent physical meaning. The radical physical assumption we have made, is to postulate, that not only T_3 but also $\mathbf{T}$ are physical quantities and take on a quantum number eigenvalue. The charge-independence of the nuclear interaction can thus also be expressed as

$$[H, \mathbf{T}] = 0$$

The formal implication is thus that H becomes 'invariant' with respect to a 'rotation in isospin space'. Those words borrowed from our experience with the angular momentum $\mathbf{J}$ have as yet little physical content as we have no picture of 'isospin space'.

For the commutation condition to be fulfilled (in analogy with the spin case) one can permit only interactions of the type

$$V(12) = A + B\tau_1 \cdot \tau_2$$

as all other products of τ_1 and τ_2 that form invariants reduce to this. A and B are then functions of $\mathbf{r}_1$, $\mathbf{r}_2$, σ_1, σ_2 etc.

With the help of the operator $\tau_1 \cdot \tau_2$ one can reproduce the effect of charge exchange between the nucleons. The process behind is, as we mentioned earlier, in the first place an exchange of π-mesons between a pair of nucleons. Let us assume that particle 1, say a neutron, emits a π^--meson. The nucleon 1 thereby becomes a proton (an alternative is a Δ^+ baryon). Nucleon 2, being a proton, absorbs the π^--meson and goes over into a neutron. What we have described is a one-pion exchange. In terms of nucleons alone we can describe this with the operator $(\tau_1)_+ (\tau_2)_-$, which is thus a component of $\tau_1 \cdot \tau_2$.

13.5 General conditions on the two-nucleon interaction

As indicated above the agents transmitting the interactions are particles other than the nucleons. We shall still assume that the internucleon interaction can be described to a certain accuracy by a potential involving only the coordinates of the two nucleons but with no reference to the intermediary agents. The degrees of freedom of the mesons are thus entirely concealed in the complications of the nucleon–nucleon potential.

The nucleon–nucleon potential is still in its details not entirely determined. However, it is being determined to better and better accuracy from scattering data. In some sense it appears fair to say that the analysis in terms of phase shifts appears completed. Given the data no potential can, however, be derived *uniquely*. It is not even clear that such a potential exists. Postulating its existence and applicability, a few requirements can be safely imposed *a priori* on the potential. Those requirements are based on simple and general invariance laws.

The coordinates on which the potential may depend are in addition to the position coordinates of the nucleons and isospin (symmetry) also the velocity and spin coordinates. In terms of these a few general conditions may be formulated.

(1) *Translational invariance* implies that the position of the origin is of no consequence. This is fulfilled by the only relevant spatial coordinate being the *relative* distance $\mathbf{r} = \mathbf{r}_1 - \mathbf{r}_2$. Similarly, the only possible momentum variable is the relative momentum $\mathbf{p} = \mathbf{p}_1 - \mathbf{p}_2$.

(2) *Rotational invariance.* The interaction cannot depend on the choice of axis of the coordinate system or the orientation of the laboratory. If vectors enter in the expression for the interaction potential V, they must appear in expressions that are scalars under rotation.

These two conditions can be expressed through the commutators

$$[V, \mathbf{p}] = 0$$

$$[V, \mathbf{J}] = 0$$

They are probably exactly fulfilled in nature in the limit of low velocities. Then come some conditions that are to a high degree approximately accurate.

(3) One is that of *reflection invariance* or parity invariance (P), by which we mean 'reflection in the origin', or $\mathbf{r}$ replaced by $-\mathbf{r}$. This reflection corresponds to a 'physical' reflection in a plane and a subsequent rotation of the reflected image 180° parallel to the plane. Expressed in more common-day language one could say that, if some physicist chose to study the outcome of all of his experiments through a mirror, he should still arrive at the same natural laws. Up until 1957 this invariance was taken for granted, surprisingly enough. Since then we have learned that in beta-decay this symmetry is actually violated. The beta-decay is, however, governed by the weak interaction. For the nucleon–nucleon interaction or *strong interaction, reflection symmetry P is assumed to hold exactly* on the basic level. Up to a factor 10^{-6} or 10^{-7} it is proven experimentally.

The replacement of $\mathbf{r}$ by $-\mathbf{r}$ through the parity operator can be expressed as

$$\mathscr{P}\mathbf{r}\mathscr{P}^{-1} = -\mathbf{r}$$

Similarly

$$\mathscr{P}\mathbf{p}\mathscr{P}^{-1} = -\mathbf{p}$$

and thus

$$\mathscr{P}(\mathbf{r} \times \mathbf{p})\mathscr{P}^{-1} = \mathbf{r} \times \mathbf{p} = \mathbf{L}$$

For the spin operator we *require* in analogy to the relation for $\mathbf{L}$ that

$$\mathscr{P}\mathbf{s}\mathscr{P}^{-1} = \mathbf{s}$$

Vectors like $\mathbf{r}$ and $\mathbf{p}$ are referred to as real vectors while $\mathbf{L}$ and $\mathbf{s}$, which are unaffected by the parity operator, are pseudovectors (or axial vectors).

Sometimes people interpret the last two equations by saying that 'seen in a mirror, the mirrored top rotates the same way as the original top'. That statement is correct provided you place the mirror under or above the top, i.e. perpendicular to its axis. If you place the mirror in a plane parallel to the mirror axis, the mirror shows a top rotating the *opposite* way. It is, however, to be remembered that $\mathscr{P}$ involves reflection in a plane plus a subsequent 180° rotation around an axis perpendicular to the plane. After the complete prescribed operation the mirrored top rotates indeed in the original sense.

(4) *Time reversal invariance.* By time reversal ($\mathscr{T}$) we mean the mathematical replacement of t with $-t$. On the more popular level we can translate this condition into one where a physicist chooses to record as films all his basic experiments and subsequently always runs the films backwards. The claim is then that he should arrive at the same set of natural laws. This is true only on the basic level. Thus we should note that on the macroscopic level there is not time reversal invariance. The difference relates to the fact that time direction and entropy are closely connected. Our world is highly ordered – although presently chaos may seem to gain. Macroscopic events therefore proceed preferentially in a direction of entropy increase. Let us repeat: *macroscopically* there is *no* time reversal invariance. On the microscopic level – and there only – there seems to be time reversal invariance, however, *only to some degree of accuracy*. We shall *require* time reversal invariance to hold *exactly for the strong interaction* exemplified in the nucleon–nucleon interaction.

Time reversal changes the sign of **p** but not **r**

$$\mathscr{T}\mathbf{r}\mathscr{T}^{-1} = \mathbf{r}$$

$$\mathscr{T}\mathbf{p}\mathscr{T}^{-1} = -\mathbf{p}$$

and consequently it follows that

$$\mathscr{T}(\mathbf{r} \times \mathbf{p})\mathscr{T}^{-1} = -\mathbf{L}$$

In complete analogy to the case for **L** we should have for **s**:

$$\mathscr{T}\mathbf{s}\mathscr{T}^{-1} = -\mathbf{s}$$

Looking at 'the film shown backwards', all motions, translational and rotational, take place backwards, according to expectations.

(5) *Charge independence.* As expressed by the isospin formulation, we shall consider the neutron and the proton to be isospin states of one and the same particle, the nucleon.

13.6 The static approximation. Central interactions and the tensor interaction

We shall first consider the approximation of no velocity-dependence, i.e. no dependence on the relative momentum **p**. This is usually termed the *static approximation*. In this case only two types of terms can occur, subject to the limitation imposed by various invariance laws, the *central forces* and the *tensor force*.

Based on the invariants formed from the vectors $\sigma_1, \sigma_2, \tau_1, \tau_2$ and using only scalar functions of **r** (= central forces) in this static approximation one ends up with the following time-honoured 'ansatz'

$$V_{\text{central}} = V_0(r) + V_\sigma(r)\sigma_1 \cdot \sigma_2 + V_\tau(r)\tau_1 \cdot \tau_2 + V_{\tau\sigma}(r)\sigma_1 \cdot \sigma_2\tau_1 \cdot \tau_2$$

This expression is often written in the following alternative forms:

(1) in terms of *exchange operators*,

$$P^\sigma = \frac{1}{2}(1 + \sigma_1 \cdot \sigma_2)$$

$$P^\tau = \frac{1}{2}(1 + \tau_1 \cdot \tau_2)$$

The P^σ-operator has the property of exchanging the spin-components of the particles 1 and 2. P^τ has apparently the same effect on the isospin variables:

$$P^\sigma \chi(S = 1) = \chi(S = 1)$$

$$P^\sigma \chi(S = 0) = -\chi(S = 0)$$

We define then an additional operator P^r, which we, without specifying its more detailed construction, posit to have the property of exchanging the position coordinates $\mathbf{r}_1$ and $\mathbf{r}_2$ for the particles 1 and 2. The generalised Pauli principle as a consequence gives rise to the relation

$$P^r P^\sigma P^\tau = -1$$

and thus

$$P^\sigma P^r = -P^\tau$$

We can then write

$$V_{\text{central}} = V_{\text{W}}(r) + V_{\text{M}}(r)P^r + V_{\text{B}}(r)P^\sigma + V_{\text{H}}(r)P^r P^\sigma$$

The different indices refer to the physicists Wigner, Majorana, Bartlett and Heisenberg, who were involved in the introduction of these terms in the early thirties. Using the operator relation above we can in the expression for V_{central} replace $P^r P^\sigma$ with $-P^\tau$.

(2) The central forces may also be expressed in terms of *projection operators*. We first introduce two projection operators in spin space:

$$P(S = 1) = \frac{1}{2}(1 + P^\sigma) = \frac{1}{4}(3 + \boldsymbol{\sigma}_1 \cdot \boldsymbol{\sigma}_2)$$

with the properties

$$P(S = 1)\chi(S = 1) = \chi(S = 1)$$

$$P(S = 1)\chi(S = 0) = 0$$

and furthermore

$$P(S = 0) = \frac{1}{2}(1 - P^\sigma) = \frac{1}{4}(1 - \boldsymbol{\sigma}_1 \cdot \boldsymbol{\sigma}_2)$$

with the properties

$$P(S = 0)\chi(S = 1) = 0$$

$$P(S = 0)\chi(S = 0) = \chi(S = 0)$$

In a way that is entirely analogous one may construct projection operators in isospin space:

$$P(T = 1) = \frac{1}{2}(1 + P^\tau)$$

$$P(T = 0) = \frac{1}{2}(1 - P^\tau)$$

We then obtain an expression for the nuclear potential that has probably been most commonly employed

$$
\begin{aligned}
V_{\text{central}} = {}^{13}V(r)P(S &= 0, T = 1, L \text{ even}) \\
+ {}^{11}V(r) = (S &= 0, T = 0, L \text{ odd}) \\
+ {}^{33}V(r) = (S &= 1, T = 1, L \text{ odd}) \\
+ {}^{31}V(r) = (S &= 1, T = 0, L \text{ even})
\end{aligned}
$$

In this terminology of the superscripts, the first figure refers to the spin degeneracy, the second to the isospin degeneracy. Thus '31' implies $S = 1$ and $T = 0$.

The central interactions were historically the first to be introduced. It was found already in the 1930s that they failed to explain the existence of a non-vanishing quadrupole moment of the deuteron. The magnitude of the quadrupole moment of the deuteron is very small indeed but clearly non-vanishing

$$Q_0 = (2.74 \pm 0.02) \times 10^{-27} \text{ cm}^2$$

Expressed in the natural unit of the deuteron cross section πR^2, the small magnitude becomes more apparent

$$\frac{Q_0}{\pi R^2} \simeq \frac{1}{200}$$

With the spin determined to be 1 and the parity pure and unadmixed, the only mixture that could enter (the ground state wave function being dominantly ${}^3S_1$) appears to be ${}^3D_1$. It turns out that the admixture of ${}^3D_1$ required by the measured quadrupole moment is $\simeq 4\%$. This in turn can be used to provide a rough determination of the magnitude of the tensor interaction term (cf. appendix 13B).

In the deuteron, with the origin at the centre-of-mass and neglecting the difference between neutron and proton masses, we have $\mathbf{r}_n + \mathbf{r}_p = 0$ and by definition $\mathbf{r}_p - \mathbf{r}_n = \mathbf{r}$ and therefore $\mathbf{r}_p = \frac{1}{2}\mathbf{r}$. If the deuteron were a pure ${}^3S_1$ state then the charge distribution or the proton distribution would be described by a Y_{00} wave function. This contains no angle-dependence

and corresponds to complete spherical symmetry and therefore implies that
$\langle Q \rangle = 0$. To reproduce the experimental measurement one is required to mix
in other states with $L \neq 0$. This can be done with a potential involving the
direction of the position vectors $\mathbf{r}_1$, $\mathbf{r}_2$ and the spin vectors $\boldsymbol{\sigma}_1$, $\boldsymbol{\sigma}_2$. It turns
out that there is only one non-trivial combination that fulfils both of the
principles of time reflection and space reflection in addition to the trivial one
of rotational invariance.

Let us consider conceivable invariants of the $\boldsymbol{\sigma}$ and the $\mathbf{r}$ operators. The
transformation of $\boldsymbol{\sigma}_1 \cdot \mathbf{r}$ under the parity and time reversal operators is easily
obtained as

$$\mathscr{P} \boldsymbol{\sigma}_1 \cdot \mathbf{r} \mathscr{P}^{-1} = -\boldsymbol{\sigma}_1 \cdot \mathbf{r}$$

$$\mathscr{T} \boldsymbol{\sigma}_1 \cdot \mathbf{r} \mathscr{T}^{-1} = -\boldsymbol{\sigma}_1 \mathbf{r}$$

Both conservation laws are thus broken. The problem is easily handled,
however. Obviously the product $(\boldsymbol{\sigma}_1 \cdot \mathbf{r})(\boldsymbol{\sigma}_2 \cdot \mathbf{r})$ preserves its sign under both
of the operators. Additionally the latter combination is also invariant under
the exchange of indices 1 and 2. This is indeed the important term of the
tensor interaction. By convention, instead of this product alone one uses
an expression where one subtracts a term proportional to $\boldsymbol{\sigma}_1 \cdot \boldsymbol{\sigma}_2$ to form
($\mathbf{e}_r = \mathbf{r}/r$)

$$S_{12} = 3(\boldsymbol{\sigma}_1 \cdot \mathbf{e}_r)(\boldsymbol{\sigma}_2 \cdot \mathbf{e}_r) - (\boldsymbol{\sigma}_1 \cdot \boldsymbol{\sigma}_2)$$

and

$$V_{\text{tensor}} = V_T(r) \cdot S_{12}$$

The operator S_{12} is chosen such that the mean value of S_{12} over all directions
vanishes. Thus the expectation value of S_{12} vanishes for an S-state, where
the wave function is isotropic in angle. Consequently,

$$\langle (\boldsymbol{\sigma}_1 \cdot \mathbf{e}_r)(\boldsymbol{\sigma}_2 \cdot \mathbf{e}_r) \rangle = \frac{1}{3} \langle \boldsymbol{\sigma}_1 \cdot \boldsymbol{\sigma}_2 \rangle$$

13.7 Velocity-independent terms. The L · S potential

We have so far not discussed the dependence on velocity, which enters
through the variables $\mathbf{p}_1$ and $\mathbf{p}_2$. Because of the limitation brought about by
the condition of translational invariance the sole coordinate is $\mathbf{p} = \mathbf{p}_1 - \mathbf{p}_2$.
The basic assumption followed is that the dependence on $\mathbf{p}$ is weak and that
a classification in terms   different powers of $\mathbf{p}$ can be considered.

We may also note that we did not allow for any $\mathbf{p}$-dependence in the
potential when we considered the separation of the Schrödinger equation into
centre-of-mass and relative coordinates. If the $\mathbf{p}$-dependence is introduced,

the separation still holds but it leads to a slightly modified kinetic energy term (so as to include a **p**-dependent mass).

The simplest invariant involving **p** is the scalar $(\mathbf{r} \times \mathbf{p}) \cdot \mathbf{S}$ (cf. the spin–orbit term of the one-particle potential, chapter 6) where $\mathbf{S} = \mathbf{s}_1 + \mathbf{s}_2$. Considering only first order powers in **p**, we should thus add a term

$$V_{LS}(\mathbf{r}) \mathbf{L} \cdot \mathbf{S}$$

Sometimes second order terms in **p** are also considered but, to a good approximation, they may in general be neglected.

13.8 The radial dependence

The analysis of the great body of scattering data is now done in terms of the mentioned central potential components, the tensor component $V_T(r)$ and the two spin–orbit components $V_{LS}(r)$. All these terms thus contain *a priori* unknown radial functions, to be determined from the fit to the scattering data. For the radial functions one now usually employs functions that are obtained from theories involving the exchange of one or several mesons (appendix 13A) in addition to some undetermined parameters to be fitted by data.

Most of the radial functions so obtained are, however, of the type indicated in fig. 13.1, thus involving an attractive long-range tail, reflecting one-pion exchange, an attractive inner minimum and finally a repulsive hard core. Sometimes, however, a few of the terms have a different behaviour, e.g. altogether repulsive.

In this elementary discussion of the nucleon–nucleon interaction, we have refrained from giving any references. For further reading, we could suggest text books e.g. by Eisenberg and Greiner (*Nuclear Theory*, vol. 3, North-Holland Publ. Comp., 1976, 1986) or by Ring and Schuck (*The Nuclear Many-Body Problem*, Springer Verlag, 1980), the *Proceedings of the International School of Physics 'Enrico Fermi'*, Course 79, 1980 (ed. A. Molinari, North-Holland Publ. Company, 1981) and the review article by Rho in *Annual Review of Nuclear and Particle Physics* (vol. 34, 1984) as well as references given in these books and papers.

Exercises

13.1   Consider two particles of masses m_1 and m_2 interacting through a potential $V = V\left(|\mathbf{r}_1 - \mathbf{r}_2|\right) = V(r)$. Find the canonical momenta $\mathbf{p}_1$ and $\mathbf{p}_2$ through the Lagrangian, $\mathscr{L} = T - V$. Introduce relative coordinates, $\mathbf{r} = (x, y, z)$, and centre-of-mass coordinates, $\mathbf{R} = (X, Y, Z)$.

Show that the corresponding momenta, $\mathbf{p}$ and $\mathbf{P}$ are given by

$$\mathbf{p} = \frac{m_2\mathbf{p}_1 - m_1\mathbf{p}_2}{m_1 + m_2} \; ; \; \mathbf{P} = \mathbf{p}_1 + \mathbf{p}_2$$

Find the Hamiltonian in the new coordinates. Specialise to the case of $m_1 = m_2 = m$.

13.2 Use the table of Clebsch–Gordan coefficients (given in appendix 6B) to derive the triplet and singlet spin wave functions. Alternatively, one may start from a general two-spin wave function, $A\alpha(1)\alpha(2) + B\alpha(1)\beta(2) + C\beta(1)\alpha(2) + D\beta(1)\beta(2)$, and determine the coefficients A, B, C and D from the requirement that S^2 and S_z should be eigenoperators (together with the normalisation condition). Do this in the triplet case when $m = 0$.

13.3 Show that $(\boldsymbol{\sigma}_1 \cdot \boldsymbol{\sigma}_2)^2 - 3 + 2(\boldsymbol{\sigma}_1 \cdot \boldsymbol{\sigma}_2) = 0$. Determine the eigenvalues of $\boldsymbol{\sigma}_1 \cdot \boldsymbol{\sigma}_2$ from this equation. The formula $(\boldsymbol{\sigma} \cdot \mathbf{A})(\boldsymbol{\sigma} \cdot \mathbf{B}) = \mathbf{A} \cdot \mathbf{B} + i\boldsymbol{\sigma} \cdot (\mathbf{A} \times \mathbf{B})$ (where $[\boldsymbol{\sigma}, \mathbf{A}] = 0$ and $[\boldsymbol{\sigma}, \mathbf{B}] = 0$), which is generally shown in elementary courses in quantum mechanics, simplifies the calculations.

13.4 Show that the triplet and singlet wave functions are eigenvectors of the operator $P^\sigma = \frac{1}{2}(1 + \boldsymbol{\sigma}_1 \cdot \boldsymbol{\sigma}_2)$ with the eigenvalues $+1$ and -1, respectively. Also show that P^σ corresponds to an exchange of the spin components, $\boldsymbol{\sigma}_1 \rightleftarrows \boldsymbol{\sigma}_2$.

13.5 Consider a simplified deuteron potential of the form

$$V(r) = f(r)(1 + a\boldsymbol{\sigma}_1 \cdot \boldsymbol{\sigma}_2)$$

where

$$f(r) = \begin{cases} -V_0, & r \leq R \\ 0, & r > R \end{cases}$$

This gives rise to a square well potential where the depths become different for the triplet and singlet wave functions. Scattering data show that $V_{\text{triplet}} \simeq 2V_{\text{singlet}}$. Determine the constant a!

13.6 (a) Derive the projection operators, $P(S = 1)$ and $P(S = 0)$ from the general expressions

$$P(S = 1) = (A + B\boldsymbol{\sigma}_1 \cdot \boldsymbol{\sigma}_2); \quad P(S = 0) = C + D\boldsymbol{\sigma}_1 \cdot \boldsymbol{\sigma}_2$$

and requiring

$$P(S = 1)\left(a\chi_0^1 + b\chi_0^0\right) = a\chi_0^1 \; ; \; P(S = 0)\left(a\chi_0^1 + b\chi_0^0\right) = b\chi_0^0 \; ;$$

$$(P(S = 1))^2 = P(S = 1) \; ; \; (P(S = 0))^2 = P(S = 0)$$

The triplet and singlet spin wave functions are denoted by χ_m^1 and χ_0^0, respectively.

(b) Carry through the expansion

$$\alpha(1)\beta(2) = c\chi_0^1 + d\chi_0^0$$

first by use of Clebsch–Gordan coefficients and then by use of the projection operators $P(S = 1)$ and $P(S = 0)$.

13.7 The central nuclear interaction can be expanded directly in terms of $\sigma_1 \cdot \sigma_2$ and $\tau_1 \cdot \tau_2$ with expansion coefficients V_0, V_σ, V_τ and $V_{\sigma\tau}$. Alternatively, one may for example use projection operators with expansion coefficients ^{31}V, ^{13}V, ^{11}V and ^{33}V. Find the relation between the two sets of coefficients.

13.8 Show that the average of the tensor force

$$V_{\text{tensor}} = V_T(r)\left[3\left(\sigma_1 \cdot e_r\right)\left(\sigma_1 \cdot e_r\right) - \sigma_1 \cdot \sigma_2\right]$$

taken over all directions of $\mathbf{r}$ vanishes.

13.9 Prove that the tensor force, V_{tensor}, gives no contribution in the single spin state. This follows for example (why?) from the relation

$$V_{\text{tensor}}\left(1 - \sigma_1 \cdot \sigma_2\right) = 0$$

which can be shown by help of the formula given in problem 13.3.

13.10 Show by direct evaluation of the commutator that $\left[S^2, V_{\text{tensor}}\right] = 0$. This commutator implies that the tensor interaction does not mix the triplet and the singlet spin wave functions. Calculate also the commutator $[\mathbf{J}, V_{\text{tensor}}]$, which must vanish for rotational invariance to be fulfilled.

13.11 Show that an $\mathbf{L} \cdot \mathbf{S}$ force has no matrix elements in a singlet spin state.

13.12 Show that the deuteron quadrupole moment can be written

$$Q = \frac{1}{10}\int\left(u(r)w(r)\sqrt{2} - \frac{1}{2}w^2(r)\right)r^2\,dr$$

in terms of the $^3S_1$ and $^3D_1$ amplitudes, $u(r)$ and $w(r)$.

13.13 Assume that the $^3S_1$ and $^3D_1$ deuteron radial amplitudes are given by

$$u(r) = \frac{1}{(1+n^2)^{1/2}}(2\kappa)^{1/2}\,e^{-\kappa r} \;;\; w(r) = \frac{n}{(1+n^2)^{1/2}}(2\kappa)^{1/2}\,e^{-\kappa r}$$

where $\kappa^{-1} = (\hbar^2/ME_B)^{1/2} \simeq 4.29$ fm (E_B is the deuteron binding energy, 2.25 MeV). Determine n to fit the experimental quadrupole moment, $Q = 2.82 \times 10^{-3}$ barn.

Appendix 13A

Theoretical basis for the phenomenological nucleon–nucleon interaction

In the present appendix we will make a very sketchy attempt to relate the largely phenomenologically determined potential in terms of the more basic 'mechanism' involving repeated scattering of different mesons of masses μ and coupling constant g between the nucleon and the exchanged meson particle. (Note that in the main text μ is also used in a different sense, namely as the reduced mass of the relative motion.)

The elementary field theoretical description of the nuclear interaction bears a close analogy to the quantum formulation of the transmission of the Coulomb interaction through the exchange of an intermediary (virtual) photon. Consider two particles having a charge. In this case the Poisson equation holds for the electromagnetic scalar field ϕ having the charge $\rho(\mathbf{r})$ as a source

$$\nabla^2 \phi = -\rho(\mathbf{r})$$

For a point charge particle we replace $\rho(r)$ by a $\delta(r)$-function at the origin. The solution to the Poisson equation is then

$$\phi(\mathbf{r}) = \frac{e}{4\pi\varepsilon_0} \frac{1}{r}$$

which is the scalar Coulomb field.

The potential energy of particle 2 of charge e moving in the electromagnetic field of particle 1 is then

$$V(r) = \frac{e^2}{4\pi\varepsilon_0 r}$$

We draw the corresponding diagram as shown in the upper part of fig. 13A.1, thereby indicating the exchange of the electromagnetic field ϕ, between particles 1 and 2, both of charge e.

In the electromagnetic case the messenger photon is of mass 0. In the

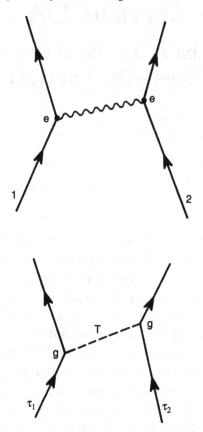

Fig. 13A.1. Diagrammatic representation of the electromagnetic interaction as exchange of a photon and the nuclear interaction as exchange of a meson.

more general case the exchanged particle may still be a scalar particle but now of mass μ. The corresponding meson field equation reads

$$\left(\nabla^2 - \mu^2\right)\phi = -g\delta\left(\mathbf{r}\right)$$

with the solution proportional to $e^{-\mu r}/r$. Obviously the electromagnetic field solution is a special case of this in the limit $\mu \to 0$. Although the field is *a scalar* in *configuration space*, it is *a vector* in *isospin space*, as the particle transmitted is an isovector particle, which may appear in one of its three states, e.g. π^+, π^0 or π^-. We therefore obtain a potential energy of the form

$$V \propto \frac{e^{-\mu r}}{r}\begin{cases} 1 & ; \ T = 0 \\ \tau_1 \cdot \tau_2 ; & T = 1 \end{cases}$$

In the first case the transmitted particle is a scalar in isospace; in the latter

Table 13A.1. *The low-mass mesons contributing to the nucleon–nucleon interaction.*

Character J^π	T	Denotion		Candidate meson	Mass (MeV)
0^-	1	pseudoscalar	isovector	π	139.6
1^-	1	vector	isovector	ρ	765
1^-	0	vector	isoscalar	ω	783
0^-	0	pseudoscalar	isoscalar	η	549
0^+	0	scalar	isoscalar	$2\pi(=\text{'}\sigma\text{'})$	279

case a vector in isospace. The corresponding diagram is drawn in the lower part of fig. 13A.1.

Without any attempt at derivation we shall just state that the pseudoscalar meson leads to an interaction

$$V(r) \propto \left[\boldsymbol{\sigma}_1 \cdot \boldsymbol{\sigma}_2 + \left(1 + \frac{3}{\mu r} + \frac{3}{\mu^2 r^2}\right) S_{12}\right] \frac{e^{-\mu r}}{r} \begin{cases} 1 \\ \boldsymbol{\tau}_1 \cdot \boldsymbol{\tau}_2 \end{cases}$$

The pseudoscalar meson field thus gives rise to a tensor force component, S_{12}. A vector meson field is finally needed to generate the nucleon–nucleon interaction terms proportional to $\mathbf{L} \cdot \mathbf{S}$ in first order. The meson fields of scalar, pseudoscalar or vector types with associated meson masses falling in the energy ranges we are considering are summarised in table 13A.1.

This appendix is only meant for supplementary reading. It does not attempt to derive the nucleon–nucleon force in a self-explanatory fashion. Instead the purpose is to outline the possible connection between the phenomenologically suggested interaction functions and the presently known lighter mesons considered as transmitter agents.

Appendix 13B

The solution of the deuteron problem with the inclusion of a tensor force

We now assume a Hamiltonian for the deuteron with a central force and a tensor force

$$H = -\frac{\hbar^2}{2\mu}\left(\frac{1}{r}\frac{\partial^2}{\partial r^2}r + \frac{\mathbf{L}^2}{r^2}\right) + V_c(r) + V_T(r)S_{12}$$

where $\mathbf{L}$ is the angular momentum operator. The wave function can be written as

$$\psi(\mathbf{r},\boldsymbol{\sigma}) = \sum_L \frac{u_L(r)}{r}\phi_{JM}^{LS}$$

where

$$\phi_{JM}^{LS} = \sum_{M_L m_S} C_{M_L M_S M}^{L\,S\,J} Y_{LM_L}\chi_{M_S}^{S}$$

Inserting this wave function 'ansatz' into the corresponding Schrödinger equation one obtains

$$\left[-\frac{\hbar^2}{2\mu}\left(\frac{d^2}{dr^2} + \frac{L(L+1)}{r^2}\right) + V_c - E\right]u_L(r) + V_T(r)$$

$$\times \sum_{L'} \left(\phi_{JM}^{LS}\,|S_{12}|\,\phi_{JM}^{L'S}\right)u_{L'}(r) = 0$$

The angular matrix element can then be evaluated using some Clebsch–Gordan technique (matrix elements with round parentheses imply angles only). The final result is given in table 13B.1 (see J.M. Eisenberg and W. Greiner, *Nuclear Theory*, vol. 3 (North-Holland, 1976, 1986)). The table reflects the obvious fact that only two L-states of the same J admix; for the deuteron case of $J = 1^+$ thus the $L = 0$ and $L = 2$ states, or the $^3S_1$ and $^3D_1$. Let us then write the wave function explicitly

Table 13B.1. *Values of the matrix elements of the tensor interaction.*

	$L' = J+1$	$L' = J$	$L' = J-1$
$L = J+1$	$-2(J+2)/(2J+1)$	0	$6[J(J+1)]^{1/2}/(2J+1)$
$L = J$	0	2	0
$L = J-1$	$6[J(J+1)]^{1/2}/(2J+1)$	0	$-2(J-1)/(2J+1)$

Table 13B.2. *Matrix elements of the tensor interaction in the special case of* $J = 1.$

	$L' = 2$	$L' = 1$	$L' = 0$
$L = 2$	-2	0	$6\sqrt{2/3} = \sqrt{8}$
$L = 1$	0	2	0
$L = 0$	$\sqrt{8}$	0	0

$$\psi(\mathbf{r},\sigma) = \frac{u(r)}{r}\frac{1}{(4\pi)^{1/2}}\chi_M^1 + \frac{w(r)}{r}\sum_{M_L M_S} C_{M_L M_S M}^{2\,1\,1} Y_{2M_L}\chi_{M_S}^1$$

where we have changed the notation to $u(r)$ and $w(r)$ instead of $u_0(r)$ and $u_2(r)$. It is evident that the $^3D_1$ wave function has only three components for each $M(M_S = -1, 0, 1)$. The listed matrix elements now take on an even simpler form as seen in table 13B.2. The fact that the bottom right corner element is zero shows that the tensor force vanishes in the $L = 0$ state.

The radial equations now simplify into

$$\frac{d^2}{dr^2}u(r) + \frac{2\mu}{\hbar^2}[E - V_c(r)]\,u(r) - \sqrt{8}\frac{2\mu}{\hbar^2}V_T(r)w(r) = 0$$

and

$$\frac{d^2}{dr^2}w(r) + \frac{2\mu}{\hbar^2}\left(E - \frac{6\hbar^2}{2\mu r^2} - V_c(r) + 2V_T(r)\right)w(r) - \sqrt{8}\frac{2\mu}{\hbar^2}V_T(r)u(r) = 0$$

The solutions are subject to the normalisation condition

$$\int \left[u(r)^2 + w(r)^2\right] dr = 1$$

13B.1 The deuteron quadrupole moment

To calculate the *electric* quadrupole moment we have to recognise that only the charged proton contributes. Let us set the neutron (n) and proton (p) masses equal in which case

$$\mathbf{r}_p = \frac{1}{2}\mathbf{r}$$

$$\mathbf{r}_n = -\frac{1}{2}\mathbf{r}$$

where we have referred the coordinates $\mathbf{r}_p$ and $\mathbf{r}_n$ to the deuteron centre of mass. Furthermore for the momenta we have

$$\mathbf{p}_p = \mathbf{p} = -\mathbf{p}_n$$

The quadrupole operator is given as

$$Q_p = 3z_p^2 - r_p^2 = \frac{1}{4}\left(3z^2 - r^2\right) = \left(\frac{\pi}{5}\right)^{1/2} r^2 Y_{20}$$

The quadrupole moment is to be evaluated for the $M = 1$ case of the $J = 1$ wave function

$$\psi_{M=1}^{J=1} = \frac{u(r)}{r}\frac{1}{(4\pi)^{1/2}}\cdot\chi_1^1$$

$$+\frac{w(r)}{r}\left[\left(\frac{1}{10}\right)^{1/2}Y_{20}\chi_1^1 - \left(\frac{3}{10}\right)^{1/2}Y_{21}\chi_0^1 + \left(\frac{3}{5}\right)^{1/2}Y_{22}\chi_{-1}^1\right]$$

The quadrupole moment has matrix elements within the $^3D_1$ component and between the $^3S_1$ and $^3D_1$. It is easy to show that (problem 13.12)

$$Q = \frac{1}{10}\int_0^\infty \left(u(r)w(r)\sqrt{2} - \frac{1}{2}w^2(r)\right) r^2\, dr$$

Thus, to lowest order, Q is linear in the relative amplitude $|w|/|u|$ of the ^3D admixture in the ^3S state.

From a comparison of Q with the empirically available quadrupole moment one obtains the squared amplitude of the D-state

$$P_D = \int w^2\, dr$$

as $P_D \simeq 5\%$ depending somewhat on the assumed radial dependences of V_c and V_T (cf. problem 13.13).

13B.2 The deuteron magnetic moment

The general expression for the magnetic moment measured in nuclear magnetons, $e\hbar/2M$, is

$$\mu = g_\ell^P \ell_p + g_s^P s_p + g_s^n s_n = \frac{1}{2}L + g_s^P s_p + g_s^n s_n$$

where the last equality follows from $g_\ell^P = 1$ and from

$$\ell_p = r_p \times p_p = \frac{1}{2}r \times p = \frac{1}{2}L$$

The g_s-factors are given as

$$g_s^P = 5.5855$$
$$g_s^n = -3.8263$$

For a pure $L = 0$ state it is apparent that

$$\mu = \frac{1}{2}\left(g_s^P + g_s^n\right)$$

while for the $^3S_1$ and $^3D_1$ (evaluated for the already explicitly written $M = 1$ component) we obtain

$$\mu = \left\langle \psi_{M=1}^{J=1} \left| \frac{1}{2}L_z + g_s^P (s_p)_z + g_s^n (s_n)_z \right| \psi_{M=1}^{J=1} \right\rangle$$
$$= \frac{1}{2}\left(g_s^P + g_s^n\right) \int u^2(r)\, dr$$
$$+ \left[\frac{1}{2}\left(\frac{3}{10} + 2 \cdot \frac{3}{5}\right) + \frac{1}{2}\left(g_s^P + g_s^n\right)\left(\frac{1}{10} - \frac{3}{5}\right)\right] \int w^2(r)\, dr$$
$$= \frac{1}{2}\left(g_s^P + g_s^n\right) - \frac{3}{4}\left(g_s^P + g_s^n - 1\right) \int w^2(r)\, dr$$

where the last equality follows from the normalisation of the wave function.
  The experimental value of the deuteron magnetic moment is

$$\mu_{exp} = 0.8574 = \frac{1}{2}\left(g_s^P + g_s^n\right) - 0.0222$$

from which we get

$$P_D = \int w^2(r)\, dr = -\frac{\frac{3}{4}\left[\mu_{exp} - \frac{1}{2}\left(g_s^P + g_s^n\right)\right]}{\left(g_s^P + g_s^n - 1\right)} = 0.039$$

in fair agreement with the estimate from the quadrupole moment.

14

The pairing interaction

In the empirical study of odd-A spectra it appears that the level density observed at low energies is well described as due to the excitation of the odd particle alone. In even–even spectra on the other hand, the general rule is that no states, with exception of members of ground state rotational bands, are found below 1 MeV of excitation energy. This is the famous even–even 'energy gap' of a magnitude of the order 2Δ, where Δ is the odd–even energy difference. The latter is defined as the average difference between the mass parabolae connecting separately odd and separately even nuclei, respectively (problem 3.5).

These facts are well described in nuclear physics by the nuclear pairing theory. The corresponding formalism goes back to solid-state physics, where it was developed to describe the phenomenon of superconductivity and the corresponding energy gap in the electronic level density of superconducting metals.

Let us first consider a single spherical subshell (ℓ, j) with a degeneracy $2\Omega = 2j + 1$. If one places two particles (1,2) into this shell, one may obtain good angular momentum states by the use of Clebsch–Gordan coupling coefficients

$$\Psi_M^I(1,2) = \sum_{m,m'} \psi_m^j(1)\psi_{m'}^j(2)C_{mm'M}^{jjI}$$

Without symmetrisation conditions the resulting states have $I = 0, 1 \ldots 2j$. If the two particles are of the same kind, however, only the even values of those I-values are permissible, or $I = 0, 2 \ldots 2j - 1$. This can be seen from the fact that to fulfil the Pauli principle we must require the two-particle wave function to be antisymmetric (chapter 13)

$$\Psi_M^I(1,2) = -\Psi_M^I(2,1)$$

290

Using

$$\Psi_M^I(2,1) = \sum_{m,m'} \psi_m^j(2)\psi_{m'}^j(1)C_{mm'M}^{jjI}$$

and renaming m and m' as m' and m, respectively, and furthermore exploiting the following symmetry of the Clebsch–Gordon coefficients (appendix 6B):

$$C_{mm'M}^{jj'I} = (-1)^{j+j'-I}C_{m'mM}^{j'jI}$$

it is easy to see that

$$\Psi_M^I(2,1) = (-1)^{2j-I}\Psi_M^I(1,2)$$

The requirement of antisymmetry is thus equivalent to $2j-I$ being an odd number. As $2j$ is an *odd* number, it follows that *only even I are allowed.*†

For these two particles all I-values allowed by the Pauli principle are degenerate in energy as long as no interaction, apart from the common potential, is considered. We shall now assume that there is added such a residual interaction of the simplest possible kind ('residual' refers to the interaction that is not included in the common potential).

The simplest interaction we can think of may be a zero-range potential or a δ-interaction. This interaction affects only particles with identical positions in space.

$$V_{\text{res}}(1,2) = -\kappa\delta\,(\mathbf{r}_1 - \mathbf{r}_2) = -\kappa\delta\,(x_1 - x_2)\,\delta\,(y_1 - y_2)\,\delta\,(z_1 - z_2)$$

$$= -\kappa\frac{\delta\,(r_1 - r_2)}{r_1^2}\delta\,(\Omega_1 - \Omega_2)$$

We simplify additionally by saying that ℓ is so large that spin $\frac{1}{2}$ is really negligible and characterise the orbitals only by ℓ. We can now show by direct calculations (see problem 14.1) that

$$\langle \ell\ell I M \,|V_{\text{res}}(1,2)|\,\ell\ell I M \rangle \propto -\kappa\left(C_{000}^{I\ell\ell}\right)^2$$

Actually, if spin is included, the formula above generalises to

$$\langle jjI M \,|V_{\text{res}}(1,2)|\,jjI M \rangle \propto -\kappa\left(C_{0\frac{1}{2}\frac{1}{2}}^{Ijj}\right)^2$$

It is interesting to evaluate the energy 'spectrum' of states with such an interaction included (cf. problem 14.2 and fig. 14.1). For large ℓ-values with the energy of $I^\pi = 0^+$ being E_0, the other spins correspond to the following energies in units of E_0:

---

† Note, however, that for two identical particles placed in *different* subshells, j_1 and j_2, all I between $|j_1 - j_2|$ and $j_1 + j_2$ are permissible.

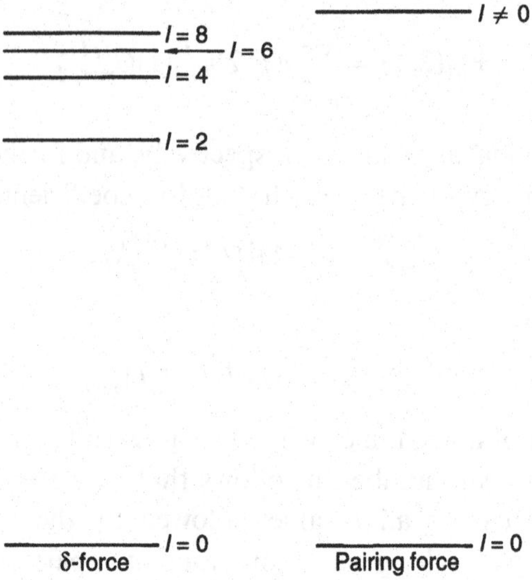

Fig. 14.1. Energy spectrum of a pair of identical particles in a spherical $j = 9/2$ subshell interacting through a δ-force and schematic pairing force, respectively.

$$E_2/E_0 = 1/4, \quad E_4/E_0 = 9/64, \quad E_6/E_0 = 25/256$$

For two particles, the spectrum is thus of a type where one combination of occupation amplitudes (corresponding to $I = 0$) fully exploits the interaction while the other permissible wave functions barely take advantage of the two-body interaction. Two things therefore suggest themselves:

(1) to analyse how this $I = 0$ wave function looks
(2) to replace the δ-interaction with an even simpler force – *the pairing force* – characterised by the even simpler two-body spectrum illustrated in fig. 14.1. In this spectrum the $I = 0$ state is depressed by E_0, while $E_{I\neq0} = 0$.

14.1 Creation and annihilation operators

Let us first investigate the ground-state wave function. We had

$$\Psi^{I=0}(1,2) = \sum_m \psi_m^j(1)\psi_{-m}^j(2)C_{m-m0}^{jj0}$$

We may now use a further symmetry of the Clebsch–Gordan coefficients (which follows from the relations given in appendix 6B), namely that

$$C_{\alpha\beta\lambda}^{abc} = \left(\frac{2c+1}{2b+1}\right)^{1/2}(-1)^{a-\alpha}C_{\alpha-\lambda-\beta}^{acb}$$

or

$$C_{m-m0}^{jj0} = \left(\frac{1}{2j+1}\right)^{1/2} (-1)^{j-m} C_{m0m}^{j0j} = \left(\frac{1}{2j+1}\right)^{1/2} (-1)^{j-m}$$

We may include the phase in the definition of a *conjugate* state $\psi_{\bar{m}}^{j}$ such that

$$\psi_{\bar{m}}^{j} = (-1)^{j-m} \psi_{-m}^{j}$$

(this definition of a conjugate state was introduced already in chapter 11). Let us call $2j+1$ the subshell degeneracy 2Ω. We furthermore let the sum run over positive m-values solely. The $I = 0$ wave function then takes the form

$$\Psi^{I=0}(1,2) = \frac{1}{\sqrt{\Omega}} \sum_{m>0} \psi_m^j(1) \psi_{\bar{m}}^j(2)$$

The state above is a particular state where the particles 1 and 2 always fill conjugate (time-reversed) pairs of orbitals, m and $\bar{m}$, with all amplitudes equal and with a particular relative phase, which can be defined as $+1$, once the $\bar{m}$ state is properly defined.

We might worry about the Pauli principle. Actually the Clebsch–Gordan coefficients take care of that for us in the two-particle case by constructing the states as either symmetric or antisymmetric. To handle more particles we shall need the second-quantisation method based on the use of creation and annihilation operators a_m^+ and a_m. Let us describe the situation of one particle occupying the orbital ψ_m^j by defining

$$a_m^+ |0\rangle \equiv |jm\rangle \equiv \psi_m^j$$

and analogously:

$$a_{\bar{m}}^+ |0\rangle \equiv (-)^{j-m} |j-m\rangle$$

Here $|0\rangle$ denotes the 'vacuum state' i.e. a situation with no particles occupying the orbitals. With $a_m^+ |0\rangle$ one particle is 'created' in the level $|jm\rangle$.

The Hermitian conjugate operator a_m 'annihilates' or 'absorbs' a particle in the orbital m. Obviously the 'vacuum' state must be subject to the condition

$$a_m |0\rangle = 0$$

for any m, as no particles can be 'absorbed' from the vacuum. Furthermore, as each orbital can be occupied by *one* particle $a_m^+ |m\rangle = 0$.

In the general case, the wave function of two particles occupying the orbitals $|\nu\rangle$ and $|\mu\rangle$,

$$\Psi(1,2) = \frac{1}{\sqrt{2}} \left[\psi_\mu(1)\psi_\nu(2) - \psi_\nu(1)\psi_\mu(2) \right]$$

is written as

$$\Psi(1,2) = a_\mu^+ a_\nu^+ |0\rangle$$

in the second-quantisation formalism. The requirement of antisymmetry, $P^{12}\Psi = -\Psi$, is equivalent to $P^{\mu\nu}\Psi = -\Psi$ and is thus taken care of by requiring $a_\mu^+ a_\nu^+ = -a_\nu^+ a_\mu^+$, i.e.

$$a_\mu^+ a_\nu^+ + a_\nu^+ a_\mu^+ = \left\{a_\mu^+, a_\nu^+\right\} = 0$$

and similarly for the Hermitian conjugate operator

$$\{a_\mu, a_\nu\} = 0$$

If one creation and one annihilation operator are mixed we find for example

$$\langle 0 | a_\mu a_\nu^+ | 0 \rangle = \int \psi_\mu^* \psi_\nu \, d^3 r = \delta_{\mu\nu}$$

$$\langle 0 | a_\nu^+ a_\mu | 0 \rangle = 0$$

and similarly if $|0\rangle$ is replaced by $|v\rangle$:

$$\langle v | a_\mu a_\nu^+ | v \rangle = 0$$

$$\langle v | a_\nu^+ a_\mu | v \rangle = \langle v | a_\nu^+ | 0 \rangle \delta_{\nu\mu} = \delta_{\nu\mu}$$

These equations are simultaneously fulfilled if

$$a_\mu a_\nu^+ + a_\nu^+ a_\mu = \{a_\mu, a_\nu^+\} = \delta_{\nu\mu}$$

The so-called Fermion anticommutator relations, which we have given above, are valid for a general many-particle wave function and the second-quantisation formalism is then a very efficient way to take care of the symmetries. For example, a three-particle wave function is given as

$$a_\mu^+ a_\nu^+ a_\omega^+ |0\rangle = a_\mu^+ a_\nu^+ |\omega\rangle = a_\mu^+ |v\omega\rangle = |\mu v\omega\rangle$$

with the antisymmetry $|\mu v\omega\rangle = -|v\mu\omega\rangle$ being equivalent to $\left\{a_\mu^+, a_\nu^+\right\} = 0$.

The operator $a_\nu^+ a_\nu$ has the eigenvalue 1 if the state $|v\rangle$ is occupied and the value 0 if $|v\rangle$ is empty. Thus, the eigenvalue of the operator (with summation over all v)

$$\mathcal{N} = \sum a_\nu^+ a_\nu$$

is the total number of particles, i.e. $\mathcal{N}$ is naturally referred to as the number operator. In the second-quantisation formalism, the *diagonal form* of the Hamiltonian is

$$\mathcal{H} = \sum_v e_v a_\nu^+ a_\nu$$

The single-particle energies e_v are given as $e_v = \langle v|H|v \rangle$. Furthermore, $\langle v|H|\mu \rangle = 0$ for $v \neq \mu$ and thus we can write

$$\mathcal{H} = \sum_{v,\mu} \langle v|H|\mu \rangle \, a_v^+ a_\mu$$

One now realises that this is the general form of a *one-particle* operator

$$\mathcal{V}_1 = \sum_{i,j} \langle i|V_1|j \rangle \, b_i^+ b_j$$

where b_i^+ is the creation operator in an arbitrary basis,

$$b_i^+ = \sum_v c_{vi} a_v^+$$

For two particles in the orbitals μ and v, which interact via a potential V_2, we write in the second-quantisation language

$$\mathcal{V}_2 = a_v^+ a_\mu^+ a_\mu a_v \, \langle v\mu|V_2|v\mu \rangle$$

For a wave function $|v\mu \rangle$, this gives the correct matrix elements:

$$\langle v\mu|\mathcal{V}_2|v\mu \rangle = \langle v\mu|V_2|v\mu \rangle = (V_2)_{v\mu}$$

For a many-particle wave function, the interaction generalises to

$$\mathcal{V}_2 = \sum_{v>\mu} a_v^+ a_\mu^+ a_\mu a_v \, \langle v\mu|V_2|v\mu \rangle = \frac{1}{2} \sum_{v,\mu} a_v^+ a_\mu^+ a_\mu a_v \, \langle v\mu|V_2|v\mu \rangle$$

If we finally expand to an arbitrary basis, we find the general expression for a *two-particle* operator

$$\mathcal{V}_2 = \frac{1}{2} \sum_{i,j,k,l} b_i^+ b_j^+ b_l b_k \, \langle ij|V_2|kl \rangle$$

Obviously this operator is capable of annihilating a pair of particles in orbitals k and l and 'recreating' them in i and j. The operator is thus associated with a diagram of the type shown in fig. 14.2.

The probability of the scattering of two particles from the orbitals k and l and into i and j is thus proportional to the square of the matrix element $\langle ij|V_2|kl \rangle$.

14.2 The pairing interaction (for degenerate configurations)

The pairing interaction H^G is characterised by having non-vanishing matrix elements of constant magnitude $-G$ for time conjugate pairs of states $(m, \bar{m})$

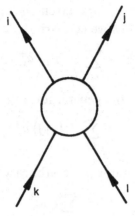

Fig. 14.2. A two-particle operator illustrated through a diagram with the pair of particles in the orbitals k and l being annihilated but being recreated in the orbitals i and j.

and $(m', \bar{m}')$, while all other matrix elements vanish. We have thus

$$H^G = -G \sum \sum a_m^+ a_{\bar{m}}^+ a_{\bar{m}'} a_{m'}$$

where we have introduced the convention that the sums run solely over positive m-values. Note first that this operator only involves orbitals occupied by pairs of particles. Considering the space spanned by paired two-particle states $a_m^+ a_{\bar{m}}^+ |0\rangle$, we can write out the entire matrix representation of H^G as

$$H^G = -G \begin{pmatrix} 1 & 1 & 1 & \dots \\ 1 & 1 & 1 & \dots \\ 1 & 1 & 1 & \dots \\ \dots & \dots & \dots & \dots \end{pmatrix}$$

a matrix of Ω rows and Ω columns ($\Omega = (2j+1)/2$). We could actually *a priori* find an eigenstate to this Hamiltonian H^G namely

$$\Psi_0 = \frac{1}{\sqrt{\Omega}} \begin{pmatrix} 1 \\ 1 \\ 1 \\ \cdot \\ \cdot \end{pmatrix}$$

such that

$$H^G \Psi_0 = -G \Omega \Psi_0$$

This is of course the straightforward vector representation of our $I = 0$ two-particle wave function

$$\Psi_0 = \frac{1}{\sqrt{\Omega}} \sum a_m^+ a_{\bar{m}}^+ |0\rangle$$

This is usually defined as a seniority-zero ($v = 0$) state. We may also conventionally denote this as

$$\Psi_0 = \frac{1}{\sqrt{\Omega}} A^+ |0\rangle$$

where

$$A^+ = \sum a_m^+ a_{\bar{m}}^+$$

We have then

$$H^G = -G A^+ A$$

It is immediately obvious that H^G is a negative definite operator

$$\langle \Psi | H^G | \Psi \rangle = -G \int |A\Psi|^2 \, d\tau \leq 0$$

Furthermore, the sum of eigenvalues E_λ equals the matrix trace (sum of diagonal elements)

$$\sum E_\lambda = -G\Omega$$

As all eigenvalues have to be negative or zero and as $\sum E_\lambda = E_0$, all E_λ with $\lambda \neq 0$ have to be zero. In fact we can readily find all the eigenfunctions as

$$\Psi_\lambda = B_\lambda^+ |0\rangle = \sum c_m^\lambda a_m^+ a_{\bar{m}}^+ |0\rangle$$

All these $\Omega - 1$ states have by definition seniority 2 ($v = 2$). The seniority number v is the number of particles not coupled two-and-two to spin zero. The wave functions can be conventionally orthogonalised by requiring

$$\sum_m c_m^\lambda c_m^{\lambda'} = \delta_{\lambda\lambda'}$$

Finally (as the eigenvalues are zero)

$$H^G \Psi(n = 2, v = 2) = 0$$

corresponds to the condition (which also assures orthogonality with the $v = 0$ state)

$$\sum_m c_m^\lambda = 0$$

One can now actually also construct wave functions in the four- and ultimately n-particles case for the so-called degenerate model (i.e. particles in a subshell of Ω degenerate orbitals).

We shall now assert that the lowest energy, spin 0, seniority 0 solution of the four-particle case corresponds to an energy $-G2(\Omega - 1)$ and the wave function (unnormalised)

$$\Psi(n = 4, v = 0) \propto \left(A^+\right)^2 |0\rangle$$

The basis space of this problem contains $\frac{1}{2}\Omega(\Omega-1)$ basis states as only pairs are considered and only one pair can be placed in each orbital.

Above the $v = 0$ state there is an energy gap and then a set of $\Omega-1$ states of seniority $v = 2$:

$$\Psi(n = 4, v = 2) \propto A^+ B_\lambda^+ |0\rangle$$

Obviously there are $\Omega-1$ of these states. Their energy is $-G(\Omega-2)$ as shown below. Finally there remain of the $\frac{1}{2}\Omega(\Omega - 1)$ four-particle states $\frac{1}{2}\Omega(\Omega - 3)$ that have energy zero and seniority 4 ($v = 4$). They can be written out as

$$\Psi(n = 4, v = 4) = C_\beta^+ |0\rangle$$

in terms of a four-particle operator C_β^+.

To prove that the wave functions we have given in the four-particle case are really eigenfunctions of H^G, we need the commutators (problem 14.4)

$$[A, A^+] = \Omega - \mathcal{N}; \quad \left[H^G, A^+\right] = -G(\Omega - \mathcal{N} + 2)A^+$$

where $\mathcal{N}$ is the particle number operator

$$\mathcal{N} = \sum \left(a_m^+ a_m + a_{\bar{m}}^+ a_{\bar{m}}\right)$$

Applied to the two-particle wave function,

$$H^G A^+ |0\rangle = A^+ H^G |0\rangle - G(\Omega - \mathcal{N} + 2)A^+ |0\rangle = -\Omega G A^+ |0\rangle$$

we get the energy eigenvalue, $-\Omega G$, which was already derived above.

In the general case, it is found that (problem 14.5)

$$\left[H^G, (A^+)^p\right] = -Gp(\Omega - \mathcal{N} + p + 1)\left(A^+\right)^p$$

which for the special case of $p = 2$ leads to

$$H^G \left(A^+\right)^2 |0\rangle = \left(A^+\right)^2 H^G |0\rangle - 2G(\Omega - \mathcal{N} + 3)\left(A^+\right)^2 |0\rangle$$
$$= -2G(\Omega - 1)\left(A^+\right)^2 |0\rangle$$

We also find

$$H^G A^+ B_\lambda^+ |0\rangle = A^+ H^G B_\lambda^+ |0\rangle - G(\Omega - \mathcal{N} + 2)A^+ B_\lambda^+ |0\rangle$$

and thus the eigenvalues $-G(\Omega - 2)$ for the four-particle $v = 2$ states. With the trace of the total matrix being $-\frac{1}{2}G\Omega(\Omega - 1)$ it is now easy to find that the $v = 4$ states have zero energy.

One can now easily write down the energy of the $v = 0$ (lowest) state of any n-particle system, which state involves $p = n/2$ pairs

$$E(n, v = 0) = -G\frac{n}{4}(2\Omega - n + 2)$$

while the $v = 2$ states which have $p = (n/2) - 1$ correspond to the energy

$$E(n, v = 2) = -G\frac{n - 2}{4}(2\Omega - n)$$

The gap between the $v = 0$ and $v = 2$ states thus remains equal to $G\Omega$ independently of n. Note, however, that the total pair correlation energy given by $E(n, v = 0)$ is to leading order proportional to the number of pairs $n/2$ and to the degeneracy Ω. For larger n-values there is a considerable second-order term of opposite sign in n. This term reproduces the effect of the Pauli principle that orbitals already occupied by the first pair are no longer available to the second pair.

Actually the largest correlation energy is obtained for a half-filled shell. For a shell with $\Omega = 5$, we obtain thus pairing energies $-5, -8, -9, -8$ and -5 in units of G for a particle number of 2, 4, 6, 8 and 10. It should, however, be noted that there is a diagonal energy equal to $-(n/2)G$ and the *true correlation energy* is thus $-4, -6, -6, -4$ and 0, again in units of G. For a completely filled shell, the Pauli principle determines the wave function entirely. There is in fact no energy arising from pure correlation.

For real nuclei this is an oversimplified description based on the availability of only one degenerate orbital. We may, however, let it simulate the empirical situation for nuclei with neutron or proton numbers between two closed shells. There the largest pair correlation energy is obtained for nuclei representing half-filled shells of neutrons and protons provided the nuclear shape remains unchanged.

14.3 Generalisation to non-degenerate configurations. The BCS formalism

The situation we now envisage is first a single-particle model of non-degenerate orbitals v, where v may alternatively be a deformed state characterised by a given value of j_3, parity and maybe an additional label or a spherical state (n, ℓ, j and m_j). In the absence of pairing we write the single-particle part of the quantised Hamiltonian as

$$H^{\mathrm{sp}} = \sum e_v \left(a_v^+ a_v + a_{\bar{v}}^+ a_{\bar{v}} \right)$$

As yet we have not included pairing. Eigenstates of H^{sp} are for example

(1) $a_\mu^+ |0\rangle$ or one orbital μ occupied by an odd particle, the associated energy eigenvalues being e_μ,

(2) $a_\mu^+ a_{\bar{\mu}}^+ |0\rangle$ two particles occupying the time-reversed orbitals ψ_μ and $\psi_{\bar{\mu}}$, the associated energy being $2e_\mu$.

We now include pairing and write

$$H = \sum e_v \left(a_v^+ a_v + a_{\bar{v}}^+ a_{\bar{v}} \right) - G \sum \sum a_\mu^+ a_{\bar{\mu}}^+ a_{\bar{v}} a_v$$

Obviously for very small level splitting the last term dominates and we should obtain the results of the previous paragraph while for large level splitting the second term becomes negligible and we expect a situation where only the lowest levels are filled. This latter case we denote as a 'sharp Fermi surface' case.

It would therefore seem natural to look for a solution based on a generalised pair operator

$$A^+ = \sum_v c_v a_v^+ a_{\bar{v}}^+$$

with c_v a diminishing function with increasing e_v. This is, however, an impractical way to approach the problem and the great inventiveness of the solid-state physicists Bardeen, Cooper and Schrieffer (BCS, 1957) is shown in the fact that they suggested a very different kind of wave function, which they applied to the theory of superconductivity. This same theory was then applied to nuclei by Bohr, Mottelson and Pines (1958), Belyaev (1959) and Migdal (1959). The new wave function is of the form (cf. appendix 14A)

$$\Psi_0 = \Psi_0^{\mathrm{BCS}} = \prod_v \left(U_v + V_v a_v^+ a_{\bar{v}}^+ \right) |0\rangle = |\tilde{0}\rangle$$

subject to the condition $U_v^2 + V_v^2 = 1$. This is the ground-state solution and we shall denote it Ψ_0^{BCS}, or the 'BCS vacuum' $|\tilde{0}\rangle$, or the 'quasiparticle vacuum'.

In this sum there are terms of the type suggested above. We can thus write

$$\Psi_0^{\mathrm{BCS}} = \prod U_v \left(1 + \frac{V_v}{U_v} a_v^+ a_{\bar{v}}^+ \right) |0\rangle$$

From this we can project out the terms containing exactly $n/2$ pairs $a_v^+ a_{\bar{v}}^+$.

These we can write

$$\Psi^{\text{Proj}}\left(p = \frac{n}{2}\right) = \left(\prod U_\nu\right)\left(\sum \frac{V_\nu}{U_\nu}a_\nu^+ a_{\bar\nu}^+\right)^{n/2}\left(\frac{n}{2}!\right)^{-1}|0\rangle$$

There are, however, also other components in the BCS wave function.

To analyse the components in the BCS wave function let us just consider the case of two levels r and s. The wave function is then

$$\Psi = (U_r + V_r a_r^+ a_{\bar r}^+)(U_s + V_s a_s^+ a_{\bar s}^+)|0\rangle$$

Multiplying this out we have

$$\Psi = \Psi_{(0)} + \Psi_{(2)} + \Psi_{(4)}$$

where

$$\Psi_{(0)} = U_r U_s$$
$$\Psi_{(2)} = V_r U_s a_r^+ a_{\bar r}^+ + U_r V_s a_s^+ a_{\bar s}^+$$
$$\Psi_{(4)} = V_r V_s a_r^+ a_{\bar r}^+ a_s^+ a_{\bar s}^+$$

We thus note that one component $\Psi_{(0)}$ contains no pairs at all, while $\Psi_{(2)}$ contains one pair alternatively occupying levels r and s, while in $\Psi_{(4)}$ both levels are occupied. Assume we want to describe one pair filling these two pairs of orbitals. Only one component of the wave function then has the proper particle number.

Thus one component of the wave function Ψ_0^{BCS} has the desirable number of particles while components with $n-2$, $n+2$, etc. are also present in the wave function. One can make sure that the correct average number of particles are present (this usually implies that the component with the correct number of particles is the largest). This is achieved through the help of the number operator

$$\mathcal{N} = \sum (a_\nu^+ a_\nu + a_{\bar\nu}^+ a_{\bar\nu})$$

This is then subtracted, multiplied by a Lagrangian multiplier λ, from the original Hamiltonian H†

$$H' = H - \lambda\mathcal{N}$$

The condition that n particles are present on the average corresponds to the relation

$$\langle\mathcal{N}\rangle = n$$

---

† There are important analogies between the auxiliary Hamiltonian, $H - \lambda\mathcal{N}$, and the cranking Hamiltonian, $H - \omega I$, which was introduced for rotating nuclei. The equation naturally leads to solutions where the 'exact' quantum numbers, particle number and angular momentum, are only preserved in an average sense. This, however, makes the solutions much more straightforward.

Let us denote the ground state wave function with Ψ_0 and the expectation values of H and H' by E_0 and E'_0, respectively. The condition that E'_0 is stationary with respect to changes in U_v and V_v that change the number of particles determines λ

$$0 = \frac{\partial E'_0}{\partial n} = \frac{\partial E_0}{\partial n} - \lambda$$

or

$$\lambda = \frac{\partial E_0}{\partial n}$$

We have now a simple understanding of λ, the 'chemical potential', within reach. This energy is the *energy of the last added particle*, the 'marginal energy' of the 'marginal particle'. To evaluate $\langle H' \rangle$ we first need the effect of $a_\mu^+ a_\mu$ on Ψ_0:

$$a_\mu^+ a_\mu \prod (U_v + V_v a_v^+ a_{\bar{v}}^+) |0\rangle$$

We thus need to commute a_μ through the operators on the right. We have then

$$[a_\mu, a_v^+ a_{\bar{v}}^+] = 0 \qquad \text{for} \quad \mu \neq v$$

and thus (by the property of the vacuum state, $a_\mu |0\rangle = 0$)

$$a_\mu a_v^+ a_{\bar{v}}^+ |0\rangle = 0 \qquad \text{for} \quad \mu \neq v$$

For $v = \mu$, we find

$$a_\mu a_\mu^+ a_{\bar{\mu}}^+ = -a_\mu^+ a_\mu a_{\bar{\mu}}^+ + a_{\bar{\mu}}^+ = a_\mu^+ a_{\bar{\mu}}^+ a_\mu + a_{\bar{\mu}}^+$$

and thus

$$a_\mu a_\mu^+ a_{\bar{\mu}}^+ |0\rangle = a_{\bar{\mu}}^+ |0\rangle$$

and finally

$$a_\mu^+ a_\mu a_\mu^+ a_{\bar{\mu}}^+ |0\rangle = a_\mu^+ a_{\bar{\mu}}^+ |0\rangle$$

It is therefore easy to prove

$$\left\langle \Psi_0 \left| a_\mu^+ a_\mu + a_{\bar{\mu}}^+ a_{\bar{\mu}} \right| \Psi_0 \right\rangle = 2V_\mu^2$$

We shall leave it to the exercises to prove that for $\mu \neq \kappa$

$$\left\langle \Psi_0 \left| a_\mu^+ a_{\bar{\mu}}^+ a_{\bar{\kappa}} a_\kappa \right| \Psi_0 \right\rangle = U_\mu V_\mu U_\kappa V_\kappa$$

while

$$\left\langle \Psi_0 \left| a_\mu^+ a_{\bar{\mu}}^+ a_{\bar{\mu}} a_\mu \right| \Psi_0 \right\rangle = V_\mu^2$$

We thus obtain ($U_\mu^2 + V_\mu^2 = 1$)

$$\left\langle \Psi_0 \left| \sum_\mu \sum_\kappa a_\mu^+ a_{\bar\mu}^+ a_{\bar\kappa} a_\kappa \right| \Psi_0 \right\rangle = \sum_{\mu\neq\kappa} \sum_\kappa U_\mu V_\mu U_\kappa V_\kappa + \sum_\mu V_\mu^2 \left(U_\mu^2 + V_\mu^2 \right)$$

$$= \sum_{all\ \mu,\kappa} U_\mu V_\mu U_\kappa V_\kappa + \sum_\mu V_\mu^4$$

and

$$E_0' = 2\sum (e_\nu - \lambda) V_\nu^2 - G\sum\sum U_\nu V_\nu U_\mu V_\mu - G\sum V_\nu^4$$

The energy E_0' is now minimised with respect to the coefficients (U_ν, V_ν). However, we have to observe the condition

$$U_\nu^2 + V_\nu^2 = 1$$

The condition of constant average number of particles,

$$\langle \mathcal{N} \rangle = \sum 2V_\nu^2 = n$$

is, on the other hand, upheld through λ. With

$$\frac{\partial U_\nu}{\partial V_\nu} = \frac{\partial}{\partial V_\nu} \left(1 - V_\nu^2\right)^{1/2} = -\frac{V_\nu}{U_\nu}$$

we now find

$$\frac{\partial E_0'}{\partial V_\nu} = 4(e_\nu - \lambda) V_\nu - 2G\left(\sum U_\mu V_\mu\right)\left(U_\nu - V_\nu \frac{V_\nu}{U_\nu}\right) - 4GV_\nu^3 = 0$$

Introducing by definition

$$\Delta = G\sum U_\nu V_\nu$$

we should note that thus $\Delta \gg G$ by a magnitude measuring the number of contributing orbitals. For realistic nuclear models this effective Ω-number, Ω_{eff}, is of the order of 4–10. In solving the stationarity equation above we take advantage of this relation between G and Δ and neglect the last term. We have thus

$$2(e_\nu - \lambda) U_\nu V_\nu = \Delta \left(U_\nu^2 - V_\nu^2\right)$$

Squaring this equation and using the normalisation condition

$$U_\nu^2 + V_\nu^2 = 1$$

we may express the relation above in a simpler form. We first define the 'quasiparticle energy' E_ν as

$$E_\nu = \left[(e_\nu - \lambda)^2 + \Delta^2\right]^{1/2}$$

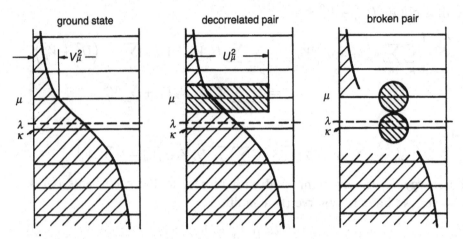

Fig. 14.3. The occupation probabilities, V_μ^2, of the orbitals around the Fermi surface λ in a BCS ground state are illustrated to the left. Furthermore, schematic illustrations are given of a decorrelated pair in orbital μ and a broken pair state with one 'odd' particle in each of the orbitals μ and κ.

We obtain

$$2U_\nu V_\nu = \frac{\Delta}{E_\nu}$$

and

$$U_\nu^2 - V_\nu^2 = \frac{e_\nu - \lambda}{E_\nu}$$

From these equations we obtain

$$V_\nu^2 = \frac{1}{2}\left(1 - \frac{e_\nu - \lambda}{E_\nu}\right)$$

$$U_\nu^2 = \frac{1}{2}\left(1 + \frac{e_\nu - \lambda}{E_\nu}\right)$$

These important factors can be interpreted as the probability of the orbitals $(\nu, \bar\nu)$ being occupied by a pair and being empty, respectively. Obviously for $e_\nu < \lambda$, where λ is the 'chemical potential', or the 'Fermi surface', the occupation probability V_ν^2 exceeds $1/2$, while the opposite is true for a level above the Fermi surface. For the complementary quantity U_ν^2 the situation is exactly the opposite. A picture of V_μ^2 for a set of levels is exhibited in fig. 14.3.

We have still not determined the relation between the 'gap parameter Δ', entering the expressions for V_ν^2 and U_ν^2 and the basic two-particle matrix element G. However, by combining the definition of Δ with the expression for $2U_\nu V_\nu$, we obtain

$$2\Delta = 2G \sum U_\nu V_\nu = G \sum \frac{\Delta}{\left[(e_\nu - \lambda)^2 + \Delta^2\right]^{1/2}}$$

This 'gap equation' has two solutions, first the 'trivial', or 'collapsed solution', corresponding to $\Delta = 0$ and secondly in most cases the one corresponding to $\Delta \neq 0$. The gap equation is generally written as

$$\frac{2}{G} = \sum \frac{1}{E_\nu} = \sum \frac{1}{\left[(e_\nu - \lambda)^2 + \Delta^2\right]^{1/2}}$$

There is still another auxiliary condition relating to the particle number

$$n = 2 \sum V_\nu^2 = \sum \left(1 - \frac{e_\nu - \lambda}{[(e_\nu - \lambda)^2 + \Delta^2]^{1/2}}\right)$$

From these two equations, both Δ and λ are determined for given single-particle energies, e_ν.

In most cases these two equations are solved by the help of a computer by numerical methods. One should note that for the case of a non-degenerate (apart from the $(\nu, \bar{\nu})$-degeneracy) level system there may not exist any non-trivial solution to the first of the auxiliary equations. In fact a minimum G-value is required in excess of G_{crit}

$$\frac{2}{G_{\text{crit}}} = \sum \frac{1}{|e_\nu - \lambda|}$$

For $G > G_{\text{crit}}$, the ground state for an even number of particles has the wave function

$$\Psi_0 = \prod (U_\nu + V_\nu a_\nu^+ a_{\bar{\nu}}^+) |0\rangle$$

The corresponding energy E_0 is obtained as the expectation value of the Hamiltonian H

$$E_0 = \langle \Psi_0 |H| \Psi_0 \rangle = E_0' + \lambda \langle \Psi_0 |\mathcal{N}| \Psi_0 \rangle$$

$$= 2 \sum e_\nu V_\nu^2 - \frac{\Delta^2}{G} - G \sum V_\nu^4$$

where $E_0' = \langle \Psi_0 |H - \lambda \mathcal{N}| \Psi_0 \rangle$ was calculated above. This should be compared with the energy of the sharp Fermi surface uncorrelated wave function

$$E(\Delta = 0) = 2 \sum_{\text{occ}} e_\nu - \frac{n}{2} G = 2 \sum_{\text{occ}} e_\nu - G \sum_\nu V_\nu^2$$

where $(n/2)G$ is the diagonal pairing energy. The pairing *correlation energy*, which is a collective energy caused by the spreading of the wave function around the Fermi surface, is thus obtained as

$$E^{\text{corr}} = 2\sum_{\nu} e_{\nu} V_{\nu}^2 - \frac{\Delta^2}{G} - 2\sum_{\text{occ}} e_{\nu} - G\left(V_{\nu}^4 - V_{\nu}^2\right)$$

The last term is small and furthermore, it was not correctly taken care of in our derivation of the V_{ν} terms. Thus, we shall in the following neglect this term.

We can now construct an excited pair or 'decorrelated pair' state. The requirement of orthogonality to the ground state leads to the following wave function:

$$\Psi_2(\mu) = \left(-V_{\mu} + U_{\mu} a_{\mu}^+ a_{\bar{\mu}}^+\right) \prod_{\nu \neq \mu} (U_{\nu} + V_{\nu} a_{\nu}^+ a_{\bar{\nu}}^+) |0\rangle$$

It is easy to prove that the excitation energy E_2^{exc} can be calculated as

$$E_2^{\text{exc}} = \langle \Psi_2 | H' | \Psi_2 \rangle - \langle \Psi_0 | H' | \Psi_0 \rangle$$

Note that H', but not H, is stationary with respect to a variation in the average particle number. If one alternatively chooses to study $\langle H \rangle$ one has to correct for changes in the average particle number (see problem 14.9).

We will leave the detailed calculations to the exercises and only give the answer (neglecting terms of the order G):

$$\langle \Psi_2 | H' | \Psi_2 \rangle - \langle \Psi_0 | H' | \Psi_0 \rangle = 2(e_{\mu} - \lambda)\left(U_{\mu}^2 - V_{\mu}^2\right) + 4\Delta U_{\mu} V_{\mu}$$

$$= 2(e_{\mu} - \lambda)\frac{(e_{\mu} - \lambda)}{E_{\mu}} + 2\Delta\frac{\Delta}{E_{\mu}} = 2E_{\mu}$$

This state we shall call a 'two-quasiparticle' state, each quasiparticle being associated with the quasiparticle energy E_{μ}. Some of the energy comes from excitation of pairs from the area around the Fermi surface and up into the level μ. In addition this pair fails to contribute to the correlation energy. The maximal loss due to this 'blocking' occuring with $e_{\nu} = \lambda$ is 2Δ. Together the single-particle and correlation energy cost is

$$2E_{\mu} = 2\left[(e_{\mu} - \lambda)^2 + \Delta^2\right]^{1/2}$$

Another way to make an excited state is to break a pair, putting one particle in orbital μ and another particle in orbital κ (fig. 14.3)

$$\Psi_2(\mu, \kappa) = a_{\mu}^+ a_{\kappa}^+ \prod_{\nu \neq \mu, \kappa} (U_{\nu} + V_{\nu} a_{\nu}^+ a_{\bar{\nu}}^+) |0\rangle$$

The corresponding excitation energy is obtained as (problem 14.11)

$$\langle\Psi_2|H'|\Psi_2\rangle - \langle\Psi_0|H'|\Psi_0\rangle = (e_\mu - \lambda)\left(1 - 2V_\mu^2\right) + (e_\kappa - \lambda)\left(1 - 2V_\kappa^2\right)$$

$$+\Delta\cdot 2\left(U_\mu V_\mu + U_\kappa V_\kappa\right) = E_\mu + E_\kappa \gtrsim 2\Delta$$

The two-quasiparticle state discussed in connection with fig. 11.10 is of this type with an excitation energy of approximately 2Δ. In an even nucleus we thus expect an energy gap of a magnitude 2Δ.

A simple special case of the 'broken-pair' problem is the 'one-quasiparticle' problem applicable to the odd-particle case. In this case the ground state wave function is of the type

$$\Psi_1 = a_\mu^+ \prod_{\mu\neq\nu}(U_\nu + V_\nu a_\nu^+ a_{\bar\nu}^+)|0\rangle$$

In solving the BCS equations connecting λ and Δ to the particle number and pairing strength G we shall adopt the convention of considering an *even-particle* vacuum of n particles where n is an *odd* number.† The wave function thus represents an average of the even neighbours on both sides. We have thus

$$n = 2\sum V_\nu^2$$

$$\frac{2}{G} = \sum\frac{1}{E_\nu}$$

with unrestricted sums (the *no-blocking case*).

For the odd-particle case, with the odd particle in the orbital μ close to the Fermi surface λ, the energy relation to the corresponding even–even quasiparticle vacuum is

$$E_1 - E_0 = \langle\Psi_1|H'|\Psi_1\rangle - \langle\Psi_0|H'|\Psi_0\rangle = E_\mu = \left[(e_\mu - \lambda)^2 + \Delta^2\right]^{1/2} \simeq \Delta$$

For $|e_\mu - \lambda| \ll \Delta$ the displacement between the odd-N mass parabola and the even-N mass parabola (of the same Z) is thus expected to be Δ_n. Similarly the odd-Z to even–even mass comparison gives a mass difference Δ_p.

14.4 The uniform model

The relation between the energy-gap parameter Δ, the coupling constant G, the number of levels involved and finally the level density can be studied in

† Because of the smaller number of paired orbitals in odd nuclei, there will be a tendency towards smaller Δ in odd than in even nuclei. Similarly, the two-quasiparticle states of even nuclei should be associated with even smaller values of Δ. None of these effects are accounted for here.

terms of the uniform model. Let us assume a case with evenly spaced doubly degenerate levels, equally many below the Fermi surface (situated at $\lambda = 0$) as above, with a level spacing of $d = \rho^{-1}$. The levels considered extend from $-S$ MeV to $+S$ MeV, and finally the pairing matrix element is assumed a constant G all over the level diagram considered. The real physical situation is that the matrix element connecting a pair at the Fermi surface with a pair far below or above diminishes with the energy distance. To simulate this roughly, the cut-off energy S is introduced.

To preserve the symmetry we also have to assume that the number of pairs is half the number of levels. In this way we assure that $\lambda = 0$. Thus only one of the BCS equations remains

$$\frac{2}{G} = \sum \frac{1}{(e_\nu^2 + \Delta^2)^{1/2}}$$

We shall furthermore assume that, in this model, sums can always be replaced by integrals

$$\frac{2}{G} = \int_{-S}^{S} \rho \, de \frac{1}{(e^2 + \Delta^2)^{1/2}} = \rho \ln \left(\frac{S + (S^2 + \Delta^2)^{1/2}}{-S + (S^2 + \Delta^2)^{1/2}} \right)$$

or

$$\exp \left(\frac{2}{G\rho} \right) = \frac{S + (S^2 + \Delta^2)^{1/2}}{-S + (S^2 + \Delta^2)^{1/2}} = \frac{\left[S + (S^2 + \Delta^2)^{1/2} \right]^2}{\Delta^2}$$

For $S \gg \Delta$, we can write

$$\frac{2S}{\Delta} \simeq \exp \left(\frac{1}{G\rho} \right)$$

and so,

$$\Delta \simeq 2S \exp \left(-\frac{1}{G\rho} \right)$$

This is a very simple and extremely useful relation between Δ, G, ρ, and the cut-off energy S. In fact the same formula appears in the solid-state problem of electron superconductivity.

For a nucleus in, say, the rare-earth region we have $\rho^{-1} \simeq 0.35$ MeV, $\Delta \approx 0.9$ MeV. Considering two shells above and two below the Fermi surface, or $S = 2\hbar\omega \approx 15$ MeV, we find that we have to use $G \approx 0.1$ MeV.

We may now calculate a 'correlation energy' defined as a difference between the correlated ($\Delta \neq 0$) state and the unpaired ($\Delta = 0$) state. Still neglecting the difference between the $G \sum V_\nu^4$ term and the diagonal pairing energy, we have

$$E(\Delta) - E(0) = \sum 2V_\nu^2 e_\nu - \frac{\Delta^2}{G} - \overset{0}{\sum} 2e_\nu$$

$$\simeq \rho \int_{-S}^{0} \left(2V^2 - 2\right) \varepsilon \, d\varepsilon + \rho \int_{0}^{S} 2V^2 \varepsilon \, d\varepsilon - \frac{\Delta^2}{G}$$

where the unpaired sum in $E(0)$ extends only up to $e = \lambda = 0$. One can
evaluate the integrals involved and obtain (problem 14.13)

$$E(\Delta) - E(0) = \rho \left[S^2 - S(S^2 + \Delta^2)^{1/2} \right] \simeq -\frac{1}{2}\rho\Delta^2$$

using $S \gg \Delta$.

The pairing correlation energy involved is indeed very small. Thus for a
deformed nucleus in the rare-earth region with $\rho \simeq 3$ MeV^{-1} and $\Delta \approx 0.9$
MeV, $E(\Delta) - E(0) \simeq 1.25$ MeV separately for the neutrons and protons, or
a total of 2.5 MeV. For near-spherical nuclei we may have an effective level
density, $\rho \simeq 5$ MeV^{-1} and $\Delta = 1$–1.5 MeV, leading to a pairing energy of
2–5 MeV for each kind of particle. For spherical nuclei at closed shells on
the other hand, the effective level density is small and, generally, $G < G_{\text{crit}}$.
Thus, the only solution of the BCS equations corresponds to $\Delta = 0$.

14.5 Application of the BCS wave function to the pure *j*-shell case

We have the BCS wave function for the $v = 0$ case (v = generalised seniority
= number of quasiparticles)

$$\Psi_0 = \prod (U_\nu + V_\nu a_\nu^+ a_{\bar\nu}^+) |0\rangle$$

For a spherical *j*-shell of degeneracy Ω, the values of V_ν and U_ν must be the
same for all the orbitals. Thus

$$V_\nu = V = \left(\frac{n}{2\Omega} \right)^{1/2}$$

$$U_\nu = U = \left(1 - \frac{n}{2\Omega} \right)^{1/2}$$

which leads to

$$\Delta = G\Omega UV = G\Omega \left[\frac{n}{2\Omega} - \left(\frac{n}{2\Omega} \right)^2 \right]^{1/2}$$

Hence the total energy is given as $(e_v = 0)$

$$E_0 = \sum e_v 2V_v^2 - \frac{\Delta^2}{G} - G\sum V_v^4 = -G\Omega^2 \frac{n}{2\Omega}\left(1 - \frac{n}{2\Omega}\right) - G\Omega\left(\frac{n}{2\Omega}\right)^2$$

$$= -G\frac{n}{4}\left(2\Omega - n + \frac{n}{\Omega}\right)$$

This should be compared with the exact formula

$$E(n, v = 0) = -G\frac{n}{4}(2\Omega - n + 2)$$

The BCS wave function thus gives a correct result up to terms of order $1/\Omega$.

14.6 Applications of the BCS theory to nuclei

The BCS equations are easily solved on a computer for an arbitrary set of single-particle orbitals. The coupling constant G is determined for example from the empirical values of the pairing gap which is set equal to the odd–even mass difference (see problem 3.5). It is furthermore necessary to make a cut-off in the single-particle orbitals and thus only include a certain number of orbitals around the Fermi surface in the BCS equations.

An estimate of the different constants is obtained from the uniform model. We thus consider a pure oscillator potential and furthermore assume $N = Z = A/2$. If the oscillator shells with principal quantum numbers smaller than N^* are filled while the N^*-shell is half-full, we get

$$A = 2N = 2Z \approx \frac{2}{3}(N^* + 1.5)^3$$

(see chapter 6 and especially problem 6.10). The number of orbitals (for protons *or* neutrons) in the N^*-shell is equal to $\frac{1}{2}(N^* + 1.5)^2$, which gives the level density

$$\rho \simeq \frac{(N^* + 1.5)^2}{2\hbar\omega_0} = \frac{1}{2}\left(\frac{3}{2}\right)^{2/3}\frac{A}{41}\quad \text{MeV}^{-1}$$

where we have eliminated N^* and inserted $\hbar\omega_0 = 41 \cdot A^{-1/3}$.

In the uniform model, we derived the formula

$$\Delta = 2S\exp(-1/G\rho)$$

which should be compared with the experimental value (fig. 3.3)

$$\Delta_n \simeq \Delta_p \simeq 12 \cdot A^{-1/2}\quad \text{MeV}$$

We make the approximation that ρ is constant over the whole pairing interval and conclude that G should vary as A^{-1} and S as $A^{-1/2}$. As an alternative to the energy interval, S, one could consider a fixed number of orbitals above and below the Fermi surface. With $\rho \propto A$, this number should be proportional to $A^{1/2}$ or equivalently to $N^{1/2}$ for neutrons and $Z^{1/2}$ for protons. A standard procedure in modified oscillator calculations is to consider $(15N)^{1/2}$ and $(15Z)^{1/2}$ orbitals, respectively, above and below the Fermi surface (Nilsson *et al.*, 1969). The following estimate is then obtained for the coupling constant:

$$G = -\left[\rho \ln\left(\frac{\Delta}{2S}\right)\right]^{-1} = -\left[\rho \ln\left(\frac{12/\sqrt{A}}{2(15 \cdot A/2)^{1/2}} \cdot \rho\right)\right]^{-1} \simeq \frac{1}{A} \cdot 18.7 \text{ MeV}$$

where the value of ρ according to the harmonic oscillator expression has been inserted.

With $N \neq Z$, one could furthermore expect that $G_n \neq G_p$. As $\Delta_n \simeq \Delta_p$, we should from the uniform model formula require that the same cut-off in energy is made for neutrons and protons and furthermore that $\rho_n G_n = \rho_p G_p$. In a similar way as in problem 6.12, we consider only first order terms in the small parameter $x = (N - Z)/A$ to obtain

$$G\binom{n}{p} = \frac{G_0}{A}\left[1 \mp \frac{1}{3}\left(\frac{N-Z}{A}\right)\right]$$

where G_0 is a constant, which is common for protons and neutrons. One furthermore finds that it is approximately correct to make the cut-off as suggested above with the number of 'paired orbitals' proportional to $\sqrt{N}$ and $\sqrt{Z}$, respectively.

In realistic calculations, one should determine the pairing constants somewhat more carefully. A possibility is to assume a pairing matrix element

$$G\binom{n}{p} = \left(g_0 \mp g_1 \frac{N-Z}{A}\right) A^{-1} \text{ MeV}$$

where the general dependence on neutron and proton number has been taken from the uniform model expression. With $(15N)^{1/2}$ or $(15Z)^{1/2}$ orbitals considered above and below the Fermi surface, the constants g_0 and g_1 have been determined from a fit of the Δ-value at the calculated ground state deformation to the odd–even mass differences for a number of nuclei in the $A = 150$–250 region (Nilsson *et al.*, 1969). The quality of the fit is illustrated for rare-earth nuclei in fig. 14.4 where the constants

$$g_0 = 19.2 \text{ MeV}; \qquad g_1 = 7.4 \text{ MeV}$$

have been used.

When calculating the nuclear energy by the shell correction method, the

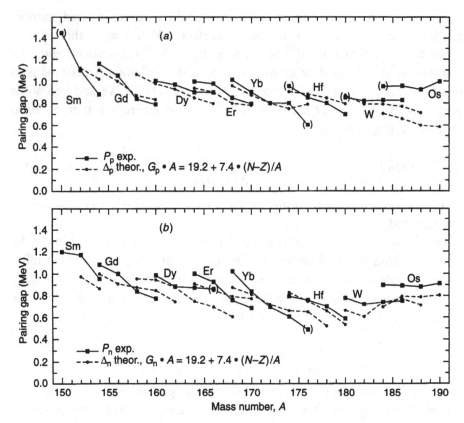

Fig. 14.4. a) Calculated pairing gap Δ_p for nuclei in the rare-earth region compared with odd–even mass differences, P_p. The latter were extracted from measured masses of even-Z and odd-Z isotones using the difference formula illustrated in problem 3.5. The theoretical pairing gaps were extracted at the calculated equilibrium deformations with the parameters given in the text (from Nilsson *et al*, 1969). b) Same as part a) but for neutrons instead, i.e. pairing gaps Δ_n compared with odd–even mass differences extracted from measured masses of odd-N and even-N isotopes.

parameters of the liquid-drop energy are fitted to reproduce the average trends. Thus, one must assume that the average pairing energy is also accounted for. This means that we should only consider the variation of the pairing energy around an average value, $\delta E_{pair} = E_{pair} - \langle E_{pair} \rangle$. The average value can be estimated from the uniform model. It comes out as

$$\langle E_{pair} \rangle = -\frac{1}{2}\left(\rho_n \Delta_n^2 + \rho_p \Delta_p^2\right) \approx -(1.15 + 1.15)\ \text{MeV} = -2.3\ \text{MeV}$$

independently of mass number, A. In calculating this value, we have used the harmonic oscillator estimate for the level density and the empirical

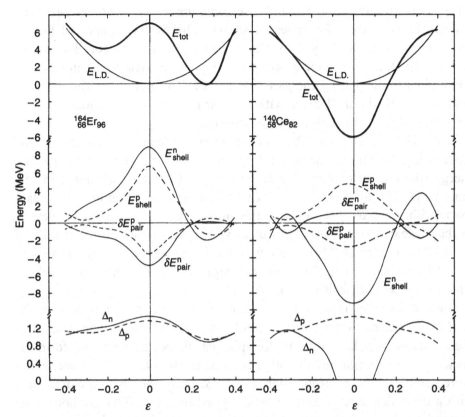

Fig. 14.5. The different energy contributions for protons (p) and neutrons (n) which build up the total energy in the shell correction approach are exhibited as a function of quadrupole deformation for ^{164}Er and ^{140}Ce. The shell energies E_{shell} were calculated by the Strutinsky method. The BCS formalism was used to calculate the pairing energies E_{pair} (where the average value, -1.15 MeV in the present approach, has been subtracted) and the pairing gap. The total energy E_{tot} is obtained as the sum of the liquid-drop energy, $E_{L.D.}$, the shell and the pairing energies.

value, $12/\sqrt{A}$ MeV, for the pairing gap. It is also possible to calculate the level density and especially its variation with energy, $\rho(e)$, by the Strutinsky method (chapter 9). The average pairing energy is then calculated numerically leading to a more accurate and systematic method to obtain $\langle E_{pair} \rangle$ (Brack et al., 1972).

In fig. 14.5, the variation of the pairing energy and its relation to the shell correction energy are illustrated as functions of quadrupole deformation, ε, for the nuclei ^{164}Er and ^{140}Ce. The energies were calculated from the single-particle orbitals of figs. 11.5 and 11.6. Let us first consider the case of ^{164}Er. From the single-particle diagram, the level density around the Fermi surface for proton number $Z = 68$ and neutron number $N = 96$, respectively, appears to be highest for spherical shape. This is also supported by the shell energy maxima of fig. 14.5. Consequently, the pairing correlations are maximal for spherical shape as manifested by the maxima in Δ_p and Δ_n and the minima in the pairing energies. It is also evident that the pairing energies tend to smooth the shell effect, with the pairing energy fluctuations being of the order 50% of the shell energy fluctuations.

In fig. 14.5, we also plot the liquid-drop energy and the total energy, the latter being the sum of the liquid drop, the shell correction and the pairing energy (the energy of for example figs. 9.3 and 9.4 was calculated in this same way). The total energy of ^{164}Er has the deepest minimum for prolate shape with $\varepsilon \approx 0.26$. The depth of this minimum gives the 'shell correction' to the nuclear mass. This is the quantity that was plotted in the middle of fig. 9.7. In the case of ^{164}Er, this correction is approximately zero as seen in fig. 14.5 (one finds that, for ^{164}Er, the depth of the minimum is very little affected by e.g. ε_4-deformations).

The case of ^{164}Er is in one way simple because the proton as well as the neutron Fermi surface is in the middle of a shell. Thus, their respective shell energies both have maxima for spherical shape and tend to deform the nucleus. The situation for the other nucleus of fig. 14.5, ^{140}Ce is somewhat more complex. The neutron number, $N = 82$, corresponds to a closed shell with a deep shell energy minimum for spherical shape. Thus, although the protons strive to make the nucleus deformed they have only a secondary effect and the total energy minimum is for spherical shape. The depth of this minimum suggests a binding energy of ^{140}Ce, which is about 6 MeV larger than the liquid-drop estimate (a comparison with experimental masses suggests that this number should rather be around 4 MeV). It is also interesting to observe that, for spherical shape and 82 neutrons, the pairing constant G is smaller than G_{crit}, corresponding to the fact that no BCS solution with $\Delta_n \neq 0$ exists. With $\Delta_n = 0$, the Fermi surface is sharp with the orbitals below the 82 gap being fully occupied ($V_\nu^2 = 1$) and those above empty ($V_\nu^2 = 0$). Furthermore, still considering only neutrons, we obtain $\delta E_{\mathrm{pair}} = -\langle E_{\mathrm{pair}} \rangle$ where, in the present approximation $\langle E_{\mathrm{pair}} \rangle = -1.15$ MeV.

We have thus demonstrated that a consistent and still relatively simple theory for the pairing correlations can be achieved through the BCS formalism. This makes it possible to get a quantitative description of phenomena such as the odd–even mass difference referred to in preceding chapters. In appendices we will demonstrate how the reduction of the nuclear moment of inertia relative to the rigid body estimate can be accounted for by the pairing theory. We will also show how a more compact description of the pairing theory is achieved through the introduction of quasiparticle operators.

Exercises

14.1 Assume a δ-interaction acting between two particles in a pure 'ℓ-shell'. Show that the matrix elements are proportional to $(C_{000}^{\ell \ell \ell})^2$.

14.2 The interaction potential between two particles in a $j = 9/2$ shell is of δ-type. Use the formula given in the text to evaluate numerically the energy of the $I = 2, 4, 6$ and 8 states in units of the $I = 0$ energy.

14.3 Consider a $j = 5/2$ shell, which has a degeneracy of $\Omega = 3$. Construct the matrix of the pairing Hamiltonian H^G if there are four paired particles in the shell. Evaluate the eigenenergies. Compare with the general formulae.

14.4 Evaluate the commutators $[A, A^+]$ and $[H^G, A^+]$.

14.5 The relation $[H^G, (A^+)^p] = -Gp(\Omega - \mathcal{N} + p + 1)(A^+)^p$ was shown for $p = 1$ in problem 14.4. Probe by induction, $p \to p + 1$, that it is true for any p.

14.6 Prove that $a_\mu^+ a_{\bar{\mu}}^+ |0\rangle$ is an eigenstate of $H^{\mathrm{sp}} = \sum_\nu e_\nu \left(a_\nu^+ a_\nu + a_{\bar{\nu}}^+ a_{\bar{\nu}} \right)$ with the eigenvalue $2e_\mu$.

14.7 Show that the BCS wave function is normalised provided $U_\mu^2 + V_\mu^2 = 1$, $\mu = 1, 2, \ldots$.

14.8 With Ψ_0 being the BCS wave function, show that $\langle \Psi_0 | a_\mu^+ a_{\bar{\mu}}^+ a_{\bar{\kappa}} a_\kappa | \Psi_0 \rangle = U_\mu V_\mu U_\kappa V_\kappa$ and $\left\langle \Psi_0 | a_\mu^+ a_{\bar{\mu}}^+ a_{\bar{\mu}} a_\mu | \Psi_0 \right\rangle = V_\mu^2$.

14.9 Let Ψ_2 be a two-quasiparticle state (a 'decorrelated pair' or a 'broken pair' state) and Ψ_0 the BCS ground state. Show that after correction for the particle number difference, $\delta N = N_2 - N_0$, we obtain the excitation energy $E_2^{\mathrm{exc}} = \langle \Psi_2 | H - \lambda \mathcal{N} | \Psi_2 \rangle - \langle \Psi_0 | H - \lambda \mathcal{N} | \Psi_0 \rangle$.

14.10 Prove that the excitation energy of a 'decorrelated pair' in the level μ is $2E_\mu$.

14.11 Prove that for a broken-pair state with one particle in the orbital μ and another in the orbital κ, the excitation energy is $E_\mu + E_\kappa$.

14.12 Calculate the excitation energy for a one-quasiparticle state with the odd particle in the orbital μ.

14.13 Prove that the correlation energy in the continuous model is approximately equal to $\frac{1}{2}\rho\Delta^2$. The level density is given by ρ and the pairing gap by Δ.

14.14 Prove that the quasiparticle operators α^+ and α satisfy the Fermion anticommutator relations if $U_\nu^2 + V_\nu^2 = 1$.

14.15 Derive the relations that are inverse to the definitions of quasiparticle operators.

14.16 Transform the Hamiltonian $H'(\Delta)$ to quasiparticle operators and find the expressions for H_{00}, H_{11} and H_{20}.

14.17 Prove that the zero-quasiparticle state, $|\Delta\rangle \propto \prod_\nu \alpha_\nu \alpha_{\bar{\nu}} |0\rangle$, is identical to the BCS wave function, $|\tilde{0}\rangle = \prod_\nu (U_\nu + V_\nu a_\nu^+ a_{\bar{\nu}}^+) |0\rangle$.

14.18 Consider the continuous model together with the linearised pairing Hamiltonian. Assume that the occupation probabilities are given by

$$V^2 = \frac{1}{2}\left(1 - \frac{e}{(e^2 + \Delta^2)^{1/2}}\right)$$

leading to a pairing energy that can be considered as a function of Δ. Derive the expression for this energy and plot it for the parameters that are approximately relevant for rare-earth nuclei: $\rho^{-1} = 0.35$ MeV, $S = 15$ MeV and $G = 0.1$ MeV. How does the function change if G is decreased to 0.06 MeV?

14.19 Show that the one-quasiparticle state, $\alpha_\mu^+ |\Delta\rangle$, is equivalent to the 'one-particle state', $a_\mu^+ \prod_{\nu \neq \mu} (U_\nu + V_\nu a_\nu^+ a_{\bar{\nu}}^+) |0\rangle$.

Appendix 14A
The quasiparticle formalism

In this appendix we shall derive the BCS wave function in an alternative way by the use of a technique based on the concept of quasiparticles. In order to make the calculation somewhat simpler the pairing interaction is linearised to a pair field. It turns out that a complete diagonalisation of the linearised Hamiltonian is equivalent to the BCS solution.

The pairing Hamiltonian can be written as

$$H^G = -GF^+F$$

where

$$F^+ = \sum_{v>0} a_v^+ a_{\bar{v}}^+$$

The summation is, as usual, only over the states v and not over the time-reversed states $\bar{v}$ ($v > 0$). H^G can now be rewritten, using identities, as follows

$$H^G = -GF^+F \equiv -G\left(F^+ - \langle F^+\rangle + \langle F^+\rangle\right)\left(F - \langle F\rangle + \langle F\rangle\right)$$
$$\equiv -G\langle F^+\rangle\left(F^+ + F\right) - G\left(F^+ - \langle F^+\rangle\right)\left(F - \langle F^+\rangle\right) + G\langle F^+\rangle^2$$

In this expression we have used that $\langle F\rangle = \langle F^+\rangle$. This is not self-evident, but it is easily verified for the BCS wave function. To proceed one assumes that the second term is small, i.e. the operators F^+ and F are approximately equal to $\langle F\rangle$. Thus, if we for the moment neglect the last constant term, we can write the linearised Hamiltonian as

$$H^\Delta = -\Delta\left(F^+ + F\right)$$

where

$$\Delta = G\langle F^+\rangle$$

The Hamiltonian H^Δ is said to represent a pair field, which simulates the pairing interaction, and Δ is the strength of the field. This pair field, or pair potential, is an approximation of the 'true' pairing Hamiltonian, H^G, in the same way as the ordinary nuclear potential in r-space is an approximation to the two-body interaction between the nucleons in the nucleus. Its main characteristic is that it describes the emission or absorption of a pair of particles. The shortcoming of the pair field is that it does not commute with the number-of-particles operator. Physically this means that the associated wave function will contain components with different number of particles.

The task is now to diagonalise the total Hamiltonian

$$H(\Delta) = \sum_v e_v \left(a_v^+ a_v + a_{\bar{v}}^+ a_{\bar{v}}\right) - \Delta \sum_v \left(a_v^+ a_{\bar{v}}^+ + a_{\bar{v}} a_v\right)$$

The main advantage of this Hamiltonian compared with the Hamiltonian studied earlier is that it is bilinear in the operators a^+ and a. The Hamiltonian can thus be diagonalised by a unitary transformation of the operators a^+ and a. This transformation was first suggested by Bogoliubov (1958) and Valatin (1958) and reads

$$\begin{cases} \alpha_v^+ = U_v a_v^+ - V_v a_{\bar{v}} \\ \alpha_v = U_v a_v - V_v a_{\bar{v}}^+ \end{cases} \quad \begin{cases} \alpha_{\bar{v}}^+ = U_v a_{\bar{v}}^+ + V_v a_v \\ \alpha_{\bar{v}} = U_v a_{\bar{v}} + V_v a_v^+ \end{cases}$$

The operators α_v^+ and α_v are denoted quasiparticle creation and annihilation operators, respectively. The signs in the formulae are understood from the equality $a_{\bar{\bar{v}}}^+ = -a_v^+$. This in turn follows from general transformation properties and can be seen from the definition of the conjugate state in a $|jm\rangle$ basis

$$a_m^+ |0\rangle \equiv |jm\rangle \ ; \quad a_{\bar{m}}^+ |0\rangle \equiv (-1)^{j-m} |j-m\rangle$$

and from the fact that an arbitrary single-particle state can be expanded in such a basis. Provided that $U_v^2 + V_v^2 = 1$, the quasiparticle operators fulfil the following (fermion) anticommutators (problem 14.14)

$$\left\{\alpha_v, \alpha_\mu\right\} = \left\{\alpha_v^+, \alpha_\mu^+\right\} = 0$$

and

$$\left\{\alpha_v, \alpha_\mu^+\right\} = \delta_{\mu v}$$

The transformation from particle to quasiparticle operators is given by (problem 14.15)

$$\begin{cases} a_v^+ = U_v \alpha_v^+ + V_v \alpha_{\bar{v}} \\ a_v = U_v \alpha_v + V_v \alpha_{\bar{v}}^+ \end{cases} \quad \begin{cases} a_{\bar{v}}^+ = U_v \alpha_{\bar{v}}^+ - V_v \alpha_v \\ a_{\bar{v}} = U_v \alpha_{\bar{v}} - V_v \alpha_v^+ \end{cases}$$

As stated above the Hamiltonian does not preserve the number of particles as a good quantum number. In order to ensure that the number of particles is right on the average a Lagrangian multiplier, $-\lambda \mathcal{N}$, is added to the Hamiltonian. Thus we want to diagonalise

$$H'(\Delta) = H(\Delta) - \lambda \mathcal{N}$$

If we denote the solution with lowest energy by $|\Delta\rangle$, then the subsidiary conditions

$$G \langle \Delta | F | \Delta \rangle = \Delta$$

$$\langle \Delta | \mathcal{N} | \Delta \rangle = n$$

where n is the number of particles, must be satisfied. The Hamiltonian $H'(\Delta)$ is now transformed using the transformation derived in problem 14.16. The result is

$$H'(\Delta) = H_{00} + H_{11} + H_{20}$$

with

$$H_{00} = \sum_{v>0} 2 \left(e_v - \lambda\right) V_v^2 - 2\Delta \sum_{v>0} U_v V_v$$

$$H_{11} = \sum_{v>0} \left[\left(e_v - \lambda\right) \left(U_v^2 - V_v^2\right) + 2\Delta U_v V_v \right] \left(\alpha_v^+ \alpha_v + \alpha_{\bar{v}}^+ \alpha_{\bar{v}} \right)$$

$$H_{20} = \sum_{v>0} \left[2 U_v V_v \left(e_v - \lambda\right) - \Delta \left(U_v^2 - V_v^2\right) \right] \left(\alpha_v^+ \alpha_{\bar{v}}^+ + \alpha_{\bar{v}} \alpha_v \right)$$

The indices of H denote the partition between creation and annihilation operators. If the Hamiltonian containing the full pairing force, H^G, had been transformed, additional terms, H_{22}, H_{31} and H_{40}, would be present.

So far the numbers U_v and V_v have been arbitrary, apart from the condition $U_v^2 + V_v^2 = 1$. It is now possible to choose U_v and V_v so that $H_{20} = 0$. This condition implies that

$$2 U_v V_v \left(e_v - \lambda\right) - \Delta \left(U_v^2 - V_v^2\right) = 0$$

The solutions, of physical interest, to this equation are

$$U_v = \left[\frac{1}{2} \left(1 + \frac{e_v - \lambda}{\left[e_v - \lambda^2 + \Delta^2\right]^{1/2}} \right) \right]^{1/2}$$

$$V_v = \left[\frac{1}{2} \left(1 - \frac{e_v - \lambda}{\left[(e_v - \lambda)^2 + \Delta^2\right]^{1/2}} \right) \right]^{1/2}$$

When $H_{20} = 0$, *the Hamiltonian is diagonal in the quasiparticle representation.* Inserting the values derived for U_v and V_v the Hamiltonian reads

$$H'(\Delta) = \sum_{v>0} 2(e_v - \lambda) V_v^2 - 2\Delta \sum_{v>0} U_v V_v + \sum_{v>0} E_v \left(\alpha_v^+ \alpha_v + \alpha_{\bar{v}}^+ \alpha_{\bar{v}} \right)$$

where we have introduced the quasiparticle energy $E_v = [(e_v - \lambda)^2 + \Delta^2]^{1/2}$. The eigenstate with lowest energy is obviously the state with no quasiparticles since $E_v > 0$, if $\Delta > 0$. If this state is denoted $|\Delta\rangle$ it must fulfil

$$\alpha_v |\Delta\rangle = \alpha_{\bar{v}} |\Delta\rangle = 0$$

for all v, i.e. $|\Delta\rangle$ is the quasiparticle vacuum. This state can be constructed as

$$|\Delta\rangle \propto \prod_v \alpha_v \alpha_{\bar{v}} |0\rangle$$

where $|0\rangle$ is the particle vacuum defined by $a_v |0\rangle = 0$ for all v. If the quasiparticles are transformed to ordinary particles, the wave function is (problem 14.17)

$$|\Delta\rangle = \prod_v \left(U_v + V_v a_v^+ a_{\bar{v}}^+ \right) |0\rangle$$

As seen by direct inspection the wave function $|\Delta\rangle$ is identical to the BCS wave function $|\tilde{0}\rangle$ derived earlier.

The parameters Δ and λ introduced in the derivation must be determined. First we have the self-consistency condition for the strength of the pair field Δ. It must satisfy

$$\Delta = G \langle \Delta | F^+ | \Delta \rangle$$

Furthermore the requirement that the expectation value of the number of particles is the desired one gives

$$\langle \Delta | \mathcal{N} | \Delta \rangle = n$$

The calculation of the expectation values is straightforward and the result is obtained as

$$\Delta = G \sum_v U_v V_v$$

$$n = 2 \sum_v V_v^2$$

These equations are identical to the equations used in the earlier treatment. As both U_v and V_v are functions of Δ and λ, the equations are coupled. In a general case they have to be solved by an iteration procedure. However, roughly speaking, one may say that the first equation determines Δ and the second one λ.

The energy of the state $|\Delta\rangle$ is now easily calculated since $|\Delta\rangle$ is the vacuum of the quasiparticles. By looking at H_{00} we see that $E(\Delta) = 2\sum V_v^2 e_v - 2\Delta^2/G$. The first part is the single-particle energy while the second term is the correlation due to the pairing interaction. As we have linearised the pairing interaction to a pair field, the pairing energy is counted twice. This is a feature that occurs quite often in many-body problems. We recall that we omitted a constant term in the linearised pairing Hamiltonian. This term compensates the double counting of the pairing energy. The energy of the ground state is thus

$$E(\Delta) = 2\sum_v e_v V_v^2 - \Delta^2/G$$

As the Hamiltonian is given as

$$H'(\Delta) = \text{constant} + \sum_{v>0} E_v \left(\alpha_v^+ \alpha_v + \alpha_{\bar{v}}^+ \alpha_{\bar{v}}\right)$$

the excitation energy of a one-quasiparticle state, $\alpha_\mu^+ |\Delta\rangle$, is trivially obtained as E_μ. Similarly, for a two-quasiparticle state $\alpha_\mu^+ \alpha_{\mu'}^+ |\Delta\rangle$ we calculate an excitation energy of $E_\mu + E_{\mu'}$.

By the quasiparticle formalism we have succeeded in reproducing in an elegant fashion the results obtained more straightforwardly in the main text. A further advantage is that the Valatin–Bogoliubov method leads itself easily to generalisations in the case of a more general two-body interaction.

Appendix 14B
The moment of inertia

It was mentioned in chapter 11 (fig. 11.2) that, because of the pairing correlations, the observed nuclear moment of inertia is much smaller than the rigid body value. In this appendix, we will derive the corresponding formula. The derivations are made within the cranking model in perturbation theory with the rotational frequency, ω, as the small parameter (Inglis, 1954).

Consider the cranking Hamiltonian for rotation around the 1-axis (chapter 12):

$$H^\omega = H_0 - \hbar\omega I_1$$

The ground state energy is given by E_0 and the corresponding (many-particle) wave function is $|\Psi_0\rangle$:

$$H_0 |\Psi_0\rangle = E_0 |\Psi_0\rangle$$

For classical rotation, the rotational energy is given by

$$E - E_0 = \frac{1}{2}\mathscr{J}\omega^2 = \frac{1}{2\mathscr{J}}I^2$$

which formula thus defines the moment of inertia, $\mathscr{J}$.

In first order perturbation theory, the cranking wave function is given by

$$|\Psi^\omega\rangle = |\Psi_0\rangle - \hbar\omega \sum_{\Psi' \neq \Psi_0} |\Psi'\rangle \frac{\langle \Psi' |I_1| \Psi_0\rangle}{E_0 - E'}$$

where Ψ' is a part of a complete set of wave functions such that $\langle \Psi' |I_1| \Psi_0\rangle \neq 0$. Furthermore,

$$H_0 |\Psi'\rangle = E' |\Psi'\rangle$$

For the non-rotating ground state, it is obvious that $\langle \Psi_0 |I_1| \Psi_0\rangle = 0$. Thus,

322

to lowest order in ω

$$\langle \Psi^\omega \, |I_1| \, \Psi^\omega \rangle = 2\hbar\omega \sum_{\Psi' \neq \Psi_0} \frac{|\langle \Psi' \, |I_1| \, \Psi_0 \rangle|^2}{E' - E_0}$$

Using second order perturbation theory, the energy in the rotating system E^ω is obtained as

$$E^\omega = E_0 - (\hbar\omega)^2 \sum_{\Psi' \neq \Psi_0} \frac{|\langle \Psi' \, |I_1| \, \Psi_0 \rangle|^2}{E' - E_0}$$

The energy E in the laboratory system is calculated as the expectation value of $H_0 = H^\omega + \hbar\omega I_1$;

$$E = E_0 + (\hbar\omega)^2 \sum_{\Psi' \neq \Psi_0} \frac{|\langle \Psi' \, |I_1| \, \Psi_0 \rangle|^2}{E' - E_0}$$

leading to the following formula for the moment of inertia:

$$\mathscr{I} = 2\hbar^2 \sum_{\Psi' \neq \Psi_0} \frac{|\langle \Psi' \, |I_1| \, \Psi_0 \rangle|^2}{E' - E_0}$$

The same formula is obtained if the relation

$$I = \mathscr{I}\omega/\hbar$$

is compared with the expectation value of I_1 derived above.

For a pure single-particle configuration (with no pairing correlation), the ground state wave function is given by

$$\Psi_0 = \prod_v a_v^+ |0\rangle$$

where the product is over all occupied orbitals (v as well as $\bar{v}$). In the second-quantisation formalism, the angular momentum operator is written as

$$I_1 = \sum_{vv'} \langle v \, |j_1| \, v' \rangle \, a_v^+ a_v$$

The only states Ψ' that have $\langle \Psi' \, |I_1| \, \Psi_0 \rangle \neq 0$ are the one-particle-one-hole states

$$\Psi' = a_{\mu'}^+ a_\mu \Psi_0$$

This state corresponds to one particle being excited from the occupied orbital μ to the empty orbital μ' and its excitation energy is given as

$$E' - E_0 = e_{\mu'} - e_\mu$$

where $e_{\mu'}$ and e_μ are the single-particle energies. It is now easy to calculate

$$\langle \Psi' |I_1| \Psi_0 \rangle = \sum_{vv'} \langle v |j_1| v' \rangle \langle \Psi_0 a_\mu^+ a_{\mu'} |a_{v'}^+ a_v| \Psi_0 \rangle = \langle \mu |j_1| \mu' \rangle$$

and thus for the moment of inertia

$$\mathscr{I} = 2\hbar^2 \sum_\mu^{\text{(occ)}} \sum_{\mu'}^{\text{(empty)}} \frac{|\langle \mu |j_1| \mu' \rangle|^2}{e_{\mu'} - e_\mu}$$

where thus the first sum runs over occupied orbitals only and the second over empty orbitals. This formula is valid for an arbitrary single-particle potential and can be applied for example to the rotating harmonic oscillator discussed in chapter 12. It is then possible to distinguish between the $\Delta N = 0$ terms (couplings of j_1 within an oscillator shell) and the $\Delta N = 2$ terms. If only the former are considered, we arrive, of course, at the $\omega = 0$ limit of the formula given in chapter 12.

We will now also incorporate pairing with the BCS function as the nuclear ground state (Belyaev, 1959; Migdal, 1959; Griffin and Rich, 1960; Nilsson and Prior, 1961). The derivation becomes somewhat more involved but gives a good insight into the calculation of general matrix elements within the quasiparticle formalism.

The first step is to express the I_1 operator in quasiparticle operators (cf. problem 14.15)

$$I_1 = \sum_{vv'} \langle \mu |j_1| v' \rangle \left(U_v \alpha_v^+ + V_v \alpha_{\bar v} \right) \left(U_{v'} \alpha_{v'} + V_{v'} \alpha_{\bar{v}'}^+ \right)$$

As the BCS ground state is the zero-quasiparticle state ($\alpha |\tilde{0}\rangle = 0$), it is evident that it is only the $\alpha^+ \alpha^+$ term ,

$$I_1 = \sum_{vv'} \langle v |j_1| v' \rangle U_v V_{v'} \left(\alpha_v^+ \alpha_{\bar{v}'}^+ + \ldots \right)$$

which contributes in the moment of inertia formula. It furthermore follows that the Ψ' which should be considered in the moment of inertia formula are the two-quasiparticle states. In the present context it is convenient to write them as

$$\Psi' = \alpha_\mu^+ \alpha_{\bar{\mu}}^+ |\tilde{0}\rangle$$

The matrix element to evaluate is

$$\langle \Psi' |I_1| \tilde{0} \rangle = \langle \tilde{0}| \alpha_{\bar{\mu}'} \alpha_\mu \sum_{vv'} \langle v |j_1| v' \rangle U_v V_{v'} \alpha_v^+ \alpha_{\bar{v}'}^+ |\tilde{0}\rangle$$

$$= \langle \mu |j_1| \mu' \rangle U_\mu V_{\mu'} - \langle \bar{\mu}' |j_1| \bar{\mu} \rangle (-1) U_{\mu'} V_\mu$$

where the (-1) in the last term depends on the sign change $|\bar{v}'\rangle = -|v'\rangle$. The two single-particle matrix elements in the formula are essentially the same. With the state $|\mu\rangle$ being given as an expansion over spherical states

$$|\mu\rangle = \sum c^{\mu}_{N\ell j\Omega} |N\ell j\Omega\rangle$$

the conjugate state is given as

$$|\bar{\mu}\rangle = \sum (-)^{j-\Omega} c^{\mu}_{N\ell j\Omega} |N\ell j - \Omega\rangle$$

The following relation is now derived

$$\langle \bar{\mu} |j_1| \bar{\mu}' \rangle = \sum\sum c^{\mu}_{N\ell j\Omega} c^{\mu'}_{N'\ell' j'\Omega'} (-1)^{j+j'-\Omega-\Omega'} \langle N'\ell'j' - \Omega' |j_1| N\ell j - \Omega\rangle$$
$$= -\langle \mu |j_1| \mu'\rangle$$

We have used the fact that the matrix element of $j_1 = (j_+ + j_-)/2$ is different from zero only if $j' = j$ and $\Omega' = \Omega \pm 1$. Furthermore, the matrix element remains the same if the sign of both Ω and Ω' is changed. Thus, as j is half-integer, the equality follows. As j_1 is a Hermitian operator with real matrix elements it also follows that

$$\langle \mu |j_1| \mu'\rangle = \langle \mu' |j_1| \mu\rangle$$

and thus for the matrix element of I_1:

$$\left\langle \Psi' |I_1| \tilde{0} \right\rangle = \langle \mu |j_1| \mu'\rangle (U_\mu V_{\mu'} - V_\mu U_{\mu'})$$

We finally insert in the moment of inertia formula to obtain

$$\mathscr{I} = 2\hbar^2 \sum_{\mu>0,\mu'} \frac{|\langle \mu |j_1| \mu'\rangle|^2}{E_\mu + E_{\mu'}} (U_\mu V_{\mu'} - V_\mu U_{\mu'})^2$$

With a full summation over both μ and μ', we had counted each state of Ψ' twice. This could have been avoided by dividing by two but, for computational reasons, we prefer instead to exclude the conjugate states in the sum over μ ($\mu > 0$).

Let us consider the case where μ is far above the Fermi surface ($V_\mu \approx 0$, $U_\mu \approx 1, E_\mu = [(e_\mu - \lambda)^2 + \Delta^2]^{1/2} \approx (e_\mu - \lambda)$) while μ' is far below ($V_{\mu'} \approx 1$, $U_{\mu'} \approx 0, E_{\mu'} \approx (\lambda - e_{\mu'})$). This leads to a denominator of $(e_\mu - e_{\mu'})$ and a UV-factor of one in agreement with the non-paired case. Similarly, one concludes that when both μ and μ' are far above or far below the Fermi surface, the UV-factor is essentially zero, and the same contribution as in the non-paired case is again obtained. It is, however, the orbitals close to the Fermi surface that give the largest contributions to the moment of inertia. For such orbitals, being partly occupied and partly empty, the

326 The moment of inertia

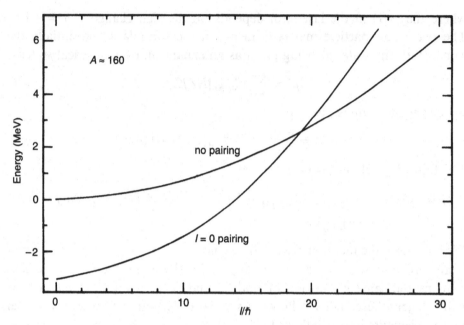

Fig. 14B.1. Schematic yrast lines for a rotational $A \approx 160$ nucleus. The 'no pairing' curve is drawn using the rigid body moment of inertia while the empirical ground state pairing energy is subtracted and the '$I = 0$ moment of inertia' is used for the '$I = 0$ pairing' curve. The figure suggests that, for high enough spin values, there will be no pairing energy in the yrast states.

UV-factor is essentially smaller than one. Furthermore, for such orbitals, the energy denominator becomes much larger than in the non-pairing case, $E_\mu + E_{\mu'} \approx 2\Delta >> e_\mu - e_{\mu'}$. This is understood from the large energy required to break up the strongly correlated BCS ground state. It is now easy to understand the general appearance of fig. 11.2 and the reduction of the moment of inertia compared with the rigid body value.

It was mentioned in chapter 12 that the pairing correlations are expected to disappear at high spins. This is referred to as Coriolis antipairing (CAP) or the Mottelson–Valatin (1960) effect and is easy to understand qualitatively. Let us take the nucleus ^{164}Er as an example. From fig. 14.5, we find that its total pairing correlation energy is around 3 MeV. Furthermore, the observed moment of inertia is $(2/\hbar^2)\mathscr{I} \approx 70$ MeV^{-1} (fig. 11.2). Assuming this moment of inertia for all spin, we get the energy versus spin as shown in fig. 14B.1. Similarly, we can draw the energy versus spin curve in the absence of pairing correlations in which case we expect to observe the rigid body moment of inertia, $(2/\hbar^2)\mathscr{I} \approx 150$ MeV^{-1}. The two curves intersect for $I \approx 20$, which is thus a very crude estimate of the spin at which the pairing correlations should disappear. A more careful analysis will show that the disappearance is gradual and is accompanied by spin alignment giving rise to two-quasiparticle states, four-quasiparticle states, etc. (fig. 11.10). The process by which the pairing correlations become less important with increasing spin I is complicated and still not understood in its detail.

Solutions to exercises

2.2   Definition ($d\tau = d^3r$):

$$\langle r^n \rangle = \frac{\int r^n \rho(r) \, d\tau}{\int \rho(r) \, d\tau}$$

Fermi distribution:

$$\rho(r) = \frac{\rho(0)}{1 + \exp \, [(r - R)/a]}$$

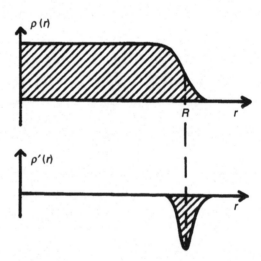

We first give a sketch of the derivation of the formula referred to in the text. For an arbitrary $f(r)$ ($f(r) < \infty$, $r = 0$) one has

$$\int_0^\infty f(r)\rho(r) \, dr = [F(r)\rho(r)]_0^\infty - \int_0^\infty F(r)\rho'(r) \, dr$$

where

$$F(r) = \int_0^r f(r') \, dr', \quad \text{i.e. } [F(r)\rho(r)]_0^\infty = F(\infty)\rho(\infty) - F(0)\rho(0) = 0$$

As $\rho'(r) \neq 0$ only in a small interval at the nuclear surface, $F(r)$ is preferably expanded around $r = R$. One further assumes $R \gg a$ and after some calculation arrives at

$$\int_0^\infty f(r)\rho(r) \, dr = \rho(0) \left[\int_0^R f(r) \, dr + \frac{\pi^2}{6} a^2 \left[\frac{df}{dr} \right]_{r=R} + \cdots \right]$$

By use of this formula it is easy to calculate

$$\int r^n \rho(r) \, d\tau = 4\pi\rho(0) \frac{R^{n+3}}{n+3} \left[1 + \frac{\pi^2}{6}(n+2)(n+3) \left(\frac{a}{R} \right)^2 + \cdots \right]$$

$$\int \rho(r) \, d\tau = 4\pi\rho(0) \frac{R^3}{3} \left[1 + \pi^2 \left(\frac{a}{R} \right)^2 + \cdots \right]$$

$$\implies \langle r^n \rangle = R^n \frac{3}{n+3} \left[1 + \frac{\pi^2}{6} \left(\frac{a}{R} \right)^2 n(n+5) + \cdots \right]$$

For the special case of $n = 2$:

$$\langle r^2 \rangle = \frac{3}{5} R^2 \left[1 + \frac{7\pi^2}{3} \left(\frac{a}{R} \right)^2 + \cdots \right]$$

2.3   The Coulomb energy is calculated as the energy required to assemble the nucleus from charges infinitely far away. The spherical symmetry is exploited and the nucleus is built up by successive addition of concentric shells.

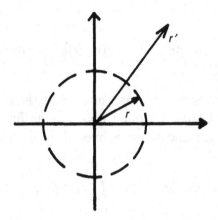

Consider first a spherical symmetric distribution of total charge Q located inside the radius r. It is then well known that the potential outside r is the same as for a point charge Q at $r = 0$.

$$V(r') = \frac{1}{4\pi\varepsilon_0} \frac{Q}{r'} ; \quad r' \geq r$$

$$r' = r : \qquad V(r) = \frac{1}{4\pi\varepsilon_0} \frac{Q}{r}$$

For the nucleus, the charge inside r is

$$Q = Q(r) = \int_0^r \rho(x) 4\pi x^2 \, dx$$

The total energy required to assemble the nucleus is

$$W = \int V \, dQ = \int_0^\infty V(r)\rho(r)4\pi r^2 \, dr$$

The Coulomb energy is

$$E_C = W = \frac{4\pi}{\varepsilon_0} \int_0^\infty r\rho(r) \, dr \int_0^r \rho(x)x^2 \, dx$$

The integral to be solved is

$$I = \int_0^\infty dr \, \rho(r)f(r) \quad \text{where} \quad \begin{cases} f(r) = r \int_0^r dx \, \rho(x)x^2 \\ \rho(r) = \rho_0 \{1 + \exp\left[(r - R)/a\right]\}^{-1} \end{cases}$$

With the formula from problem 2.2:

$$I = \rho(0) \left[\int_0^R dr \, r \int_0^r dx \, \rho(x)x^2 + \frac{\pi^2}{6} a^2 \left[\frac{d}{dr} \left(r \int_0^r x^2 \rho(x) \, dx \right) \right]_R + \ldots \right]$$

We perform a partial integration to eliminate integrals of the type $\int_0^r \ldots dx$ where r is a variable

$$I = \rho(0)\left[\left[\frac{r^2}{2}\int_0^r dx\, \rho(x)x^2\right]_0^R\right.$$

$$\left. - \int_0^R dr\, \frac{r^2}{2}\rho(r)r^2 + \frac{\pi^2}{6}a^2\left(\int_0^R x^2\rho(x)\, dx + RR^2\rho(R)\right) + \ldots\right]$$

All integrals are now of the type $\int_0^R f(r)\rho(r)\, dr$ and can be solved with a method analogous to that for $\int_0^\infty f(r)\rho(r)\, dr$ (problem 2.2):

$$\int_0^R f(r)\rho(r)\, dr = [F(r)\rho(r)]_0^R - \int_0^R F(r)\rho'(r)\, dr$$

$$= F(R)\rho(R) - F(0)\rho(0) + \sum_{m=0}^\infty \frac{1}{m!}\left(\frac{d^m F}{dr^m}\right)_R \rho(0)a^m \int_{-\infty}^0 \frac{x^m e^x\, dx}{(1+e^x)^2}$$

With the definition

$$I'_m = \int_{-\infty}^0 \frac{x^m e^x\, dx}{(1+e^x)^2} : \quad I'_0 = 1/2; \quad I'_2 = \frac{\pi^2}{6}$$

(it will later turn out that the integral I'_1 is not needed). With $F(0) = 0;\ \rho(R) = \rho(0)/2$:

$$\int_0^R f(r)\rho(r)\, dr = \rho(0)\left[F(R) + aI'_1 f(R) + \frac{\pi^2}{12}a^2\left(\frac{df}{dr}\right)_{r=R} + \ldots\right]$$

One now easily obtains

$$\int_0^R r^4\rho(r)\, dr = \rho(0)\left(\frac{R^5}{5} + aI'_1 R^4 + \frac{\pi^2}{12}a^2 4R^3 + \ldots\right)$$

$$\int_0^R r^2\rho(r)\, dr = \rho(0)\left(\frac{R^3}{3} + aI'_1 R^2 + \frac{\pi^2}{2}a^2 2R^3 + \ldots\right)$$

Up to second order in (a/R) the integral I is now given by

$$I = [\rho(0)]^2 \frac{R^5}{15}\left[1 + \frac{5}{6}\left(\frac{a}{R}\right)^2 \pi^2 + \ldots\right]$$

The value of $\rho(0)$ was already calculated in problem 2.2:

$$Ze = 4\pi\rho(0)\frac{R^3}{3}\left[1 + \left(\frac{a}{R}\right)^2 \pi^2 + \ldots\right]$$

With $E_C = (4\pi/\varepsilon_0)\, I$, we then obtain

$$E_C = \frac{4\pi}{\varepsilon_0} \frac{9(Ze)^2}{(4\pi)^2 R^6} \frac{R^5}{15} \frac{\left[1 + \frac{5}{6}\left(a/R\right)^2 \pi^2 + \ldots\right]}{\left[1 + \left(a/R\right)^2 \pi^2 + \ldots\right]^2}$$

The denominator is expanded and we get for the Coulomb energy

$$E_C = \frac{3}{5} \frac{(Ze)^2}{R} \frac{1}{4\pi\varepsilon_0}\left[1 - \frac{7}{6}\left(\frac{a}{R}\right)^2 \pi^2 + \ldots\right]$$

3.1 We study the reaction

$$\text{n} + {}^{235}\text{U} \rightarrow {}^{236}\text{U} \rightarrow {}^{91}\text{Kr} + {}^{142}\text{Ba} + 3\text{n} + \text{energy}$$

The energy is thus obtained from

$$^{235}\text{U} \rightarrow {}^{91}\text{Kr} + {}^{142}\text{Ba} + 2\text{n} + \text{energy}$$

The atomic masses are expressed in the units

$$u = m\left({}^{12}\text{C}\right)/12 = 1.661 \times 10^{-27}\text{ kg}$$

corresponding to 931.5 MeV. From the table in the text, the energy gain is now easily obtained as

$$(235.0439 - 90.9232 - 141.9165 - 2 \times 1.0087) \times 931.5\,\text{MeV} = 174\,\text{MeV}$$

This is roughly in agreement with the estimate in the text:

$$142 \times 0.7 + 91 \times 1.0 - 2 \times 7.6 \approx 175\text{ MeV}.$$

3.2 2000 MW is equivalent to

$$\frac{2000 \times 10^6}{1.60 \times 10^{-13}} \times \underbrace{60 \times 60 \times 24 \times 365}_{\text{s y}^{-1}} \text{ MeV y}^{-1} = 3.94 \times 10^{29}\text{ MeV y}^{-1}$$

(2000 MW $= 2000 \times 10^6$ J s^{-1}; 1 MeV $= 1.60 \times 10^{-13}$ J). With 174 MeV being released for each atom of ^{235}U it is now easy to calculate how many kg of ^{235}U are being 'burned' per year:

$$\frac{3.94 \times 10^{29}}{174} \times 235 \times 1.661 \times 10^{-27} = 884\text{ kg y}^{-1}$$

With the isotopic abundance of ^{235}U being 0.72%, 123×10^3 kg of natural uranium are needed per year.

3.3 $E = 8.7$ MeV; 832 kg ^{11}B y^{-1}.

3.4   Volume per nucleon: $\frac{4}{3}\pi R^3/A$. Diameter of nucleon: d.

$$\frac{4}{3}\pi \left(\frac{d}{2}\right)^3 \approx \frac{4}{3}\pi R^3/A \Rightarrow d \approx 2R/A^{1/3}$$

$$\frac{\text{volume of surface layer}}{\text{total volume}} \approx \frac{4\pi R^2 \cdot d}{(4\pi/3)\,R^3} \approx 6 \cdot A^{-1/3}$$

The nucleons in the surface layer have neighbours on five sides out of six or maybe three sides out of four.

$\Rightarrow$ Loss of energy due to surface:

$$\left(\frac{1}{4} \rightarrow \frac{1}{6}\right) \cdot 6 \cdot A^{-1/3} \cdot E_{\text{vol}} \approx A^{-1/3} \cdot E_{\text{vol}}$$

$$E_{\text{vol}} = a_{\text{vol}} \cdot A \Rightarrow E_{\text{surf}} \simeq -a_{\text{vol}} \cdot A^{2/3} \Rightarrow a_{\text{surf}} \simeq a_{\text{vol}}$$

3.5   We first briefly discuss the formulae in the text. Consider two para-bolae, $f(x)$ and $g(x) = f(x) + \Delta$. It is then easy to check that (x_0 and d arbitrary constants):

$$\Delta = \frac{1}{4}\left[g\left(x_0 + 2d\right) - 3f\left(x_0 + d\right) + 3g\left(x_0\right) - f\left(x_0 - d\right)\right]$$

Thus, for example for a constant even proton number Z, the formula in the text for Δ_n is exact if the masses for different N-values lie on two parabolae with that for odd N displaced Δ_n compared with that for even N (the difference formula suggested is of lowest order to give an exact result in the parabola approximation).

The binding energies are taken from mass tables (e.g. A.H. Wapstra and K. Bos, *At. Data and Nucl. Data Tables* **19**, 175 (1977)).

$$\left.\begin{array}{l} B\left(^{168}_{70}\text{Yb}\right) = 1362.798 \text{ MeV} \\[4pt] B\left(^{169}_{70}\text{Yb}\right) = 1369.665 \text{ MeV} \\[4pt] B\left(^{170}_{70}\text{Yb}\right) = 1378.134 \text{ MeV} \\[4pt] B\left(^{171}_{70}\text{Yb}\right) = 1384.748 \text{ MeV} \end{array}\right\} \Rightarrow \Delta_n = 0.864 \text{ MeV}$$

$$\left.\begin{array}{l} B\left(^{168}_{68}\text{Er}\right) = 1365.783 \text{ MeV} \\[4pt] B\left(^{169}_{69}\text{Tm}\right) = 1371.356 \text{ MeV} \\[4pt] B\left(^{170}_{70}\text{Yb}\right) = 1378.134 \text{ MeV} \\[4pt] B\left(^{171}_{71}\text{Lu}\right) = 1382.485 \text{ MeV} \end{array}\right\} \Rightarrow \Delta_p = 0.908 \text{ MeV}$$

$$B\left({}^{169}_{68}\text{Er}\right) = 1371.786 \text{ MeV}$$
$$B\left({}^{170}_{69}\text{Tm}\right) = 1377.948 \text{ MeV}$$
$$B\left({}^{171}_{70}\text{Yb}\right) = 1384.748 \text{ MeV}$$
$$B\left({}^{172}_{71}\text{Lu}\right) = 1389.464 \text{ MeV}$$

$$\Rightarrow \Delta_p - E_{np} = 0.681 \text{ MeV}$$

These values are thus the 'experimental' pairing parameters for ^{170}Yb and ^{171}Yb, respectively. With $\Delta_p(^{170}\text{Yb}) \simeq \Delta_p(^{171}\text{Yb})$: $E_{np} \simeq 0.2$ MeV. It is interesting to compare with the empirical relations:

$$\Delta = 12/\sqrt{A} \text{ MeV} \quad E_{np} = 20/A \text{ MeV}$$

For $A = 170$, these formulae give

$$\Delta = 0.92 \text{ MeV} \quad E_{np} = 0.12 \text{ MeV}$$

3.6 Alpha decay of ^{242}Pu: $^{242}_{94}\text{Pu} \rightarrow {}^{238}_{92}\text{U} + \alpha$. The total energy released is

$$Q_\alpha = \frac{m_\alpha v_\alpha^2}{2} + \frac{m_U v_U^2}{2}$$

The kinetic energy of the α-particle is measured:

$$E(\alpha_0) = \frac{m_\alpha v_\alpha^2}{2} = 4.903 \text{ MeV}$$

The kinetic energy of the ^{238}U nucleus is then obtained from momentum conservation:

$$m_\alpha v_\alpha = m_U v_U$$

$$\Rightarrow Q_\alpha = \left(1 + \frac{m_\alpha}{m_U}\right)\frac{m_\alpha v_\alpha^2}{2} = \left(1 + \frac{4}{238}\right) 4.903 \text{ MeV} = 4.985 \text{ MeV}$$

From mass tables:

$$Q_\alpha = m\left({}^{242}\text{Pu}\right) - m\left({}^{238}\text{U}\right) - m\left({}^{4}\text{He}\right)$$
$$= (242.058737 - 238.050785 - 4.002603)u$$
$$= 0.005349 \times 931.5 \text{ MeV} = 4.983 \text{ MeV}$$

3.7 (a) The *line of beta-stability* is defined by

$$\left(\frac{\partial m}{\partial N}\right)_{A=\text{constant}} = 0$$

We express the mass as a function of N and A:

$$m(N,A)c^2 = N(M_n - M_H)c^2 + AM_Hc^2 - a_vA + a_sA^{2/3}$$
$$+ a_C\frac{(A-N)^2}{A^{1/3}} + a_{sym}\frac{(2N-A)^2}{2A}$$

$$\frac{\partial m}{\partial N} = 0 \Rightarrow N - Z = \frac{a_CA^{2/3} - (M_n - M_H)c^2}{(2a_{sym}/A) + a_C/A^{1/3}}$$

With $A \to \infty$ this formula leads to $N \simeq A$ (one finds $Z \propto A^{1/3}$). The mass formula is fitted for nuclei with $A \lesssim 250$ and should in principle not be applied too far outside this region. However, with $A \to \infty$, we get the reasonable result that $N >> Z$. This is due to the increasing importance of the Coulomb energy.

(b) The *neutron drip line*. This is characterised by $(\partial B/\partial N)_{Z=constant} = 0$. We then express the binding energy as a function of N and Z and get the partial derivative as

$$\frac{\partial B(N,Z)}{\partial N} = a_v - \frac{2}{3}a_s(N+Z)^{-1/3} + \frac{a_C}{3}\frac{Z^2}{(N+Z)^{4/3}}$$
$$- \frac{a_{sym}(N-Z)}{N+Z} + \frac{a_{sym}}{2}\frac{(N-Z)^2}{(N+Z)^2} = 0$$

$$N + Z = A ; \ Z = A - N \Rightarrow$$

$$N^2\left(\frac{a_C}{3A^{4/3}} + \frac{2a_{sym}}{A^2}\right) - N\left(\frac{2a_CA}{3A^{4/3}} + \frac{4a_{sym}A}{A^2}\right)$$
$$+ a_v + \frac{3}{2}a_{sym} - \frac{2}{3}a_s\frac{1}{A^{1/3}} + \frac{a_C}{3}A^{2/3} = 0$$

This equation is of the form

$$N^2 - 2NA + \alpha A^2 = 0$$

$$N = A\left[(1 \overset{(+)}{\underset{-}{}} (1-\alpha)^{1/2}\right]$$

$$\alpha = \left(a_v + \frac{3}{2}a_{sym} - \frac{2}{3}a_sA^{-1/3} + \frac{a_C}{3}A^{2/3}\right) \bigg/ \left(\frac{a_C}{3}A^{2/3} + 2a_{sym}\right)$$

(c) The *proton drip line*

$$\frac{\partial B(N,Z)}{\partial Z} = 0$$

leads to

$$Z = A \left\{ (1 + \beta) \overset{(+)}{-} [(1 + \beta)^2 - \gamma]^{1/2} \right\}$$

where

$$\beta = \left(2a_C A^{2/3} \right) \Big/ \left(a_C A^{2/3} + 6a_{sym} \right)$$

$$\gamma = \left(a_v + \frac{3}{2}a_{sym} - \frac{2}{3}a_s A^{-1/3} \right) \Big/ \left(\frac{1}{3}a_C A^{2/3} + 2a_{sym} \right)$$

With the constants inserted we get the following table:

A	Beta-stability Z	Neutron drip line N	Proton drip line Z
100	43.4	67.8	49.4
200	80.3	138.1	88.1
300	113.3	209.4	121.4

4.1   In the semi-empirical mass formula, the binding energy is given by

$$B = a_v A \left(1 - \kappa_v I^2 \right) - a_s A^{2/3} \left(1 - \kappa_s I^2 \right) - a_C \frac{Z^2}{A^{1/3}}$$

where $I = (N - Z)/A$.

The binding energy is thus split up into a volume term, a surface term and a Coulomb term.

$$B(A, Z) = E_v(A) - \overset{0}{E}_s(A, Z) - \overset{0}{E}_C(A, Z)$$

In the general case of division into n equal fragments, we calculate the 'gain' in energy as

$$\Delta E = nB \left(\frac{A}{n}, \frac{Z}{n} \right) - B(A, Z)$$

$$= nE_v \left(\frac{A}{n} \right) - E_v(A) - n \overset{0}{E}_s \left(\frac{A}{n}, \frac{Z}{n} \right) - n \overset{0}{E}_C \left(\frac{A}{n}, \frac{Z}{n} \right)$$

$$+ \overset{0}{E}_s(A, Z) + \overset{0}{E}_C(A, Z)$$

$$= \overset{0}{E}_s(A, Z) + \overset{0}{E}_C(A, Z) - n \overset{0}{E}_s(A, Z) \frac{1}{n^{2/3}} - n \overset{0}{E}_C(A, Z) \frac{n^{1/3}}{n^2}$$

With $x = \overset{0}{E}_C(A, Z) / 2 \overset{0}{E}_s(A, Z)$, i.e. $\overset{0}{E}_C = 2x \cdot \overset{0}{E}_s$:

$$\Delta E = \overset{0}{E}_s \left(1 + 2x - n^{1/3} - 2xn^{-2/3} \right)$$

The division is energetically possible if $\Delta E > 0$

$$2x \left(1 - n^{-2/3}\right) - \left(n^{1/3} - 1\right) > 0$$

Special cases:

$$n = 2; \ x > 0.351 \ (A \gtrsim 90\text{--}95)$$
$$n = 3; \ c > 0.426 \ (A \gtrsim 110\text{--}120)$$

These A-values can also be graphically extracted from fig. 3.8.

4.2 (a)

$$R(\theta) = R_a \left(1 + a_2 P_2(\cos \theta)\right)$$

$$P_2(x) = \frac{1}{2} \left(3x^2 - 1\right)$$

$$R(\theta) = R_a \left(1 + \frac{3}{2} a_2 \cos^2 \theta - \frac{1}{2} a_2\right)$$

$$a_2 = 0.5 \Rightarrow R(\theta) = \frac{3}{4} R_a \left(1 + \cos^2 \theta\right)$$

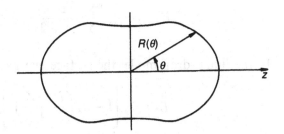

(b) The derivation is carried through in the text. We will fill in some details. The surface S is derived as

$$S = R_a^3 \int \frac{[1 + a_2 P_2(\cos \theta)]^2}{R_a} \left(1 + \frac{a_2^2 \, [(d/d\theta)P_2(\cos \theta)]^2}{[1 + a_2 P_2(\cos \theta)]^2}\right)^{1/2} d\Omega$$

To third order in a_2 (noting that $\int_{-1}^{1} P_2(\cos \theta) \, d(\cos \theta) = 0$):

$$S = 2\pi R_a^2 \int_{-1}^{1} \left\{1 + a_2^2 \left[P_2^2 + \frac{1}{2} \left(\frac{d}{d\theta} P_2\right)^2\right]\right\} d(\cos \theta)$$

With P_2 and $(d/d\theta)P_2$ inserted $(x = \cos\theta)$:

$$S = 2\pi R_a^2 \left[2 + a_2^2 \left(\frac{2}{5} + \frac{1}{2} \int_{-1}^{1} 9x^2 \left(1 - x^2\right) dx \right) \right]$$

$$= 4\pi R_a^2 \left(1 + \frac{4a_2^2}{5} + \cdots \right)$$

It now remains to determine the deformation dependence of R_a. With R_0 being the radius of a sphere with the same volume as the deformed nucleus, we get

$$\frac{4}{3}\pi R_0^3 = \int dV = \int d\Omega \int_0^{R(\theta)} r^2\, dr = \frac{1}{3} \int R^3(\theta)\, d\Omega$$

$$= \frac{R_a^3}{3} \cdot \int \left(1 + 3a_2 P_2 + 3a_2^2 P_2^2 + a_2^3 P_2^3 \right) d\Omega$$

To third order in a_2:

$$R_a = R_0 \left(1 - \frac{1}{5}a_2^2 - \frac{2}{105}a_2^3 + \cdots \right)$$

(d) The expression for the height of the fission barrier is calculated in the text:

$$E_{\text{barr}} = \overset{0}{E}_{\text{s}}\, \frac{98}{15} \cdot \frac{(1-x)^3}{(1+2x)^2} \; ; \; x < 1$$

In the liquid-drop model the surface energy $\overset{0}{E}_{\text{s}}$ is obtained as

$$\overset{0}{E}_{\text{s}} = a_{\text{s}} \left[1 - \kappa_{\text{s}} \left(\frac{N-Z}{A} \right)^2 \right] A^{2/3}$$

With $a_{\text{s}} = 17.9$ MeV and $\kappa_{\text{s}} = 1.78$ we find for ^{238}U:

$$\overset{0}{E}_{\text{s}} \left(^{238}\text{U} \right) = 624\ \text{MeV}$$

The fissility parameter is

$$x \left(^{238}\text{U} \right) = 0.769$$

and we thus find for the fission barrier:

$$E_{\text{barr}} \left(^{238}\text{U} \right) = 7.8\ \text{MeV}$$

(e) For $x < 1$ and $x > 1$, respectively, the fission barrier has the following appearance as a function of a_2:

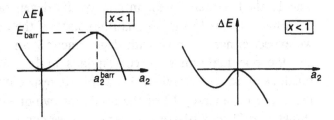

However, there are no stable oblate nuclei with $x > 1$, at least not in the liquid-drop model.

If triaxial shapes are also considered, it is found that the 'oblate minimum' is 'only' a saddle point and the nucleus can go to fission without passing any energy barrier.

5.1   (a) Spherical shape; $\omega_x = \omega_y = \omega_z = \omega_0$:

$$E = \hbar\omega_0 \left(n_x + n_y + n_z + 3/2\right)$$

We get the following table (because of the two possible spin directions, the degeneracy becomes a factor 2 larger than that obtained directly from the oscillator).

$E/\hbar\omega_0$	n_x	n_y	n_z	Degeneracy	Magic numbers
3/2	0	0	0	2	2
5/2	1	0	0		
	0	1	0	6	8
	0	0	1		
7/2	2	0	0		
	1	1	0		
	1	0	1	12	20
	0	2	0		
	0	1	1		
	0	0	2		
9/2	3	0	0		
	⋮	⋮	⋮	20	40
	⋮	⋮	⋮		
	⋮	⋮	⋮		

If the solution is instead carried through in spherical coordinates, we obtain ℓ-shells, $(n+1)\ell = 1s$, 2s, 1p, etc. Their degeneracy is $2(2\ell+1)$. With ℓ constant, the energy must increase with increasing n and with n constant, the energy must increase with increasing ℓ. The 1s shell is then lowest in energy. Its degeneracy is 2 and it corresponds thus to the no-quantum state of the oscillator. Next, we expect either 2s or 1p with degeneracies 2 and 6, respectively, i.e. the one-quantum state corresponds to 1p. With 1s and 1p filled, 2s or 1d with degeneracies 2 and 10, respectively, must come next. The degeneracy 12 of the oscillator two-quantum state then shows that 2s and 1d are degenerate. If we continue, we find that also 2p and 1f are degenerate etc. The general formula is derived in section 6:

$$E = (2n + \ell + 3/2)\,\hbar\omega_0$$

(cf. also fig. 6.1).

(b) $\omega_x = \omega_y = 2\omega_z \Rightarrow E = \hbar\omega_z (2n_x + 2n_y + n_z + 5/2)$

$E/\hbar\omega_z$	n_x	n_y	n_z	Degeneracy	Magic numbers
5/2	0	0	0	2	2
7/2	0	0	1	2	4
9/2	1	0	0		
	0	1	0	6	10
	0	0	2		
11/2	1	0	1		
	0	1	1	6	16
	0	0	3		
13/2	2	0	0		
	1	1	0		
	0	2	0	12	28
	1	0	2		
	0	1	2		
	0	0	4		
15/2	2	0	1		
$\vdots$	$\vdots$	$\vdots$	$\vdots$	12	40
	$\vdots$	$\vdots$	$\vdots$		

(cf. fig. 8.1).

6.2   The eigenfunctions of the harmonic-oscillator potential are given by

$$\psi_{n\ell m} = C\rho^\ell F\left(-n, \ell + 3/2; \rho^2\right) e^{-\rho^2/2} Y_{\ell m}(\theta, \phi)$$

The Kummer function F has the form

$$F(a, c; z) = 1 + \frac{a}{c}\frac{z}{1} + \frac{a(a+1)}{c(c+1)}\frac{z^2}{2!} + \cdots$$

The principal quantum number N is defined as

$$N = 2n + \ell$$

$\underline{N = 0, \ell = 0 \Rightarrow n = 0}$

$$F\left(-0, 3/2, \rho^2\right) = 1$$

$$\psi_{000} = C_0 e^{-\rho^2/2} Y_{00} = C_0 \exp\left(-\frac{M\omega_0}{2\hbar}r^2\right) Y_{00}$$

As the $Y_{\ell m}$-functions are already normalised ($Y_{00} = 1/(4\pi)^{1/2}$) we only need to normalise the radial integral:

$$1 = C_0^2 \int_0^\infty \exp\left(-\frac{M\omega_0}{\hbar}r^2\right) r^2 \, dr$$

We use the formula

$$\int_0^\infty x^n e^{-\alpha^2 x^2} \, dx = \frac{\Gamma((n+1)/2)}{2\alpha^{n+1}}$$

which can either be found in tables or derived by partial integration. The gamma function is given by

$$\Gamma(n+1) = n\Gamma(n) \; ; \; \Gamma(1/2) = \sqrt{\pi}$$

We thus find for the normalisation integral:

$$1 = C_0^2 \frac{\Gamma(3/2)}{2(M\omega_0/\hbar)^{3/2}} \Rightarrow C_0^2 = \frac{4(M\omega_0/\hbar)^{3/2}}{\sqrt{\pi}}$$

$$\Rightarrow \psi_{000} = 2\left(\frac{M\omega_0}{\hbar}\right)^{3/4}\left(\frac{1}{\pi}\right)^{1/4} \exp\left(-\frac{M\omega_0}{2\hbar}r^2\right) Y_{00}$$

$\underline{N = 3, \ell = 1 \Rightarrow n = 1}$

$$\psi_{11m} = C_1\left(\frac{M\omega_0}{\hbar}\right)^{1/2} r\left(1 - \frac{2}{5}\frac{M\omega_0}{\hbar}r^2\right)\exp\left(-\frac{M\omega_0}{2\hbar}r^2\right) Y_{1m}$$

$$\int |\psi_{11m}|^2 \, d\tau = 1 \Leftrightarrow$$

$$1 = C_1^2 \cdot \frac{M\omega_0}{\hbar} \int r^2\left(1 - \frac{2}{5}\frac{M\omega_0}{\hbar}r^2\right)^2 \exp\left(-\frac{M\omega_0}{\hbar}r^2\right) r^2 \, dr$$

$$\Rightarrow C_1 = \left(\frac{20}{3}\right)^{1/2} \left(\frac{M\omega}{\hbar}\right)^{3/4} \left(\frac{1}{\pi}\right)^{1/4}$$

The other examples are now easily carried through in a similar way.

6.4

$$j_\ell(\xi) = \xi^\ell \left(-\frac{1}{\xi}\frac{\mathrm{d}}{\mathrm{d}\xi}\right)^\ell \frac{\sin \xi}{\xi} \Longrightarrow$$

$$j_0(\xi) = \frac{\sin \xi}{\xi}$$

$$j_1(\xi) = -\frac{\mathrm{d}}{\mathrm{d}\xi}\left(\frac{\sin \xi}{\xi}\right) = \frac{\sin \xi}{\xi^2} - \frac{\cos \xi}{\xi}$$

$$j_2(\xi) = \xi^2 \left(\frac{1}{\xi}\frac{\mathrm{d}}{\mathrm{d}\xi}\right)\left(\frac{1}{\xi}\frac{\mathrm{d}}{\mathrm{d}\xi}\right)\frac{\sin \xi}{\xi} = \xi\frac{\mathrm{d}}{\mathrm{d}\xi}\left(-\frac{\sin \xi}{\xi^3} + \frac{\cos \xi}{\xi^2}\right)$$

$$= \frac{3\sin \xi}{\xi^3} - \frac{3\cos \xi}{\xi^2} - \frac{\sin \xi}{\xi}$$

$$j_3(\xi) = \ldots = \frac{15\sin \xi}{\xi^4} - \frac{15\cos \xi}{\xi^3} - \frac{6\sin \xi}{\xi^2} + \frac{\cos \xi}{\xi}$$

The radial functions of the infinite square well potential are given by

$$R_\ell = N_\ell j_\ell(kr)$$

With R being the radius of the well we have $j_\ell(kR) = 0$. For $\ell = 0$:

$$R_0 = N_0 \frac{\sin kr}{kr} \ ; \ kR = n\pi$$

Normalisation:

$$1 = N_0^2 \int_0^R \left(\frac{\sin kr}{kr}\right)^2 r^2 \, \mathrm{d}r = \frac{N_0^2}{k^2} \cdot \frac{R}{2}$$

$$\Rightarrow N_0 = k \cdot \left(\frac{2}{R}\right)^{1/2} = \frac{n\pi}{R}\left(\frac{2}{R}\right)^{1/2}$$

$$\Rightarrow R_0 = \left(\frac{2}{R}\right)^{1/2} \frac{\sin (n\pi r/R)}{r}$$

$$\ell = 1: \quad R_1 = N_1 \left(\frac{\sin kr}{(kr)^2} - \frac{\cos kr}{kr}\right)$$

Normalisation:

$$1 = N_1^2 \int_0^R \left(\frac{\sin^2 kr}{(kr)^4} - \frac{2\sin kr \cos kr}{(kr)^3} + \frac{\cos^2 kr}{(kr)^2}\right) r^2 \, \mathrm{d}r$$

$$\Leftrightarrow 1 = \frac{N_1^2}{k^3} \int_0^{kR} \left(\frac{\sin^2 x}{x^2} - \frac{2 \sin x \cos x}{x} + \cos^2 x \right) dx$$

$$\left. \begin{array}{l} \displaystyle \int_0^{kR} \frac{\sin^2 x}{x^2} dx = \left(-\frac{1}{x} \sin^2 x \right)_0^{kR} + \int_0^{kR} \frac{2 \sin x \cos x}{x} dx \\[2em] \displaystyle \int_0^{kR} \cos^2 x \, dx = \int_0^{kR} \frac{1 + \cos 2x}{2} dx = \frac{kR}{2} + \frac{\sin 2kR}{4} \end{array} \right\} \Rightarrow$$

$$1 = N_1^2 \frac{1}{k^3} \left(\frac{kR}{2} + \frac{\sin 2kR}{4} - \frac{1}{kR} \sin^2 kR \right)$$

If we also use $j_1(kR) = 0$ ($\sin kR = kR \cos kR$), the expression for N_1 can be somewhat simplified and we finally arrive at

$$R_1 = k^{3/2} \left(\frac{kR}{2} - \frac{1}{4} \sin 2kR \right)^{-1/2} \left(\frac{\sin kr}{(kr)^2} - \frac{\cos kr}{kr} \right)$$

Here, the possible values of k are obtained as the zeros of $j_1(kR)$, see table 6A.2 (and problem 6.5).

6.6 For a particle with mass μ, charge e and spin S, the magnetic moment **m** is given by

$$\mathbf{m} = \frac{e}{\mu} \mathbf{S}$$

According to the formulas from special relativity, the electric and magnetic field vectors, **E** and **B**, are transformed according to

$$\mathbf{E}'_\parallel = \mathbf{E}_\parallel \qquad \mathbf{E}'_\perp = \frac{1}{(1 - v^2/c^2)^{1/2}} (\mathbf{E}_\perp + \mathbf{v} \times \mathbf{B})$$

$$\mathbf{B}'_\parallel = \mathbf{B}_\parallel \qquad \mathbf{B}'_\perp = \frac{1}{(1 - v^2/c^2)^{1/2}} \left(\mathbf{B}_\perp - \frac{\mathbf{v} \times \mathbf{E}}{c^2} \right)$$

Thus, a particle with velocity **v** in the electrostatic field $\mathbf{E} = -(\partial\phi/\partial r)\mathbf{e}_r$ will also experience a magnetic field ($v = |\mathbf{v}| \ll c$):

$$\mathbf{B} = -\frac{1}{c^2} \mathbf{v} \times \mathbf{E} = -\varepsilon_0 \mu_0 \mathbf{v} \times \mathbf{E} = \varepsilon_0 \mu_0 \mathbf{v} \times \mathbf{e}_r \frac{\partial\phi}{\partial r}$$

With the orbital angular momentum $\mathbf{L} = \mathbf{r} \times \mathbf{p} = \mu \mathbf{r} \times \mathbf{v}$:

$$\mathbf{B} = -\varepsilon_0 \mu_0 \frac{1}{\mu r} \frac{\partial\phi}{\partial r} \mathbf{L}$$

For a magnetic moment **m** in a magnetic field **B**, the energy is given by

$$\xi_{\text{magn}} = -\mathbf{m} \cdot \mathbf{B}$$

With **m** and **B** inserted and $V = e\phi$:

$$\xi_{\text{magn}} = \frac{\varepsilon_0 \mu_0}{\mu^2} \frac{1}{r} \frac{\partial V}{\partial r} \mathbf{L} \cdot \mathbf{S}$$

Note that this is an incomplete derivation. A more correct treatment gives corrections of approximately a factor 2.

6.7   For this problem and also for some of the following problems we must calculate sums of the type

$$\sum_{N'=0}^{N} (N')^p$$

where p is an integer that will never be greater than 4. Such sums may be calculated for example by use of the Euler–Maclaurin summation formula:

$$\sum_{k=0}^{n} f(x_k) = \frac{1}{h} \int_{x_0}^{x_n} f(x)\,dx + \frac{1}{2}\left[f(x_0) + f(x_n)\right]$$
$$+ \frac{h}{12}\left[f^{(1)}(x_n) - f^{(1)}(x_0)\right]$$
$$- \frac{h^3}{720}\left[f^{(3)}(x_n) - f^{(3)}(x_0)\right]$$
$$+ \frac{h^5}{30240}\left[f^{(5)}(x_n) - f^{(5)}(x_0)\right] - \cdots$$

Here $x_k = x_0 + k \cdot h$ and the nth derivative of $f(x)$ is denoted by $f^{(n)}(x)$. We now get

$$\sum_{N'=0}^{N} N' = \int_0^N x\,dx + \frac{1}{2}(0+N) + \frac{1}{12}(1-1) = \frac{N}{2}(N+1)$$
$$\sum_{N'=0}^{N} N'^2 = \cdots = \frac{N}{6}(2N+1)(N+1)$$
$$\sum_{N'=0}^{N} N'^3 = \cdots = \frac{1}{4}N^2(N+1)^2$$
$$\sum_{N'=0}^{N} N'^4 = \cdots = \frac{N}{30}(N+1)(2N+1)\left(3N^2 + 3N - 1\right)$$

To solve the problem we first show that the number of degenerate levels within one N-shell is $(N+1)(N+2)$.

We have $N = 2n + \ell$ or $\ell = N - 2n$.

For ℓ fixed, there are $2\ell + 1$ different orbitals (with $m_\ell = -\ell$, $-\ell + 1, \ldots, 0, \ldots, \ell - 1, \ell$). Furthermore, the spin can take on two values and assuming that N is even we thus find the degeneracy:

$$2\sum_\ell (2\ell + 1) = 2\sum_{n=0}^{N/2} [2(N - 2n) + 1]$$

$$= 2(2N + 1)\left(\frac{N}{2} + 1\right) - 8\sum_0^{N/2} n$$

$$= \ldots = (N + 1)(N + 2)$$

For N odd we get

$$2\sum_{n=0}^{N/2 - \frac{1}{2}} [2(N - 2n) + 1] = \ldots = (N + 2)(N + 1)$$

The expectation value of ℓ^2 is $\ell(\ell + 1)$. The degeneracy $2(2\ell + 1)$ then leads to

$$\sum_{\text{one shell}} \left\langle \ell^2 \right\rangle = 2\sum_\ell (2\ell + 1)\ell(\ell + 1)$$

First assume that N is even. The sum may be performed over $n = (N - \ell)/2$ as above but we prefer to perform it over $\ell/2 \, (= v)$ which also takes the values $0, 1, \ldots N/2$:

$$\sum_{\text{one shell}} \left\langle \ell^2 \right\rangle = 4\sum_{v=0}^{N/2} (4v + 1)v(2v + 1)$$

$$= 4\sum \left(8v^3 + 6v^2 + v\right) = \frac{N}{2}(N + 1)(N + 2)(N + 3)$$

With the degeneracy $(N + 1)(N + 2)$, we obtain

$$\left\langle \ell^2 \right\rangle_N = N(N + 3)/2$$

For odd N, ℓ takes the values $1, 3, \ldots N$. If the summation is carried out over $\mu = (\ell + 1)/2$, $\mu = 1, 2, \ldots, (N + 1)/2$, the calculations become easy. The same result as for even N is obtained.

6.8   The virial theorem: $2\langle T \rangle = \langle \mathbf{r} \cdot \nabla V \rangle$

$$V = \frac{1}{2}M\omega_0^2 r^2 \Rightarrow \mathbf{r} \cdot \nabla V = M\omega_0^2 r^2 = 2V$$

and thus for a harmonic oscillator:

$$\langle T \rangle = \langle V \rangle$$

The eigenvalue equation

$$\langle i|H|i \rangle = \langle H \rangle = \langle T + V \rangle = \hbar\omega_0 \left(N_i + \frac{3}{2} \right)$$

now leads to

$$\left\langle i \left| r^2 \right| i \right\rangle = \langle r_i^2 \rangle = \frac{2}{M\omega_0^2}\langle V \rangle = \frac{\hbar}{M\omega_0}\left(N_i + \frac{3}{2} \right)$$

6.9   With the shells from $N' = 0$ up to $N' = N$ being filled and with a shell degeneracy of $(N' + 1)(N' + 2)$, see problem 6.7, we find for the number of particles (equal number of protons and neutrons):

$$A = \sum_{N'=0}^{N} 2(N' + 1)(N' + 2) = \sum_{0}^{N} 2\left(N'^2 + 3N' + 2 \right)$$

$$= \frac{N}{3}(2N + 1)(N + 1) + 6\frac{N}{2}(N + 1) + 4(N + 1)$$

$$= \ldots = \frac{2}{3}(N + 1)(N + 2)(N + 3)$$

The value of $\langle r^2 \rangle$ for a specific orbital was calculated in problem 6.8,

$$\left\langle r^2 \right\rangle_{N'\ell} = \frac{\hbar}{M\omega_0}(N' + 3/2)$$

$$\Rightarrow \sum \langle r^2 \rangle = \frac{\hbar}{M\omega_0} \sum_{N'=0}^{N} 2\,(N' + 3/2)\,(N' + 1)\,(N' + 2) = \ldots$$

$$= \frac{\hbar}{M\omega_0}\frac{1}{2}(N + 1)(N + 2)^2(N + 3) = A\left\langle r^2 \right\rangle$$

One finds that the approximate expressions are constructed in such a way that not only the coefficient of highest but also that of next highest order in N are the same:

$$A = \frac{2}{3}\left(N^3 + 6N^2 + \ldots \right)$$

$$A\left\langle r^2 \right\rangle = \frac{1}{2}\frac{\hbar}{M\omega_0}\left(N^4 + 8N^3 + \ldots \right)$$

We thus expect the approximate expressions to be quite accurate also for not very high N-values.

6.10   We first perform the sums exactly and thus assume that the shells with principal quantum number $\leq N$ are full and the shell with principal quantum number $N + 1$ is half full.

Degeneracy in the '$(N + 1)$' shell: $(N + 2)(N + 3)$.

Radius in the '$(N + 1)$' shell: $\langle r^2 \rangle = (N + 5/2)\hbar/M\omega_0$.

Assuming that the number of neutrons and protons is equal and using the results from problem 6.9 we get

$$A = \frac{2}{3}(N+1)(N+2)(N+3) + \frac{1}{2}(N+2)(N+3) \cdot 2 = \ldots$$

$$= \frac{2}{3}(N+2)(N+3)\left(N+\frac{5}{2}\right) \approx \frac{2}{3}\left[\left(N+\frac{1}{2}\right)+2\right]^3$$

$$\sum \langle r^2 \rangle = \frac{\hbar}{M\omega_0}\frac{1}{2}(N+1)(N+2)^2(N+3)$$

$$+ \frac{\hbar}{M\omega_0}\frac{1}{2}(N+2)(N+3)\left(N+\frac{5}{2}\right) \cdot 2$$

$$= \ldots = \frac{\hbar}{M\omega_0}\frac{1}{2}(N+2)(N+3)\left(N^2+5N+7\right)$$

$$\approx \frac{\hbar}{M\omega_0}\cdot\frac{1}{2}\left[\left(N+\frac{1}{2}\right)+2\right]^4$$

We conclude that, if the approximate expressions are generalised in a natural way, replacing N by $N + 1/2$, they are about as accurate when the last shell is half full as when it is completely filled.

6.11   We first let $r \leq R$. Integrate Maxwell's equation,

$$\operatorname{div} \mathbf{D} = \rho$$

over a sphere with radius r:

$$\int_V \nabla \cdot \mathbf{D}\, d\tau = \int_V \rho \, d\tau \Rightarrow \oint_S \mathbf{D} \cdot d\mathbf{S} = Q_r$$

where Q_r is the charge within the sphere, i.e. $Q_r = (r/R)^3 Ze$. Furthermore, due to the spherical symmetry, $\mathbf{D} = D(r)\mathbf{e}_r$.

$$\mathbf{D} = \frac{Q_r}{4\pi r^2}\mathbf{e}_r = \frac{Zer}{4\pi R^3}\mathbf{e}_r$$

$$\Rightarrow \mathbf{E} = \frac{1}{4\pi\varepsilon_0}Ze\frac{r}{R^3}\mathbf{e}_r$$

$$\mathbf{E} = -\nabla V = -\frac{\partial V}{\partial r}\mathbf{e}_r \Rightarrow V = -\frac{1}{4\pi\varepsilon_0}Ze\frac{1}{2}\frac{r^2}{R^3} + C$$

For $r \geq R$ we have

$$V = \frac{1}{4\pi\varepsilon_0}\frac{Ze}{r}$$

The potential must be continuous at $r = R$ and this leads to

$$C = \frac{1}{4\pi\varepsilon_0}\frac{3}{2}\frac{Ze}{R}$$

$$V(r) = \frac{1}{4\pi\varepsilon_0}\frac{Ze}{R}\left[-\frac{1}{2}\left(\frac{r}{R}\right)^2 + \frac{3}{2}\right] \; ; \; (r \leq R)$$

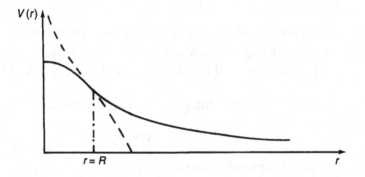

6.12   Assume that for the protons the shells with principal quantum number $N \leq N_Z$ and for the neutrons those with $N \leq N_N$, are filled (if the last shell is only partly filled this will correspond to N_N or N_Z non-integer, see problem 6.10).

The number of neutrons is:

$$N = \frac{1}{3}(N_N + 2)^3$$

The sum of $\langle r^2 \rangle$ for the neutrons is

$$\sum \langle r^2 \rangle = \frac{1}{4} \frac{\hbar}{M\omega_0^N} (N_N + 2)^4$$

The neutron radius is

$$\langle r^2 \rangle_N = \frac{1}{N} \sum \langle r^2 \rangle = \frac{\hbar}{M\omega_0^N} \frac{3}{4} (N_N + 2) = \frac{\hbar}{M\omega_0^N} \frac{3}{4} (3N)^{1/3}$$

With the number of protons being Z we find in an analogous way

$$\langle r^2 \rangle_Z = \frac{\hbar}{M\omega_0^Z} \frac{3}{4} (3Z)^{1/3}$$

We now define

$$x = \frac{N - Z}{A}$$

i.e. $N = (A/2)(1 + x)$ and $Z = (A/2)(1 - x)$

$$\langle r^2 \rangle_Z = \langle r^2 \rangle_N \Rightarrow \frac{\omega_0^N}{\omega_0^Z} = \frac{(1 + x)^{1/3}}{(1 - x)^{1/3}}$$

It is now natural to make the following ansatz:

$$\omega_0^N = \omega_0 (1 + x)^{1/3}$$

$$\omega_0^Z = \omega_0 (1 - x)^{1/3}$$

where ω_0^Z and ω_0^N are functions of N and Z or equivalently of x and A. This means that also ω_0 might depend on these variables, $\omega_0(x, A)$. Either ω_0^N or ω_0^Z is now inserted in the expression for the corresponding radius:

$$\langle r^2 \rangle_N = \langle r^2 \rangle = \frac{\hbar(3/4)}{M\omega_0(x, A)(1 + x)^{1/3}} \left(\frac{3A}{2} (1 + x) \right)^{1/3}$$

As $\langle r^2 \rangle$ only depends on A, this must also be the case for ω_0, i.e. $\omega_0 = \omega_0(A)$:

$$\omega_0^N = \omega_0(A)(1 + x)^{1/3} \approx \omega_0(A) \left(1 + \frac{1}{3} x \right)$$

$$\omega_0^Z = \omega_0(A)(1 - x)^{1/3} \approx \omega_0(A) \left(1 - \frac{1}{3} x \right)$$

6.13   For j_z it is trivial to show that

$$j_z (Y_{\ell\ell}\alpha) = (\ell_z + s_z) Y_{\ell\ell}\alpha = (\ell + 1/2) Y_{\ell\ell}\alpha$$

Thus, $Y_{\ell\ell}\alpha$ is an eigenfunction of j_z with the eigenvalue $\ell + 1/2$. To apply j^2 to $Y_{\ell\ell}\alpha$ we rewrite j^2 in a similar way as j_z above:

$$j^2 = (\ell + s)^2 = \ell^2 + s^2 + 2\ell \cdot s$$

where

$$2\ell \cdot s = 2\ell_x s_x + 2\ell_y s_y + 2\ell_z s_z = \ell_+ s_- + \ell_- s_+ + 2\ell_z s_z$$

Thus

$$2\ell \cdot s (Y_{\ell\ell}\alpha) = (\ell_+ s_- + \ell_- s_+ + 2\ell_z s_z)(Y_{\ell\ell}\alpha)$$

$$= \left(0 + 0 + 2 \cdot \ell \cdot \frac{1}{2}\right)(Y_{\ell\ell}\alpha) = \ell(Y_{\ell\ell}\alpha)$$

Finally for j^2:

$$j^2 (Y_{\ell\ell}\alpha) = \left(\ell(\ell + 1) + \frac{1}{2} \cdot \frac{3}{2} + \ell\right)(Y_{\ell\ell}\alpha)$$

$$= (\ell + 1/2)(\ell + 3/2)(Y_{\ell\ell}\alpha)$$

6.14   We want to show that

$$|\ell\ 1/2\ j = \ell + 1/2\ m\rangle = \left(\frac{\ell + m + 1/2}{2\ell + 1}\right)^{1/2} Y_{\ell m - 1/2}\, \alpha$$

$$+ \left(\frac{\ell - m + 1/2}{2\ell + 1}\right)^{1/2} Y_{\ell m + 1/2}\, \beta$$

In problem 6.13 it was shown that

$$|\ell\ 1/2\ j = \ell + 1/2\ \ell + 1/2\rangle = Y_{\ell\ell}\alpha$$

We now apply $j_- = \ell_- + s_-$ '$j - m$' times to this equation. We use the formula $j_- |j m\rangle = [(j + m)(j - m + 1)]^{1/2}|j\ m - 1\rangle$ and the analogous formulae for ℓ_- and s_-.

It is then easy to calculate

$$(j_-)^{j-m} |\ell\ 1/2\ j = \ell + 1/2\ \ell + 1/2\rangle$$
$$= [2j(2j - 1) \ldots (j + 1 + m)]^{1/2} [(j - m)!]^{1/2} |\ell\ 1/2\ \ell + 1/2\ m\rangle$$

In a similar way $(\ell_- + s_-)^{j-m}$ is applied to the right side of the equation above:

$$(\ell_- + s_-)^{j-m} Y_{\ell\ell}\alpha = \left(\ell_-^{j-m} + (j-m)\ell_-^{j-1-m}s_-\right) Y_{\ell\ell}\alpha$$

$$= [(2j-1)(2j-2)\ldots(j+m)]^{1/2}[(j-m)!]^{1/2} Y_{\ell m-1/2}\,\alpha$$

$$+(j-m)[(2j-1)(2j-2)\ldots(j+m+1)]^{1/2}[(j-m-1)!]^{1/2} Y_{\ell m+1/2}\,\beta$$

A comparison of these two expressions leads to

$$(2j)^{1/2}\left|\ell\,1/2\,\ell+1/2\,m\right) = (j+m)^{1/2}\,Y_{\ell m-1/2}\,\alpha + (j-m)^{1/2}\,Y_{\ell m+1/2}\,\beta$$

We finally insert $j = \ell + 1/2$:

$$\left|\ell\,1/2\,\ell+1/2\,m\right) = \left(\frac{\ell+m+1/2}{2\ell+1}\right)^{1/2} Y_{\ell m-1/2}\,\alpha$$

$$+ \left(\frac{\ell-m+1/2}{2\ell+1}\right)^{1/2} Y_{\ell m+1/2}\,\beta$$

The coefficients are the Clebsch–Gordan coefficients $C_{m-\frac{1}{2}\,\frac{1}{2}\,m}^{\ell\,\frac{1}{2}\,\ell+\frac{1}{2}}$ and $C_{m+\frac{1}{2}\,-\frac{1}{2}\,m}^{\ell\,\frac{1}{2}\,\ell+\frac{1}{2}}$, respectively.

6.16

$$C_{101}^{101} = \left[\frac{2!\,0!\,0!\,2!\,0!}{3!\,0!\,2!\,0!\,0!}\right]^{1/2} \sum_H \frac{(-1)^{H+0}\sqrt{3}}{(0-H)!(2-H)!} \cdot \frac{(2-H)!\,H!}{H!\,H!}$$

where it is only for $H = 0$ that no factorials of negative arguments occur.

$$\implies C_{101}^{101} = 1$$

Similarly,

$$C_{112}^{112} = \ldots = 1$$

6.17

$$C_{m-\frac{1}{2}\,\frac{1}{2}\,m}^{\ell\,\frac{1}{2}\,\ell+\frac{1}{2}} = \left(\frac{(2\ell)!\,1!\,0!\,\left(\ell+m+\frac{1}{2}\right)!\,\left(\ell-m+\frac{1}{2}\right)!}{(2\ell+2)!\,\left(\ell-m+\frac{1}{2}\right)!\,\left(\ell+m-\frac{1}{2}\right)!\,0!\,1!}\right)^{1/2}$$

$$\times \sum_H \frac{(-1)^{H+1}(2\ell+2)^{1/2}}{(1-H)!\,\left(\ell+m+\frac{1}{2}-H\right)!} \cdot \frac{\left(\ell+m+\frac{1}{2}-H\right)!\,\left(\ell-m+\frac{1}{2}+H\right)!}{H!\,\left(H+\ell-m-\frac{1}{2}\right)!}$$

$$= \left(\frac{\left(\ell + m + \frac{1}{2} \right)}{(2\ell + 1)(2\ell + 2)} \right)^{1/2}$$

$$\left[(2\ell + 2)^{1/2} \left(\ell - m + \frac{3}{2} \right) - (2\ell + 2)^{1/2} \left(\ell - m + \frac{1}{2} \right) \right] = \left(\frac{\ell + m + \frac{1}{2}}{2\ell + 1} \right)^{1/2}$$

In table 6B.1, we find

$$\left(\frac{1}{2} \, j_2 \, \frac{1}{2} \, m_2 | j_2 + \frac{1}{2} \, m \right) = C^{\frac{1}{2} \, j_2 \, j_2 + \frac{1}{2}}_{\frac{1}{2} \, m_2 \, m} = \left(\frac{j_2 + \frac{1}{2} + m}{2 j_2 + 1} \right)^{1/2}$$

The symmetry relation for interchange of 'j_1' and 'j_2' gives a phase factor, $(-1)^{j_1 + j_2 - j}$, which equals $(-1)^0$ in our special case. The calculated value is thus consistent with table 6B.1.

7.1   Formula (j in units of $\hbar$)

$$\langle jm|\mathbf{t}|jm' \rangle = \frac{\langle jm|\mathbf{j}|jm' \rangle \langle jm'|\mathbf{t} \cdot \mathbf{j}|jm' \rangle}{j(j+1)}$$

This formula is a special case of the well-known Wigner–Eckart theorem, which is discussed in many books on quantum mechanics. For $m = m'$:

$$\langle jm|\mathbf{t}|jm \rangle = \frac{\langle jm|\mathbf{j}|jm \rangle \langle jm|\mathbf{t} \cdot \mathbf{j}|jm \rangle}{j(j+1)}$$

Interpretation: 'Only the projection of the vector $\mathbf{t}$ on $\mathbf{j}$ gives any contribution to the expectation value above.'

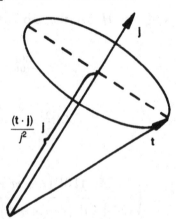

The magnetic moment is calculated from the formula

$$\mu = \langle jj | g_s s_z + g_\ell \ell_z | jj \rangle = \langle jj | g_\ell j_z + (g_s - g_\ell) s_z | jj \rangle$$
$$= g_\ell j + (g_s - g_\ell) \langle jj | s_z | jj \rangle$$

We use the formula above to calculate $\langle jj|s_z|jj\rangle$ and we then need

$$\mathbf{j}\cdot\mathbf{s} = \boldsymbol{\ell}\cdot\mathbf{s}+s^2 = \frac{1}{2}\left(j^2-\ell^2-s^2\right)+s^2 = \frac{1}{2}\left(j^2-\ell^2+s^2\right)$$

$$\implies \langle jj|s_z|jj\rangle = \frac{\left\langle jj\left|\frac{1}{2}\left(j^2-\ell^2+s^2\right)\right|jj\right\rangle}{\langle jj|j^2|jj\rangle}\langle jj|j_z|jj\rangle$$

$$= \frac{1}{2}\frac{j(j+1)-\ell(\ell+1)+s(s+1)}{j(j+1)}j = \pm\frac{1}{2\ell+1}j\,; \quad j=\ell\pm\frac{1}{2}$$

$$\implies \mu = j\left(g_\ell\pm(g_s-g_\ell)\frac{1}{2\ell+1}\right)\,; \quad j=\ell\pm\frac{1}{2}$$

7.2 The quadrupole moment Q is defined by

$$Q = Q_{33} = \int\left(3z^2-r^2\right)\frac{\rho}{e}\,\mathrm{d}^3r$$

For spheroidal shape with half-axes a and b, the integral is easily solved through the introduction of the 'ellipsoidal coordinates':

$$\begin{cases} x/a = r\sin\theta\cos\varphi \\ y/a = r\sin\theta\sin\varphi \\ z/b = r\cos\theta \end{cases} \implies \frac{\mathrm{d}(x,y,z)}{\mathrm{d}(r,\theta,\varphi)} = a^2br^2\sin\theta \implies$$

$$Q = \frac{\rho}{e}\int_{r=0}^{1}\int_{\theta=0}^{\pi}\int_{\varphi=0}^{2\pi}\left(2b^2\cos^2\theta-a^2\sin^2\theta\right)r^2a^2br^2\sin\theta\,\mathrm{d}\theta\,\mathrm{d}\varphi\,\mathrm{d}r$$

$$= \frac{\rho}{e}a^2b\frac{1}{5}\cdot2\pi\int_{\cos\theta=-1}^{1}\left[\left(2b^2+a^2\right)\cos^2\theta-a^2\right]\mathrm{d}(\cos\theta)$$

$$= \frac{\rho}{e} a^2 b \frac{2\pi}{5} 2 \left(\frac{2b^2 - 2a^2}{3} \right)$$

$$\rho = \frac{Ze}{\frac{4}{3}\pi a^2 b} \Rightarrow Q = \frac{2Z}{5} \left(b^2 - a^2 \right)$$

We first calculate the full quadrupole tensor in the system of principal axes and then make a coordinate rotation by an angle β. Define the quadrupole tensor:

$$Q_{ij} = \frac{\rho}{e} \int \left(3x_i x_j - r^2 \delta_{ij} \right) d\tau$$

In another coordinate system with $x_i' = \sum_k A_{ik} x_k$:

$$Q_{ij}' = \frac{\rho}{e} \int \left(3 \sum_k \sum_\ell A_{ik} x_k A_{j\ell} x_\ell - r^2 \delta_{ij} \right) d\tau$$

$$= \frac{\rho}{e} \sum_{k\ell} A_{ik} \left(\int \left(3x_k x_\ell - r^2 \delta_{k\ell} \right) d\tau \right) \left(A^{\mathrm{T}} \right)_{\ell j}$$

where we have inserted $\delta_{ij} = \sum_{k\ell} A_{ik} \delta_{k\ell} \left(A^{\mathrm{T}} \right)_{\ell j}$

$$\Rightarrow Q_{ij}' = \sum_{k\ell} A_{ik} Q_{k\ell} \left(A^{\mathrm{T}} \right)_{\ell j} \quad \text{or} \quad Q' = A Q A^{\mathrm{T}}$$

We now calculate the tensor Q: It is easy to see that $Q_{ij} = 0$ for $i \neq j$. Furthermore $Q_{11} + Q_{22} + Q_{33} = \mathrm{Tr}(Q) = 0$. As $Q_{11} = Q_{22}$ we thus have $Q_{11} = Q_{22} = -\frac{1}{2} Q_{33}$.

The symmetry axis (the z-axis) is rotated an angle β for example by the transformation

$$A = \begin{pmatrix} 1 & 0 & 0 \\ 0 & \cos\beta & -\sin\beta \\ 0 & \sin\beta & \cos\beta \end{pmatrix}$$

$$\Longrightarrow Q' = \frac{1}{2} Q_{33}$$

$$\times \begin{pmatrix} 1 & 0 & 0 \\ 0 & \cos\beta & -\sin\beta \\ 0 & \sin\beta & \cos\beta \end{pmatrix} \begin{pmatrix} -1 & 0 & 0 \\ 0 & -1 & 0 \\ 0 & 0 & 2 \end{pmatrix} \begin{pmatrix} 1 & 0 & 0 \\ 0 & \cos\beta & \sin\beta \\ 0 & -\sin\beta & \cos\beta \end{pmatrix}$$

$$Q' = Q_{33}' = Q_{33} \left(\frac{3}{2} \cos^2\beta - \frac{1}{2} \right)$$

7.4 The Coulomb potential for a homogeneously charged sphere was calculated in problem 6.11 (the 'step function' is denoted by θ):

$$V_C(r) = \frac{Ze}{4\pi\varepsilon_0}\left(\frac{3R^2 - r^2}{2R^3}\theta(R-r) + \frac{1}{r}\theta(r-R)\right)$$

We treat $V_C(r)$ as a perturbation.

The wave function is assumed to be of 'harmonic oscillator type':

$$\phi(n\ell sj\Omega) = R_{n\ell}(r)\sum_{\Lambda\Sigma}\langle\ell s\,\Lambda\Sigma|\ell sj\Omega\rangle Y_{\ell\Lambda}\chi_{s\Sigma}$$

with $R_{n\ell}(r) = C_{n\ell}\rho^\ell F\left(-n, \ell + 3/2; \rho^2\right)e^{-\rho^2/2}$; $\rho = (M\omega/\hbar)^{1/2}r$ where $C_{n\ell}$ is the normalisation constant. The Kummer function is defined by

$$F(a,c;z) = 1 + \frac{a}{c}\frac{z}{1} + \frac{a(a+1)}{c(c+1)}\frac{z^2}{2!} + \cdots$$

We want to compare an $1g_{7/2}$ and a $2d_{5/2}$ proton:

$1g_{7/2}$: $\ell = 4$, $n = 0$; the radial function is given by $R_{04}(r)$ with $F\left(0, 11/2; \rho^2\right) = 1$.

$2d_{5/2}$: $\ell = 2$, $n = 1$; the radial function is given by $R_{12}(r)$ with $F\left(-1, 7/2; \rho^2\right) = 1 - \frac{2}{7}\rho^2$.

In first order perturbation theory, the energy shift due to the Coulomb potential is given by

$$\Delta E(1g_{7/2}) = \int_0^\infty R_{04}^*(r)eV_C(r)R_{04}(r)r^2\,dr$$

$$\Delta E(2d_{5/2}) = \int_0^\infty R_{12}^*(r)eV_C(r)R_{12}(r)r^2\,dr$$

It is now straightforward to calculate the energy shifts. The calculations are, however, quite lengthy and we will only estimate the answers. At the centre the Coulomb potential is given by

$$V_C(0) = \frac{3}{2}\frac{Ze}{4\pi\varepsilon_0}\cdot\frac{1}{R}$$

and at the nuclear surface $V_C(R) = \frac{2}{3}V_C(0)$. For a wave function centred somewhere in between the shift should be of the order

$$\Delta E \simeq \frac{Ze^2}{4\pi\varepsilon_0}\cdot\frac{1}{R}\cdot 1.3$$

We have $R = r_0 \cdot A^{1/3}$ and let us put that $Z = A/2 \Rightarrow \Delta E \simeq \frac{1.3}{2}(e^2/4\pi\varepsilon_0 r_0)A^{2/3} \cong 0.8 \cdot A^{2/3}$ MeV. With $A = 100$, $\Delta E \simeq 17$ MeV. Furthermore, the $1g_{7/2}$ wave function is 'closer' to the nuclear surface than the $2d_{5/2}$ wave function and we thus expect

$$\Delta E(2d_{5/2}) > \Delta E(1g_{7/2})$$

where a difference of maybe 1 MeV could be expected for $A \cong 100$. We also observe that $\Delta E \propto A^{2/3}$, which means that the Coulomb effects are not very important for small A but become more and more important with increasing A.

7.5   We want to calculate the quadrupole moment for a particle outside closed shells and which has a charge e ($e = 1$ for a proton, $e = 0$ for a neutron):

$$Q = \langle \ell\, 1/2\, jj | Q^{\mathrm{op}} | \ell 1/2\, jj \rangle$$

(a) $j = \ell + 1/2$: $|\ell 1/2 jj\rangle = Y_{\ell\ell}\alpha \Rightarrow \langle \ell\, 1/2\, j\, j | Y_{20} | \ell\, 1/2\, j\, j \rangle = \int Y_{\ell\ell}^*(\Omega) Y_{20}(\Omega) Y_{\ell\ell}(\Omega)\, d\Omega$. The addition theorem gives

$$Y_{20} Y_{\ell\ell} = \sum_L \left(\frac{5(2\ell+1)}{4\pi(2L+1)} \right)^{1/2} C_{0\ell\ell}^{2\ell L} C_{000}^{2\ell L} Y_{L\ell}$$

We also use the orthogonality

$$\int Y_{\ell\ell}^*(\Omega) Y_{L\ell}(\Omega)\, d\Omega = \delta_{L\ell}$$

$$\Rightarrow \langle \ell\, 1/2\, jj | Y_{20} | \ell\, 1/2\, jj \rangle = \left(\frac{5}{4\pi} \right)^{1/2} C_{0\ell\ell}^{2\ell\ell} C_{000}^{2\ell\ell}$$

The Clebsch–Gordan coefficients are taken from table 6B.1:

$$C_{000}^{2\ell\ell} = \langle 2\,\ell\, 0\, 0 | \ell\, 0 \rangle$$
$$= \frac{-2\ell(\ell+1)}{[(2\ell-1)2\ell(2\ell+2)(2\ell+3)]^{1/2}} = -\left(\frac{\ell(\ell+1)}{(2\ell-1)(2\ell+3)} \right)^{1/2}$$

$$C_{0\ell\ell}^{2\ell\ell} = \frac{2\left(3\ell^2 - \ell(\ell+1)\right)}{[(2\ell-1)2\ell(2\ell+2)(2\ell+3)]^{1/2}} = \left(\frac{(2\ell-1)\ell}{(\ell+1)(2\ell+3)} \right)^{1/2}$$

$$C_{0\ell\ell}^{2\ell\ell} C_{000}^{2\ell\ell} = -\frac{\ell}{2\ell+3} = -\frac{j-1/2}{2j+2} \quad \text{for} \quad j = \ell + 1/2$$

$$Q = -e \left(\frac{16\pi}{5} \right)^{1/2} \langle r^2 \rangle_{N\ell} \left(\frac{5}{4\pi} \right)^{1/2} \frac{j-1/2}{2j+2}$$

$$Q = -e\left\langle r^2\right\rangle_{N\ell} \frac{2j-1}{2j+2}; \quad j = \ell + 1/2$$

(b) $j = \ell - 1/2$. We have

$$|\ell\, 1/2\, jj\rangle = \sum_{m_\ell m_s} \langle \ell\, 1/2\, m_\ell m_s|\ell\, 1/2\, jj\rangle\, |\ell\, 1/2\, m_\ell m_s\rangle$$

Thus, with a change in the notation:

$$\Longrightarrow |\ell\, 1/2\, j = \ell - 1/2\, m = j\rangle$$
$$= C_{\ell\,-1/2\,\ell-1/2}^{\ell\,1/2\,\ell-1/2} Y_{\ell\ell}\beta + C_{\ell-1\,1/2\,\ell-1/2}^{\ell\,1/2\,\ell-1/2} Y_{\ell\ell-1}\alpha$$
$$= aY_{\ell\ell}\beta + bY_{\ell\ell-1}\alpha$$

We now get for the matrix element:

$$\langle \ell\, 1/2\, j = \ell - 1/2\, j|Y_{20}|\ell\, 1/2\, j = \ell - 1/2\, j\rangle$$

$$= \langle a\, Y_{\ell\ell}\beta + bY_{\ell\ell-1}\alpha|Y_{20}| a Y_{\ell\ell}\beta + bY_{\ell\ell-1}\alpha\rangle$$

$$= |a|^2 \left(Y_{\ell\ell}|Y_{20}| Y_{\ell\ell}\right) + |b|^2 \left(Y_{\ell\ell-1}|Y_{20}| Y_{\ell\ell-1}\right)$$

We calculated above that

$$\langle Y_{\ell\ell}|Y_{20}| Y_{\ell\ell}\rangle = -\left(\frac{5}{4\pi}\right)^{1/2} \frac{\ell}{2\ell+3}$$

The second term is obtained as

$$\left(Y_{\ell\ell-1}|Y_{20}| Y_{\ell\,\ell-1}\right) = \int Y_{\ell\ell-1}^* Y_{20} Y_{\ell\ell-1}\, d\Omega$$

$$= \left(\frac{5}{4\pi}\right)^{1/2} C_{0\,\ell-1\,\ell-1}^{2\,\ell\,\ell} C_{0\,0\,0}^{2\,\ell\,\ell} = \cdots = -\left(\frac{5}{4\pi}\right)^{1/2} \frac{\ell-3}{2\ell+3}$$

where we have used the addition theorem for $Y_{20} Y_{\ell\ell-1}$ and orthogonality for the sperical harmonics:

$$\langle \ell\, 1/2\, jj|Y_{20}|\ell\, 1/2\, jj\rangle = -\left(\frac{5}{4\pi}\right)^{1/2} \left(|a|^2 \frac{\ell}{2\ell+3} + |b|^2 \frac{\ell-3}{2\ell+3}\right)$$

It remains to determine $a = C_{\ell-1/2\,\ell-1/2}^{\ell\,1/2\,\ell-1/2}$ and $b = C_{\ell-1\,1/2\,\ell-1/2}^{\ell\,1/2\,\ell-1/2}$. We explore the symmetry $\left|C_{m_1 m_2 m}^{j_1 j_2 j}\right| = \left|C_{m_2 m_1 m}^{j_2 j_1 j}\right|$ and obtain from tables $a^2 = 2\ell/(2\ell+1)$ and $b^2 = 1/(2\ell+1)$. Thus, for the quadrupole moment:

$$Q = -e\left(\frac{16\pi}{5}\right)^{1/2} \left\langle r^2\right\rangle_{N\ell} \left(\frac{5}{4\pi}\right)^{1/2} \frac{\ell-1}{2\ell+1} = -e\left\langle r^2\right\rangle_{N\ell} \frac{2j-1}{2j+2}$$

7.6   The magnetic moment is calculated in the nucleon state having $s = s_z = \frac{1}{2}$. Thus, of the two-quark states having $s = 1$ we need to consider those having $s_z = 0$ and $s_z = 1$:

$$|s = 1 \; s_z = 1\rangle_{2q} = |11\rangle_{2q} = \left|\frac{1}{2}\frac{1}{2}\right\rangle_{q_1} \left|\frac{1}{2}\frac{1}{2}\right\rangle_{q_2}$$

$$|10\rangle_{2q} = \sum_{m_1 m_2} C^{\frac{1}{2}\frac{1}{2}1}_{m_1 m_2 0} \left|\frac{1}{2}m_1\right\rangle_{q_1} \left|\frac{1}{2}m_2\right\rangle_{q_2}$$

$$= \frac{1}{\sqrt{2}} \left(\left|\frac{1}{2}\frac{1}{2}\right\rangle_{q_1} \left|\frac{1}{2}-\frac{1}{2}\right\rangle_{q_2} + \left|\frac{1}{2}-\frac{1}{2}\right\rangle_{q_1} \left|\frac{1}{2}\frac{1}{2}\right\rangle_{q_2} \right)$$

We have

$$C^{\frac{1}{2}\frac{1}{2}1}_{\frac{1}{2}-\frac{1}{2}0} = C^{\frac{1}{2}\frac{1}{2}1}_{-\frac{1}{2}\frac{1}{2}0} = \frac{1}{\sqrt{2}}$$

$$\left|\frac{1}{2}\frac{1}{2}\right\rangle_{3q} = \sum_{M m_3} C^{1\frac{1}{2}\frac{1}{2}}_{M m_3 \frac{1}{2}} |1M\rangle_{2q} \left|\frac{1}{2}m_3\right\rangle_{q_3}$$

$$= \left(\frac{2}{3}\right)^{1/2} \left(\left|\frac{1}{2}\frac{1}{2}\right\rangle_{q_1} \left|\frac{1}{2}\frac{1}{2}\right\rangle_{q_2} \left|\frac{1}{2}-\frac{1}{2}\right\rangle_{q_3} \right)$$

$$- \left(\frac{1}{6}\right)^{1/2} \left(\left|\frac{1}{2}\frac{1}{2}\right\rangle_{q_1} \left|\frac{1}{2}-\frac{1}{2}\right\rangle_{q_2} \left|\frac{1}{2}\frac{1}{2}\right\rangle_{q_3} + \left|\frac{1}{2}-\frac{1}{2}\right\rangle_{q_1} \left|\frac{1}{2}\frac{1}{2}\right\rangle_{q_2} \left|\frac{1}{2}\frac{1}{2}\right\rangle_{q_3} \right)$$

The total magnetic moment operator is

$$\mu = \mu_0 \left[Q_1 (s_z)_1 + Q_2 (s_z)_2 + Q_3 (s_z)_3 \right]$$

$$\left\langle \frac{1}{2}\frac{1}{2} |\mu| \frac{1}{2}\frac{1}{2} \right\rangle_{3q} = \mu_0 \left\{ Q_1 \cdot \frac{1}{2} \left[\frac{2}{3} + \left(\frac{1}{6} - \frac{1}{6} \right) \right] + Q_2 \cdot \frac{1}{2} \left(\frac{2}{3} + 0 \right) \right.$$

$$\left. + Q_3 \frac{1}{2} \left(-\frac{2}{3} + \frac{1}{6} + \frac{1}{6} \right) \right\}$$

$$= \mu_0 Q_1 \frac{1}{3} + \mu_0 Q_2 \frac{1}{3} - \mu_0 Q_3 \frac{1}{6} = \frac{\mu_0}{6} (2Q_1 + 2Q_2 - Q_3)$$

For the proton $Q_1 = Q_2 = \frac{2}{3}e$; $Q_3 = -\frac{1}{3}e \Rightarrow \mu_p = \mu_0 e/2$.
For the neutron $Q_1 = Q_2 = -\frac{1}{3}e$; $Q_3 = \frac{2}{3}e \Rightarrow \mu_n = -\mu_0 e/3$, i.e.

$$\mu_n = -\left(\frac{2}{3}\right) \mu_p$$

The measured ratio is $g_s^n / g_s^p = -(3.83/5.56) = -0.69$.

8.1   The normalised wave function is given by

$$|N\ell sj\Omega\rangle = R_{n\ell}(r)\sum_{\Lambda\Sigma}\langle\ell s\Lambda\Sigma|\ell sj\Omega\rangle Y_{\ell\Lambda}\chi_{s\Sigma}$$

where it is assumed that

$$\langle R_{n\ell}(r)|R_{n\ell}(r)\rangle = \int |R_{n\ell}(r)|^2\, r^2\, \mathrm{d}r = 1$$

We want to calculate the diagonal matrix elements of

$$-M\omega_0^2\frac{2}{3}\varepsilon r^2 P_2 = -M\omega_0^2\frac{2}{3}\varepsilon\left(\frac{4\pi}{5}\right)^{1/2} r^2 Y_{20}(\theta,\varphi)$$

Consider first the angular part:

$$\langle\ell sj\Omega|Y_{20}(\theta,\varphi)|\ell sj\Omega\rangle$$

$$= \left\langle \sum_{\Lambda\Sigma}\langle\ell s\Lambda\Sigma|\ell sj\Omega\rangle Y_{\ell\Lambda}\chi_{s\Sigma}\right|Y_{20}\left|\sum_{\Lambda'\Sigma'}\langle\ell s\Lambda'\Sigma'|\ell sj\Omega\rangle Y_{\ell\Lambda'}\chi_{s\Sigma'}\right\rangle$$

$$= \sum_{\Lambda\Sigma}|\langle\ell s\Lambda\Sigma|\ell sj\Omega\rangle|^2 \left\langle Y_{\ell\Lambda}\chi_{s\Sigma}\right|\left(\frac{5}{4\pi}\right)^{1/2} C^{2\ell\ell}_{0\Lambda\Lambda}C^{2\ell\ell}_{000}\left|Y_{\ell\Lambda}\chi_{s\Sigma}\right\rangle$$

$$= \sum_{\Lambda\Sigma}|\langle\ell s\Lambda\Sigma|\ell sj\Omega\rangle|^2 \left(\frac{5}{4\pi}\right)^{1/2}\langle 2\ell 0\Lambda|2\ell\ell\Lambda\rangle\langle 2\ell 00|2\ell\ell 0\rangle$$

where, in a similar way as in problem 7.5, we have first used orthogonality for the spin functions and the addition theorem for spherical harmonics (noting that the 'm_ℓ quantum numbers' should add to zero) and then orthogonality for the spherical harmonics.

In order to find numerical values of the Clebsch–Gordan coefficients in table 6B.1, we need the symmetry relation

$$\langle\ell_1\ell_2 m_1 m_2|\ell_1\ell_2 LM\rangle = (-1)^{\ell_1+\ell_2-L}\langle\ell_2\ell_1 m_2 m_1|\ell_2\ell_1 LM\rangle;\quad (\ell_2 = 1/2)$$

For a fixed value of Ω, there are two terms in the sum, namely $\Sigma = 1/2, \Lambda = \Omega - 1/2$ and $\Sigma = -1/2, \Lambda = \Omega + 1/2$. Furthermore, it turns out that the calculations in the two cases $j = \ell \pm 1/2$ can be carried out simultaneously:

$$\left(\frac{4\pi}{5}\right)^{1/2}\langle N\ell j\Omega|Y_{20}|N\ell j\Omega\rangle$$

$$= \frac{1}{2\ell+1}\left((\ell\pm\Omega+1/2)\frac{2\left[3(\Omega-1/2)^2-\ell(\ell+1)\right][-2\ell(\ell+1)]}{(2\ell-1)2\ell(2\ell+2)(2\ell+3)}\right.$$

$$+ (\ell \mp \Omega + 1/2) \frac{2 \left[3(\Omega + 1/2)^2 - \ell(\ell+1) \right] \left[-2\ell(\ell+1) \right]}{(2\ell-1)2\ell(2\ell+2)(2\ell+3)} \Bigg)$$

$$\cdots = -\frac{(\ell + 1/2 \mp 1)6\Omega^2 - 2(\ell + 1/2)(\ell + 3/2)(\ell - 1/2)}{8(\ell - 1/2)(\ell + 1/2)(\ell + 3/2)}$$

$$= \begin{Bmatrix} -\dfrac{(j-1)6\Omega^2}{8(j-1)j(j+1)} - \dfrac{1}{4} \\ -\dfrac{(j+2)6\Omega^2}{8j(j+1)(j+2)} - \dfrac{1}{4} \end{Bmatrix} = -\frac{3\Omega^2 - j(j+1)}{4j(j+1)} \begin{cases} j = \ell + 1/2 \\ j = \ell - 1/2 \end{cases}$$

We thus obtain the desired relation:

$$\left\langle N\ell j\Omega \left| - M\omega_0^2 \frac{2}{3}\varepsilon r^2 P_2 \right| N\ell j\Omega \right\rangle = \langle r^2 \rangle M\omega_0^2 \frac{\varepsilon}{6} \frac{3\Omega^2 - j(j+1)}{j(j+1)}$$

8.4   First consider the pure oscillator Hamiltonian,

$$H_{\mathrm{osc}} = \frac{-\hbar^2}{2M}\Delta + \frac{M}{2}\left[\omega_\perp^2 \left(x^2 + y^2 \right) + \omega_z^2 z^2 \right] = T + V_{\mathrm{osc}}$$

We want to calculate $[H_{\mathrm{osc}}, j_z] = [H_{\mathrm{osc}}, \ell_z] + [H_{\mathrm{osc}}, s_z] = [H_{\mathrm{osc}}, \ell_z]$ where

$$\ell_z = \frac{\hbar}{\mathrm{i}}\left(x\frac{\partial}{\partial y} - y\frac{\partial}{\partial x} \right) = \frac{\hbar}{\mathrm{i}}\frac{\partial}{\partial \varphi}$$

We also find that ℓ_z in the stretched system,

$$(\ell_z)_t = \frac{\hbar}{\mathrm{i}}\left(\xi\frac{\partial}{\partial \eta} - \eta\frac{\partial}{\partial \xi} \right) = \frac{\hbar}{\mathrm{i}}\frac{\partial}{\partial \varphi_t}$$

is identical to ℓ_z (and $\varphi_t \equiv \varphi$).

The kinetic energy is rewritten in cylindrical coordinates (see section on asymptotic wave functions, chapter 8):

$$T = -\frac{1}{2}\hbar\omega_\perp \left(\frac{1}{\rho}\frac{\partial}{\partial \rho}\rho\frac{\partial}{\partial \rho} + \frac{1}{\rho^2}\frac{\partial^2}{\partial \varphi^2} \right) - \frac{1}{2}\hbar\omega_z\frac{\partial^2}{\partial \zeta^2}$$

and it is immediately obvious that $[T, \ell_z] = 0$. (Indeed, we have the general relation $[T, \ell_z] = 0$ as should be obvious from the introduction of chapter 6.) Furthermore, V_{osc} is independent of φ and thus

$$[V_{\mathrm{osc}}, \ell_z] = 0$$

Indeed, as long as V stays axially symmetric, i.e. by definition independent of φ, $V = V(\rho, \zeta)$, we must have $[V, \ell_z] = 0$.

It remains to calculate

$$\left[\ell^2, j_z\right] = \left[\ell^2, \ell_z + s_z\right] = 0$$

$$[\ell \cdot s, j_z] = \frac{1}{2}\left[j^2 - \ell^2 - s^2, j_z\right] = \ldots = 0$$

$$\Rightarrow [H_{\text{MO}}, j_z] = 0$$

Consider now:

$$[\ell \cdot s, \ell_z] = [\ell_x, \ell_z]\, s_x + [\ell_y, \ell_z]\, s_y$$
$$= -i\hbar \ell_y s_x + i\hbar \ell_x s_y = i\hbar \left(\ell_x s_y - \ell_y s_x\right) \neq 0$$

i.e. ℓ_z is not generally a preserved quantum number if an $\ell \cdot s$ term is present in the Hamiltonian.

8.5 In the spherical case

$$H_0 = \frac{1}{2}\hbar\omega_0\left(-\Delta_\xi + \rho^2\right) - \hbar\omega_0\left[2\kappa\ell \cdot s + \mu'\left(\ell^2 - \left\langle \ell^2\right\rangle_N\right)\right]$$

For the spherical eigenvectors, $|N\ell j\Omega\rangle$, the energies are

$$E_j \begin{Bmatrix} j = \ell + 1/2 \\ j = \ell - 1/2 \end{Bmatrix} = \langle N\ell j\Omega | H_0 | N\ell j\Omega\rangle$$

$$= \hbar\omega_0\left[(N + 3/2) - \kappa\begin{Bmatrix} \ell \\ -(\ell+1) \end{Bmatrix} - \mu'\left(\ell(\ell+1) - \frac{N(N+3)}{2}\right)\right]$$

$$|22\,5/2\,\Omega\rangle = d_{5/2} \quad E_{5/2} = 3.28\hbar\omega_0$$

$$|22\,3/2\,\Omega\rangle = d_{3/2} \quad E_{3/2} = 3.78\hbar\omega_0$$

$$|20\,1/2\,\Omega\rangle = s_{1/2} \quad E_{1/2} = 3.60\hbar\omega_0$$

For small deformations:

$$H = H_0 + \varepsilon h'; \quad \varepsilon h' = -\hbar\omega_0 \frac{2}{3}\varepsilon\rho^2 P_2(\cos\theta)$$

where

$$\langle N\ell j\Omega | \varepsilon h' | N\ell j\Omega'\rangle = \frac{1}{6}\hbar\omega_0\varepsilon\left\langle \rho^2\right\rangle \frac{3\Omega^2 - j(j+1)}{j(j+1)}\delta_{\Omega\Omega'}$$

$$\left\langle \left| \rho^2 \right| \right\rangle = N + 3/2; \quad \text{i.e.} \quad \left\langle \rho^2 \right\rangle = 7/2 \quad \text{for} \quad N = 2$$

$$\Rightarrow \langle \varepsilon h' \rangle = \frac{7}{12} \hbar \omega_0 \varepsilon \left(\frac{3\Omega^2}{j(j+1)} - 1 \right)$$

j	5/2			3/2		1/2
Ω	5/2	3/2	1/2	3/2	1/2	1/2
$\langle h' \rangle / \hbar \omega_0$	0.67	−0.13	−0.53	0.47	−0.47	0

For large deformations:

$$E_{\text{osc}}(n_z, n_\perp) = \hbar \omega_z \left(n_z + 1/2 \right) + \hbar \omega_\perp \left(n_\perp + 1 \right)$$

$$= \hbar \omega_0 \left(N + \frac{3}{2} + (n_\perp - 2n_z) \frac{\varepsilon}{3} \right)$$

First order perturbation for the $\ell \cdot \mathbf{s}$ and ℓ^2-terms:

$$\langle Nn_z \Lambda \Sigma | \ell \cdot \mathbf{s} | Nn_z \Lambda \Sigma \rangle = \Lambda \Sigma$$

$$\langle Nn_z \Lambda \Sigma | \ell^2 - \langle \ell^2 \rangle_N | Nn_z \Lambda \Sigma \rangle = \Lambda^2 + 2n_\perp n_z + 2n_z + n_\perp - \frac{N(N+3)}{2}$$

Nn_z	Λ	Σ	Ω	H_{osc}	'$\ell \cdot \mathbf{s}$'	'ℓ^2'	Total $\varepsilon = 0.75$
22	0	1/2	1/2	$3.5 - 4\varepsilon/3$	0	−1	2.52
21	1	1/2	3/2	$3.5 - \varepsilon/3$	1/2	1	3.13
	1	−1/2	1/2		−1/2	1	3.33
20	2	1/2	5/2	$3.5 + 2\varepsilon/3$	1	1	3.78
	2	−1/2	3/2		−1	1	4.18
	0	1/2	1/2		0	−3	4.06

In the last column, $\kappa = 0.1$ and $\mu' = 0.02$ have been inserted (assuming that the $\ell \cdot \mathbf{s}$ and ℓ^2-terms are multiplied by the ε-dependent frequency, $\hbar \omega_0$). Note that for each Ω, there is one 'conjugate' state with negative Ω (and inverted signs for Λ and Σ) having the same energy.

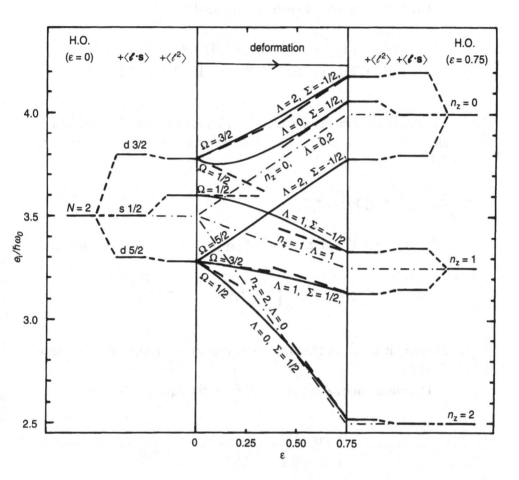

In the figure, the levels at spherical shape have been drawn to the left and those at $\varepsilon = 0.75$ to the right. The addition of first the $\ell \cdot s$ and then the ℓ^2-terms is illustrated when approaching the middle of the figure. Note that the calculations are exact for spherical shape but not for $\varepsilon = 0.75$. The perturbation terms as functions of ε are then drawn by dashed lines and the levels at the two perturbation limits are connected, noting that levels with the same Ω are coupled and therefore cannot intersect. As there is only one orbital with $\Omega = 5/2$, the calculations are exact (in both perturbation limits) in this case. The pure oscillator levels are drawn as functions of deformation by thin dot–dashed lines. As usual, all energies are given in $\hbar\omega_0$ units.

8.6   The diagonalisation is carried out in an $|Nn_z\Lambda\Sigma\rangle$ basis with basis vectors $|211\ 1/2\rangle$ and $|202 - 1/2\rangle$.

The following matrix elements are needed:

$$\begin{bmatrix} \langle 202 - 1/2 |H_{MO}| 202 - 1/2\rangle & \langle 202 - 1/2 |H_{MO}| 211\ 1/2\rangle \\ \langle 211\ 1/2 |H_{MO}| 202 - 1/2\rangle & \langle 211\ 1/2 |H_{MO}| 211\ 1/2\rangle \end{bmatrix}$$

The operator formalism is used to calculate the non-diagonal matrix elements of $\ell \cdot s$ and ℓ^2 (the diagonal ones were given in problem 8.5):

$$\langle 211\ 1/2 |\ell \cdot s| 202 - 1/2\rangle =$$

$$\langle n'_z = 1\ r' = 1\ s' = 0\ \Sigma' = 1/2\ |\ell \cdot s|\ n_z = 0\ r = 2\ s = 0\ \Sigma = -1/2\rangle$$

$$= -\frac{1}{\sqrt{2}}\ [(n_z + 1)r]^{1/2} = -1$$

Further, it is found that the ℓ^2-term does not couple the two basis vectors.

The eigenvalues λ are obtained from the equation ($\kappa = 0.1$):

$$\begin{vmatrix} 3.38 - \frac{1}{3}\varepsilon - \lambda & -0.2 \\ -0.2 & 3.68 + \frac{2}{3}\varepsilon - \lambda \end{vmatrix} = 0$$

$$\Rightarrow \lambda_{1,2} = 3.53 + \frac{\varepsilon}{6} \pm \left(0.0625 + 0.15\varepsilon + \varepsilon^2/4\right)^{1/2}$$

or

$$\lambda_{1,2} = 3.53 + \frac{\varepsilon}{6} \pm \frac{1}{2}\left[(\varepsilon + 0.3)^2 + 0.16\right]^{1/2}$$

That is, the two orbitals 'come closest together' for $\varepsilon = -0.3$, see figure.

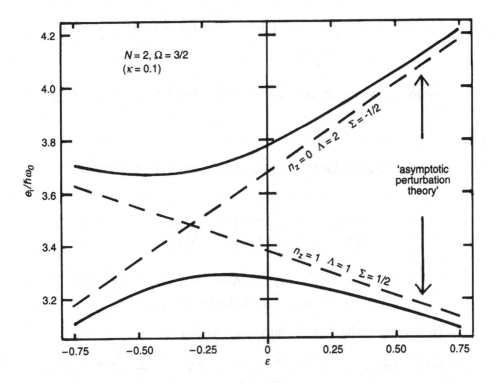

8.7 The formula $a^+|n\rangle = C_n|n+1\rangle$ leads to $\langle n|aa^+|n\rangle = (C_n)^2$. Now apply the two relations $[a, a^+] = 1$ and $a^+a|n\rangle = n|n\rangle$:

$$\langle n|aa^+|n\rangle = \langle n|a^+a + 1|n\rangle = n+1$$

$\Rightarrow (C_n)^2 = n + 1$, which means that $C_n = (n + 1)^{1/2}$ is one possible choice.

8.8

$$[R, R^+] = \frac{1}{2}\left[(a_x - ia_y), \left(a_x^+ + ia_y^+\right)\right] = \frac{1}{2}\left[a_x, a_x^+\right] + \frac{1}{2}\left[a_y, a_y^+\right] = 1 \quad \text{etc.}$$

8.9

$$\ell_z = \frac{1}{i}\left(\xi\frac{\partial}{\partial\eta} - \eta\frac{\partial}{\partial\xi}\right)$$

$$= \frac{1}{2i}\left[\left(a_x + a_x^+\right)\left(a_y - a_y^+\right) - \left(a_y + a_y^+\right)\left(a_x - a_x^+\right)\right]$$

$$= \frac{1}{i}\left(a_x^+a_y - a_xa_y^+\right)$$

$$\left.\begin{array}{l} R^+ = (1/\sqrt{2})\left(a_x^+ + ia_y^+\right) \\ S^+ = (1/\sqrt{2})\left(a_x^+ - ia_y^+\right) \end{array}\right\} \Rightarrow \left\{\begin{array}{l} a_x^+ = (1/\sqrt{2})\left(R^+ + S^+\right) \\ a_y^+ = -(i/\sqrt{2})\left(R^+ - S^+\right) \end{array}\right.$$

$$\Rightarrow \ell_z = \frac{1}{2i}\left[(R^+ + S^+)\, i(R - S) - (R + S)(-i)\,(R^+ - S^+)\right]$$

$$= \ldots = \frac{1}{2}\,(R^+ R - S^+ S)$$

where we have used commutation rules like $[R, R^+] = 1$, $[R^+, S] = 0$ etc.

8.10

$$A(B|A)) = (BA + B)|A\rangle = Ba|A\rangle + B|A\rangle = (a+1)B|A$$

8.11

$$\left.\begin{matrix} n_\perp = r + s \\ \Lambda = r - s \end{matrix}\right\} \Rightarrow \begin{cases} r = (n_\perp + \Lambda)/2 \\ s = (n_\perp - \Lambda)/2 \end{cases}$$

$$\langle |\ell \cdot \mathbf{s}| \rangle = \langle n_z + 1\ n_\perp - 1\ \Lambda - 1\ \Sigma + 1 |\ell \cdot \mathbf{s}|\, n_z\ n_\perp\ \Lambda\ \Sigma\rangle$$
$$= \langle n_z + 1\ r - 1\ s\ \Sigma + 1 |\ell \cdot \mathbf{s}|\, n_z\ r\ s\ \Sigma\rangle$$

In chapter 8, the $\ell \cdot \mathbf{s}$ operator was derived as

$$\ell \cdot \mathbf{s} = (R^+ R - S^+ S)\, s_z - \frac{1}{\sqrt{2}}\,(a_z R^+ - a_z^+ S)\, s_- - \frac{1}{\sqrt{2}}\,(a_z^+ R - a_z S^+)\, s_+$$

The only term that contributes is $(-1/\sqrt{2})a_z^+ R s_+$;

$$a_z^+ |n_z\rangle = (n_z + 1)^{1/2}\, |n_z + 1\rangle$$

$$R\,|r\rangle = \sqrt{r}\,|r - 1\rangle$$

$$s_+\,|\Sigma = -1/2\rangle = |\Sigma = 1/2\rangle$$

$$\Rightarrow \langle |\ell \cdot \mathbf{s}| \rangle = \frac{-1}{\sqrt{2}}\,(n_z + 1)^{1/2}\,\sqrt{r} = -\frac{1}{2}\,[(n_z + 1)\,(n_\perp + \Lambda)]^{1/2}$$

8.12

$$x^2 = \frac{\hbar}{M\omega_x}\xi^2 = \frac{1}{2}\frac{\hbar}{M\omega_x}\,(a_x + a_x^+)\,(a_x + a_x^+)$$

$$= \frac{1}{2}\frac{\hbar}{M\omega_x}\,(a_x^+ a_x + a_x a_x^+ + a_x a_x + a_x^+ a_x^+)$$

$$= \frac{\hbar}{M\omega_x}\left(a_x^+ a_x + \frac{1}{2} + \frac{1}{2}\,(a_x a_x + a_x^+ a_x^+)\right)$$

$$\Rightarrow \langle n_x |x^2| n_x\rangle = \frac{\hbar}{M\omega_x}\left(n_x + \frac{1}{2}\right)$$

$$\Rightarrow \langle n_x n_y n_z \left| r^2 \right| n_x n_y n_z \rangle$$

$$= \frac{\hbar}{M\omega_x}(n_x + 1/2) + \frac{\hbar}{M\omega_y}(n_y + 1/2) + \frac{\hbar}{M\omega_z}(n_z + 1/2)$$

Spherical shape corresponds to $\omega_x = \omega_y = \omega_z = \omega_0$:

$$\langle \left| r^2 \right| \rangle = \frac{\hbar}{M\omega_0}(n_x + n_y + n_z + 3/2) = \frac{\hbar}{M\omega_0}(N + 3/2)$$

9.1   With the number of protons and neutrons equal, the total degeneration for an N-shell is $2(N+1)(N+2)$ (cf. problem 6.7). Thus

$$A = \sum_{N=0}^{N^*} 2(N+1)(N+2) + 2\rho(N^*+2)(N^*+3)$$

$$= \frac{2}{3}(N^*+1)(N^*+2)(N^*+3) + 2\rho(N^*+2)(N^*+3)$$

$$= \frac{2}{3}(N^*+2)(N^*+3)(N^*+1+3\rho)$$

$$E = \sum_{N=0}^{N^*} 2(N+1)(N+2)(N+3/2)\hbar\omega_0$$
$$+ 2\rho(N^*+2)(N^*+3)(N^*+5/2)\hbar\omega_0 =$$

$$2\hbar\omega_0 \left[\frac{1}{4}(N^*+1)(N^*+2)^2(N^*+3) + \rho(N^*+2)(N^*+3)\left(N^*+\frac{5}{2}\right)\right]$$

We now want to express N^* as a function of A (and ρ). The following equation is obtained:

$$N^{*3} + (6+3\rho)N^{*2} + (11+15\rho)N^* + 6 + 18\rho - 3A/2 = 0$$

Through the substitution $N^* = x - (6+3\rho)/3 = x - 2 - \rho$, the term of second order is eliminated:

$$x^3 + \left(-3\rho^2 + 3\rho - 1\right)x + 2\rho^3 - 3\rho^2 + \rho - 3A/2 = 0$$

The equation $x^3 + 3px + 2q = 0$ has one and only one real root if $D = q^2 + p^3 > 0$. In our case

$$D = \left(\frac{2\rho^3 - 3\rho^2 + \rho - 3A/2}{2}\right)^2 + \left(\frac{-3\rho^2 + 3\rho - 1}{3}\right)^3$$

We have $0 \le \rho \le 1$ and if we confine ourselves to A-values of practical interest, let us say $A \geq 4$, it is trivial to conclude that $D > 0$. It is possible to find the exact solution of any third order equation but, in this case, it is easier to make an ansatz

$$N^* = \alpha A^{1/3} + \beta + \gamma A^{-1/3} + \ldots$$

This expression is inserted in the equation. The requirement that the coefficient of the A-term should vanish gives

$$\alpha^3 - 3/2 = 0 \Rightarrow \alpha = (3/2)^{1/3}$$

The $A^{2/3}$-term is

$$3\alpha^2\beta + (6 + 3\rho)\alpha^2 = 0 \Rightarrow \beta = -(2+\rho)$$

Finally, γ is obtained from the $A^{1/3}$-term, leading to the following solution:

$$N^* = \left(\frac{3}{2}\right)^{1/3} A^{1/3} - (2+\rho) + \frac{1}{3}\left(\frac{2}{3}\right)^{1/3}\left(3\rho^2 - 3\rho + 1\right)A^{-1/3} + \ldots$$

which, inserted in the expression for the energy E, gives

$$E = \tilde{E} + E_{\text{shell}}$$

where $\tilde{E}$ and E_{shell} are given in the text of the problem (in the present approach, the division of the $A^{1/3}$-term between $\tilde{E}$ and E_{shell} is somewhat arbitrary).

It is instructive to draw E_{shell} and compare it with figs. 3.9 and 9.7:

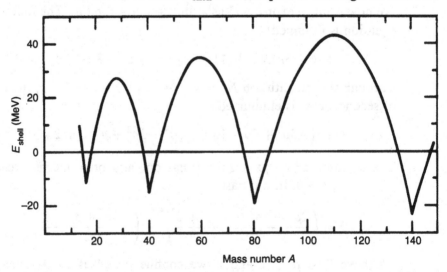

9.2

$$\rho^3 Y_{30} = \frac{1}{4}\left(\frac{7}{4\pi}\right)^{1/2}\left(5\zeta^3 - 3\rho^2\zeta\right) = \frac{1}{4}\left(\frac{7}{4\pi}\right)^{1/2}\left[2\zeta^3 - 3\left(\xi^2 + \eta^2\right)\zeta\right]$$

Now introduce $\zeta = (1/\sqrt{2})\,(a_z^+ + a_z)$ to obtain

$$\zeta^3 = \frac{1}{2\sqrt{2}}\left[(a_z^+)^3 + 3a_z^+ a_z^+ a_z + 3a_z^+ + 3a_z^+ a_z a_z + 3a_z + (a_z)^3\right]$$

where the operators have been put in so called normal order, i.e. creation operators before annihilation operators ($[a_z, a_z^+] = 1$).

Similarly:

$$\left(\xi^2 + \eta^2\right)\zeta = (1/\sqrt{2})\,(1 + R^+R + S^+S + R^+S^+ + RS)\,(a_z^+ + a_z)$$

where

$$a_x^+ = (1/\sqrt{2})\,(R^+ + S^+); \quad [R, R^+] = 1, \quad \text{etc.}$$

We now obtain the selection rules for the ζ^3-term:

$$\Delta n_z = n_z' - n_z = \pm 1, \pm 3; \quad \Delta r = \Delta s = 0; \quad \text{i.e.} \quad \Delta n_\perp = \Delta\Lambda = 0$$

and for the $\left(\xi^2 + \eta^2\right)\zeta$-term:

$$\Delta n_z = \pm 1; \ \Delta r = \Delta s = 0, \pm 1; \quad \text{i.e.} \quad \Delta n_\perp = 0, \pm 2, \ \Delta\Lambda = 0$$

(the selection rule $m' + \mu = m$ of the matrix element $\langle Y_{\ell m} | Y_{\lambda\mu} | Y_{\ell'm'}\rangle$ makes it evident from the beginning that $\Delta\Lambda = 0$).

Expressed in n_z and $n_\perp$, the harmonic oscillator levels are given by (see chapter 8)

$$E_{\text{osc}} = (n_\perp + n_z + 3/2)\,\hbar\omega_0 + (1/3)\varepsilon\hbar\omega_0\,(n_\perp - 2n_z)$$

$$\Rightarrow \delta E = (\Delta n_\perp(1 + \varepsilon/3) + \Delta n_z(1 - 2\varepsilon/3))\,\hbar\omega_0$$

Thus, at large positive ε and for the orbitals that couple through the $\rho^3 Y_{30}$-term, δE is smallest for $\Delta n_\perp = 0$, $\Delta n_z = 1$:

$$\langle n_\perp\ n_z + 1\ \Lambda\,\big|\rho^3 Y_{30}\big|\ n_\perp n_z\Lambda\rangle$$

$$\propto \frac{1}{\sqrt{2}}\langle r\ s\ n_z + 1\,|3a_z^+ a_z^+ a_z + 3a_z^+ - 3\,(1 + R^+R + S^+S)\,a_z^+\,|\,r\ s\ n_z\rangle$$

$$= \frac{3\,(n_z - n_\perp)}{\sqrt{2}}\langle rsn_z + 1\,|a_z^+|\,rsn_z\rangle = 3\,(n_z - n_\perp)\,[(n_z + 1)/2]^{1/2}$$

For $N = n_z + n_\perp$ fixed, i.e. essentially for fixed mass number, this matrix element is largest for n_z being maximum and $n_\perp$ being minimum.

This corresponds e.g. to the coupling between $[Nn_z\Lambda] = [550]$ and $[660]$ orbitals. The matrix element is, however, also quite large for n_z minimum and $n_\perp$ maximum, e.g. couplings between $[Nn_z\Lambda] = [404]$ and $[514]$. These orbitals are illustrated in fig. 9.5. They are important for the formation of 'mass asymmetric fission barriers' not only because the matrix elements are large but also because they have a high degeneracy and because ε_4 helps to make them approach each other at large ε.

9.3

$$G = \alpha_0 + \alpha_1 e + \alpha_2 e^2 + \alpha_3 e^3 + \dots$$

The smearing function is

$$f(u) = \frac{1}{\gamma\sqrt{\pi}} f_{\text{corr}}(u)e^{-u^2}; \quad u = \frac{e - e'}{\gamma}$$

We require that

$$G(e') = \int_{-\infty}^{\infty} G(e)f\left(\frac{e - e'}{\gamma}\right) de \Leftrightarrow$$

$$G(e') = \int_{-\infty}^{\infty} G(\gamma u + e')\gamma f(u) \, du = \int_{-\infty}^{\infty} G(\gamma u + e')\frac{1}{\sqrt{\pi}} f_{\text{corr}}(u)e^{-u^2} \, du$$

(a)

$$G = \alpha_0 + \alpha_2 e^2; \quad f_{\text{corr}}(u) = a_0 + a_2 u^2$$

$$\alpha_0 + \alpha_2 e'^2$$

$$= \int \left(\alpha_0 + \alpha_2\gamma^2 u^2 + 2\alpha_2\gamma u e' + \alpha_2 e'^2\right)\frac{1}{\sqrt{\pi}}\left(a_0 + a_2 u^2\right)e^{-u^2} \, du$$

$$= \left(\alpha_0 + \alpha_2 e'^2\right)\left(a_0 + \frac{1}{2}a_2\right) + \alpha_2\gamma^2\left(\frac{1}{2}a_0 + \frac{3}{4}a_2\right)$$

$$\Rightarrow \begin{cases} a_0 + \frac{1}{2}a_2 = 1 \\ \frac{1}{2}a_0 + \frac{3}{4}a_2 = 0 \end{cases} \Rightarrow \begin{cases} a_0 = 3/2 \\ a_2 = -1 \end{cases}$$

We have used the general solution

$$\int_{-\infty}^{\infty} x^n e^{-x^2} \, dx = \begin{cases} \sqrt{\pi} & \text{for } n = 0 \\ 1 \times 3 \times 5 \dots (n-1)\sqrt{\pi}/2^{n/2} & \text{for } n = 2, 4, 6, \dots \\ 0 & \text{for } n \text{ odd} \end{cases}$$

The smearing function is more peaked than a pure Gaussian and becomes negative before it goes to zero.

(b) With G being a function of order 3, we obtain

$$\alpha_0 + \alpha_1 e' + \alpha_2 e'^2 + \alpha_3 e'^3 =$$
$$\left(\alpha_0 + \alpha_1 e' + \alpha_2 e'^2 + \alpha_3 e'^3 \right) \left(a_0 + \frac{a_2}{2} \right) + \gamma^2 \left(\alpha_2 + 3\alpha_3 e' \right) \left(\frac{a_0}{2} + \frac{3a_2}{4} \right)$$

leading to the same equations for a_0 and a_2 as above.

(c) From the special cases studied it is straightforward to see the pattern that develops. Thus, in the general case:

$$
\begin{pmatrix}
1 & \frac{1}{2} & \frac{3}{2^2} & \frac{3\cdot5}{2^3} & \cdots \\
\frac{1}{2} & \frac{3}{2^2} & \frac{3\cdot5}{2^3} & \frac{3\cdot5\cdot7}{2^4} & \cdots \\
\frac{3}{2^2} & \frac{3\cdot5}{2^3} & \frac{3\cdot5\cdot7}{2^4} & \frac{3\cdot5\cdot7\cdot9}{2^5} & \cdots \\
\frac{3\cdot5}{2^3} & \cdots & & & \\
\vdots & & & &
\end{pmatrix}
\cdot
\begin{pmatrix}
a_0 \\
a_2 \\
a_4 \\
a_6 \\
\vdots
\end{pmatrix}
=
\begin{pmatrix}
1 \\
0 \\
0 \\
0 \\
\vdots
\end{pmatrix}
$$

The regular pattern makes it rather straightforward to find the solutions for polynomials of order 4, 6, ...

10.1 The action integral

$$K = \frac{2}{\hbar} \int_0^1 \left[2k\mu \cdot 0.096 \, \overset{0}{E_s} \, a_2^2 (1 - a_2) \right]^{1/2} R_0 \, \mathrm{d}a_2$$

The integral to be calculated is

$$I = \int_0^1 \left[a_2^2 (1 - a_2) \right]^{1/2} \mathrm{d}a_2$$

The substitution $a_2 = \sin^2 \theta$ leads to

$$I = s \int_0^{\pi/2} \sin^3 \theta \cos^2 \theta \, \mathrm{d}\theta = -2 \int_0^{\pi/2} \left(1 - \cos^2 \theta \right) \cos^2 \theta \, \mathrm{d}(\cos \theta)$$
$$= -2 \int_1^0 \left(1 - x^2 \right) x^2 \, \mathrm{d}x = \ldots = 4/15$$

With a surface coefficient of the mass formula, $a_s \approx 17$ MeV, we obtain for ^{238}U that $E_s \approx 650$ MeV. For division into two equal parts, the reduced mass μ is 1/4 of the mass of the mother nucleus, $\mu = MA/4$. With $\hbar^2/(Mr_0^2) = 28.8$ MeV and with $R_0 = r_0 A^{1/3}$ we obtain

$$\frac{2R_0}{\hbar} \left(2k\mu \cdot 0.096 \, \overset{0}{E_s} \right)^{1/2} \approx 200 \sqrt{k}$$

$$\left. \begin{array}{l} t_{1/2} \approx \ln 2 \cdot 10^{-21} \exp\left(53\sqrt{k}\right) \text{ s} \\ t_{1/2}^{\text{exp}} \approx 10^{16} \text{ years} \approx 3 \times 10^{23}\text{s} \end{array} \right\} \Rightarrow k \approx 3.75$$

$\Delta B_r = 0.1 B_r \Rightarrow t_{1/2}$ changes by approximately a factor of 10^2.

10.2   From fig. 4.2, we find that when r is changed by about $0.05R_0$ away from the ground state minimum, V is increased by about 3 MeV:

$$V = V_0 + \frac{1200}{R_0^2}(r - d_0)^2 \text{ MeV} \quad (R_0 = 1.2 \cdot A^{1/3} \text{ fm})$$

With the mass being m, the H.O. potential is generally written

$$V = \frac{1}{2}m\omega^2 r^2$$

(a)

$$m = \frac{MA}{4} \Rightarrow \frac{MA\omega^2}{8} = \frac{1200}{R_0^2}$$

$$\Rightarrow \hbar\omega = \left(\frac{8 \cdot 1200}{A \cdot A^{2/3}} \frac{\hbar^2}{M r_0^2}\right)^{1/2} \approx 5.46 \text{ MeV}$$

The lowest vibrational state for a one-dimensional oscillator is at

$$E_0 = \frac{1}{2}\hbar\omega_0 \approx 2.7 \text{ MeV}$$

(b)

$$m = \frac{10MA}{4} \Rightarrow E_0 \approx 0.9 \text{ MeV}$$

The β-vibrational state of ^{240}Pu is observed at 0.86 MeV. For $\hbar\omega_0 = 1.7$ MeV, the vibrational frequency becomes $\omega_0 = 2.7 \times 10^{21}\text{s}^{-1}$.

10.3   The solution to this problem can be found in elementary books on quantum mechanics.

10.4   (a) $^{264}108 \rightarrow {}^{260}106 + \alpha$

$$\left. \begin{array}{l} B(^{264}108) = 1928.3 \text{ MeV} \\ B(^{260}106) = 1910.3 \text{ MeV} \\ B(\alpha) \quad\;\; = 28.3 \text{ MeV} \end{array} \right\} \Rightarrow E_\alpha = 10.3 \text{ MeV}$$

$$\Rightarrow t_{1/2} \approx 10^{-12} \text{ years}$$

Sol

(b) $^{294}110 \rightarrow {}^{290}108 + \alpha$

$$E_\alpha = 2089.6 + 28.3 - 2110.8 = 7.1 \text{ MeV} \Rightarrow t_{1/2} \approx 10^0 \text{ years}$$

(c) $^{284}_{98}\text{Cf} \rightarrow {}^{280}_{96}\text{Cm} + \alpha$

$$E_\alpha = 1993.5 + 28.3 - 2021.3 = 0.5 \text{ MeV} \Rightarrow t_{1/2} \approx 10^{156} \text{ years}$$

It is also interesting to see how $t_{1/2}$ varies with E_α. For example, in the case of $^{294}110 \rightarrow {}^{290}108 + \alpha$, the Taagepera–Nurmia formula gives

E_α (MeV)	4.1	6.1	7.1	8.1	10.1
$t_{1/2}$ (years)	10^{20}	10^5	10^0	10^{-4}	10^{-11}

11.1   Denote the longer axis by b and shorter by a. With a mass AM of the nucleus, we obtain for the rigid moment of inertia (rotation around the x-axis):

$$\mathscr{I}_{\text{rig}} = \frac{AM \int_V \left(y^2 + z^2\right) \mathrm{d}^3 r}{\int_V \mathrm{d}^3 r}$$

The integrals are solved as in problem 7.2:

$$\mathscr{I}_{\text{rig}} = \frac{MA \left[(4\pi/15)b^2 a \left(a^2 + b^2\right)\right]}{(4\pi/3)b^2 a} = \frac{MA}{5}\left(a^2 + b^2\right)$$

For a central sphere of radius a:

$$\mathscr{I}_{\text{rig}}^{\text{sphere}} = \frac{2MA}{5}a^2$$

Thus, for the moment of inertia in the two-fluid model:

$$\mathscr{I}_{\text{rig}}^{\text{TF}} = \mathscr{I}_{\text{rig}} - \mathscr{I}_{\text{rig}}^{\text{sphere}} = \frac{MA}{5}\left(b^2 - a^2\right)$$

$$\frac{\mathscr{I}_{\text{rig}}^{\text{TF}}}{\mathscr{I}_{\text{rig}}} = \frac{(b/a)^2 - 1}{(b/a)^2 + 1} = \begin{cases} 0.26 \; ; & b = 1.3a \\ 0.60 \; ; & b = 2a \end{cases}$$

$b = 1.3a$ is a typical value in the ground state of a deformed nucleus while $b \simeq 2a$ in the fission isomers of the actinides or in the superdeformed high-spin states, discussed in chapter 12.

11.2   With $E_I = \left(\hbar^2/2\mathscr{J}\right) I(I+1)$ we get

$$\Delta E_I = E_{I+2} - E_I = \frac{\hbar^2}{2\mathscr{J}}(4I+6)$$

For ^{178}Hf

I	E_I (keV)	$(\Delta E_I/4I + 6)$ (keV)
0	0	15.5
2	93.2	15.3
4	306.8	14.8
6	632.5	14.2
8	1059	

$$\Rightarrow \frac{\hbar^2}{2\mathscr{J}} \simeq 15 \text{ keV}$$

The quadrupole moment of a spheroid with half-axes a and b is according to problem 7.2

$$Q = \frac{2}{5}Z\left(b^2 - a^2\right)$$

To take account of volume conservation to first order, we define

$$\begin{cases} b = R_0\left(1 + \frac{2}{3}\varepsilon\right) \\[2mm] a = R_0\left(1 - \frac{1}{3}\varepsilon\right) \end{cases} ; \quad R_0 = 1.2 \cdot A^{1/3} \text{ fm}$$

(this ε is to first order the same as ε of the M.O. potential)

$$Q = \frac{2}{5}Z R_0^2 \cdot 2\varepsilon + O\left(\varepsilon^2\right) \Rightarrow \varepsilon = 0.29$$

$$\Longrightarrow \frac{\hbar^2}{2\mathscr{J}_{\text{rig}}} = \frac{5\hbar^2}{2MAr_0^2A^{2/3}2(1+\varepsilon/3)} \approx 6 \text{ keV}; \quad \left(\frac{\hbar^2}{Mr_0^2} = 28.8 \text{ MeV}\right)$$

11.3   The orbital with $j = 5/2$ and $j_x = 5/2$ is given by ($\ell = 2$)

$$Y_{22}(\hat{x})\alpha_x$$

where α_x is the wave function having spin up in the x-direction. The transformation of α_x is trivial:

$$\alpha_x = \frac{1}{\sqrt{2}} (\alpha_z + \beta_z)$$

where α_z and β_z have spin up and spin down in the z-direction. The spherical harmonics are given by

$$Y_{22}(\hat{x}) = \frac{1}{r^2} \left(\frac{15}{32\pi} \right)^{1/2} \left(y^2 + 2iyz - z^2 \right)$$

$$Y_{20}(\hat{z}) = \frac{1}{r^2} \left(\frac{5}{16\pi} \right)^{1/2} \left(2z^2 - x^2 - y^2 \right)$$

$$Y_{21}(\hat{z}) + Y_{2-1}(\hat{z}) = -\frac{2}{r^2} \left(\frac{15}{8\pi} \right)^{1/2} (iyz)$$

$$Y_{22}(\hat{z}) + Y_{2-2}(\hat{z}) = \frac{1}{r^2} \left(\frac{15}{8\pi} \right)^{1/2} \left(x^2 - y^2 \right)$$

(the symmetry requires that $Y_{2\mu}(\hat{z})$ and $Y_{2-\mu}(\hat{z})$ enter with the same amplitude).

After some straightforward calculations

$$Y_{22}(\hat{x}) = -\left(\frac{3}{8} \right)^{1/2} Y_{20}(\hat{z}) - \frac{1}{2} (Y_{21}(\hat{z}) + Y_{2-1}(\hat{z})) - \frac{1}{4} (Y_{22}(\hat{z}) + Y_{2-2}(\hat{z}))$$

We now expand:

$$|j = 5/2 \ j_x = 5/2\rangle = \sum_{j_z} C_{j_z} |j = 5/2 \ j_z\rangle$$

Combination with the spin wave function leads to

$$\left| C_{\pm 1/2} \right|^2 = 5/16 \qquad \left| C_{\pm 3/2} \right|^2 = 5/32 \qquad \left| C_{\pm 5/2} \right|^2 = 1/32$$

Compare these numbers with those for recoupling of a $j = 13/2$ shell, which were given without proof in the text. Note that neither ℓ_z nor s_z are good quantum numbers but $\ell = 2$ and $j = 5/2$ are preserved.

11.4 With the kinetic energy of the projectile being T MeV, we obtain for its velocity (which defines the x-direction)

$$v_{1x} = \left(\frac{2T}{A_1 M} \right)^{1/2}$$

We must go to the centre-of-mass system. In this system, the velocities are

$$v'_{1x} = \frac{A_2}{A_1 + A_2} v_{1x}; \qquad v'_{2x} = -\frac{A_1}{A_1 + A_2} v_{1x}$$

The angular momentum with respect to the centre of mass is thus $\left(R_2 = r_0 A_2^{1/3}\right)$

$$\begin{aligned}
I &= MA_1 \frac{A_2}{A_1 + A_2}\left(\frac{3R_2}{4}\right)\frac{A_2}{A_1 + A_2}v_{1x} \\
&\quad + MA_2 \frac{A_1}{A_1 + A_2}\left(\frac{3R_2}{4}\right)\frac{A_1}{A_1 + A_2}v_{1x} \\
&= (2M)^{1/2}\frac{3r_0}{4}\sqrt{T}\frac{\sqrt{A_1 A_2^{4/3}}}{A_1 + A_2}
\end{aligned}$$

where $[A_2/(A_1 + A_2)]\,3R_2/4$ and $[A_1/(A_1 + A_2)]\,3R_2/4$ are the perpendicular distances to the respective velocity vectors. With $\hbar^2/Mr_0^2 = 28.8$ MeV, this leads to

$$I = \frac{3}{4}\hbar\left(\frac{2T}{28.8}\right)^{1/2}\cdot\frac{\sqrt{A_1 A_2^{4/3}}}{A_1 + A_2}$$

(a) $A_1 = 4$, $A_2 = 160$, $T = 60$ MeV $\Rightarrow I \simeq 16\hbar$
(b) $A_1 = 40$, $A_2 = 124$, $T = 180$ MeV $\Rightarrow I \simeq 63\hbar$

The kinetic energy from the motion of the centre of mass is

$$2T_{\mathrm{CM}} = (A_1 + A_2)M\left(\frac{A_1}{A_1 + A_2}\right)^2 v_1^2 = \frac{A_1^2}{A_1 + A_2}Mv_1^2$$

The rest of the kinetic energy is transformed to excitation energy (including rotational energy):

$$2T^{\mathrm{exc}} = A_1 Mv_1^2 - \frac{A_1^2}{A_1 + A_2}Mv_1^2 = A_1 Mv_1^2 \frac{A_2}{A_1 + A_2}$$

In case (a) $\frac{1}{2}A_1 Mv_1^2 = 60$ MeV and $T^{\mathrm{exc}} = 58.5$ MeV.
The total excitation energy is ($B =$ binding energy)

$$\begin{aligned}
E^{\mathrm{exc}} &= B\left(^{164}\mathrm{Er}\right) - B\left(^{160}\mathrm{Dy}\right) - B\left(^4\mathrm{He}\right) + T^{\mathrm{exc}} \\
&= (1336.5 - 1309.5 - 28.3 + 58.5)\ \mathrm{MeV} = 57\ \mathrm{MeV}
\end{aligned}$$

For $I = 16\hbar$, this corresponds to about 54 MeV above the yrast line (see fig. 11.11). Thus, if six neutrons are emitted each carrying their binding energy of about 8 MeV and in addition a small amount of kinetic energy but no angular momentum, an $I = 16$ state close to the yrast line of $^{158}\mathrm{Er}$ is reached.

Case (b) $A_1 M v_1^2 = 180$ MeV; $T^{\text{exc}} = 136.1$ MeV

$$E^{\text{exc}} = B\left(^{164}\text{Er}\right) - B\left(^{124}\text{Sm}\right) - B\left(^{40}\text{Ar}\right) + T^{\text{exc}}$$
$$= (1336.5 - 1050.0 - 343.8 + 136.1)\text{MeV} = 79 \text{ MeV}$$

With an estimated yrast energy of 35 MeV for $I = 63\hbar$, this suggests the possibility of a $(^{40}\text{Ar}, 5n)$ or $(^{40}\text{Ar}, 4n)$ reaction.

12.1 Coordinate transformation:

$$x = x_1$$
$$y = x_2 \cos \omega t - x_3 \sin \omega t$$
$$z = x_2 \sin \omega t + x_3 \cos \omega t$$

The velocities are then given by

$$\dot{x} = \dot{x}_1$$
$$\dot{y} = (\dot{x}_2 - \omega x_3) \cos \omega t - (\dot{x}_3 + \omega x_2) \sin \omega t$$
$$\dot{z} = (\dot{x}_3 + \omega x_2) \cos \omega t + (\dot{x}_2 - \omega x_3) \sin \omega t$$

The kinetic energy is:

$$T = \frac{1}{2}m\left(\dot{x}^2 + \dot{y}^2 + \dot{z}^2\right) = \ldots =$$
$$= \frac{1}{2}m\left[\dot{x}_1^2 + \dot{x}_2^2 + \dot{x}_3^2 - 2\omega\left(\dot{x}_2 x_3 - \dot{x}_3 x_2\right) + \omega^2\left(x_2^2 + x_3^2\right)\right]$$

The Lagrangian is given by $\mathscr{L} = T - V$ where V is the potential energy, $V(x_1, x_2, x_3)$. The generalised momenta are

$$p_1 = \frac{\partial \mathscr{L}}{\partial \dot{x}} = m\dot{x}_1$$
$$p_2 = \frac{\partial \mathscr{L}}{\partial \dot{x}_2} = m\dot{x}_2 - m\omega x_3$$
$$p_3 = \frac{\partial \mathscr{L}}{\partial \dot{x}_3} = m\dot{x}_3 + m\omega x_2$$

The Hamiltonian is given by

$$H(x_i, p_i) = \sum_i \dot{x}_i \frac{\partial \mathscr{L}}{\partial \dot{x}_i} - \mathscr{L}$$
$$= \sum_i \frac{1}{2}m\dot{x}_i^2 - \frac{1}{2}m\omega^2\left(x_2^2 + x_3^2\right) + V(x_1, x_2, x_3)$$
$$= T + m\omega\left[x_3\left(\dot{x}_2 - \omega x_3\right) - x_2\left(\dot{x}_3 + \omega x_2\right)\right] + V(x_1, x_2, x_3)$$

$$= \sum \frac{p_i^2}{2m} + \omega(x_3 p_2 - x_2 p_3) + V(x_1, x_2, x_3)$$

$$= \sum \frac{p_i^2}{2m} + V(x_1, x_2, x_3) - \omega \ell_1$$

where ℓ_1 is the angular momentum operator. This is thus equivalent to the cranking Hamiltonian, $h^\omega = h - \omega \ell_1$.

12.2   We use the same formalism as for the rotating potential and define

$$\Sigma_k = \sum_{occ} \left\langle v \left| a_k^+ a_k + \frac{1}{2} \right| v \right\rangle ; \quad k = 1, 2, 3$$

with summation over occupied orbitals. The energy is given by

$$E = \hbar \omega_1 \Sigma_1 + \hbar \omega_2 \Sigma_2 + \hbar \omega_3 \Sigma_3$$

and minimisation under the constraint of volume conservation, $\omega_1 \omega_2 \omega_3 = \left(\overset{0}{\omega_0} \right)^3$, leads to

$$\omega_1 \Sigma_1 = \omega_2 \Sigma_2 = \omega_3 \Sigma_3 = \text{constant}$$

The expectation value, $\langle x_k^2 \rangle$ is given by

$$\langle x_k^2 \rangle = \sum_{occ} \left\langle v \left| \frac{\hbar}{2M\omega_k} [2a_k^+ a_k + 1 - (a_k^+ a_k^+ + a_k a_k)] \right| v \right\rangle$$

$$= \sum \frac{\hbar}{M\omega_k} \left(n_k + \frac{1}{2} \right)_v = \frac{\hbar}{M\omega_k} \Sigma_k = \text{constant} \frac{\hbar}{M\omega_k^2}$$

The shape of an equipotential surface is ellipsoidal with half-axes a, b and c proportional to $1/\omega_i$. Thus

$$\langle x_1^2 \rangle = \int_V x_1^2 \, d^3 r = \frac{4\pi}{15} abc \, a^2 \propto \frac{1}{5} V \cdot \frac{1}{\omega_1^2}$$

where the integral is solved as in problem 7.2.

Equivalent expressions are found for $\langle x_2^2 \rangle$ and $\langle x_3^2 \rangle$. Consequently, in both cases $\langle x_k^2 \rangle \propto \left(1/\omega_k^2 \right)$.

12.3   The parameters ε and γ are defined as

$$\omega_1 = \omega_0(\varepsilon, \gamma) \left[1 - \frac{2}{3} \varepsilon \cos \left(\gamma + \frac{2\pi}{3} \right) \right]$$

$$\omega_2 = \omega_0(\varepsilon, \gamma) \left[1 - \frac{2}{3} \varepsilon \cos \left(\gamma - \frac{2\pi}{3} \right) \right]$$

$$\omega_3 = \omega_0(\varepsilon, \gamma)\left(1 - \frac{2}{3}\varepsilon\cos\gamma\right)$$

with the self-consistent frequencies

$$\overset{0}{\omega}_0\left(\Sigma_1\tilde{\Sigma}_2\tilde{\Sigma}_3\right)^{1/3}/\Sigma_1 = \omega_0\left(1 + \frac{1}{3}\varepsilon\cos\gamma + \frac{\varepsilon}{\sqrt{3}}\sin\gamma\right) \qquad \text{(E12.1)}$$

$$\overset{0}{\omega}_0\left(\Sigma_1\tilde{\Sigma}_2\tilde{\Sigma}_3\right)^{1/3}/\tilde{\Sigma}_2 = \omega_0\left(1 + \frac{1}{3}\varepsilon\cos\gamma - \frac{\varepsilon}{\sqrt{3}}\sin\gamma\right) \qquad \text{(E12.2)}$$

$$\overset{0}{\omega}_0\left(\Sigma_1\tilde{\Sigma}_2\tilde{\Sigma}_3\right)^{1/3}/\tilde{\Sigma}_3 = \omega_0\left(1 - \frac{2}{3}\varepsilon\cos\gamma\right) \qquad \text{(E12.3)}$$

We thus have three equations to determine ε, γ and $\omega_0(\varepsilon, \gamma)$.

$$(1)+(2)+(3) \Rightarrow \overset{0}{\omega}_0\left(\Sigma_1\tilde{\Sigma}_2\tilde{\Sigma}_3\right)^{1/3}\left(\Sigma_1^{-1} + \tilde{\Sigma}_2^{-1} + \tilde{\Sigma}_3^{-1}\right) = 3\omega_0 \quad \text{(E12.4)}$$

This is the volume conservation condition and gives $\omega_0/\overset{0}{\omega}_0$ as a function of ε and γ:

$$(1)-(2) \Rightarrow \varepsilon\sin\gamma = \frac{\sqrt{3}}{2}\frac{\overset{0}{\omega}_0}{\omega_0}\left(\Sigma_1\tilde{\Sigma}_2\tilde{\Sigma}_3\right)^{1/3}\left(\Sigma_1^{-1} - \tilde{\Sigma}_2^{-1}\right) \quad \text{(E12.5)}$$

$$(1)+(2)-2\times(3) \Rightarrow \varepsilon\cos\gamma = \frac{\overset{0}{\omega}_0}{2\omega_0}\left(\Sigma_1\tilde{\Sigma}_2\tilde{\Sigma}_3\right)^{1/3}\left(\Sigma_1^{-1} + \tilde{\Sigma}_2^{-1} - 2\tilde{\Sigma}_3^{-1}\right)$$
$$\text{(E12.6)}$$

Eq. (5) is now divided by eq. (6):

$$\gamma = \arctan\frac{\sqrt{3}\left(\Sigma_1^{-1} - \tilde{\Sigma}_2^{-1}\right)}{\left(\Sigma_1^{-1} + \tilde{\Sigma}_2^{-1} - 2\tilde{\Sigma}_3^{-1}\right)}$$

Eqs. (5) and (6) are squared and added. The value of $\omega_0/\overset{0}{\omega}_0$ is taken from eq. (4):

$$\varepsilon = \frac{3\left(\Sigma_1^{-2} + \tilde{\Sigma}_2^{-2} + \tilde{\Sigma}_3^{-2} - \Sigma_1^{-1}\tilde{\Sigma}_2^{-1} - \tilde{\Sigma}_2^{-1}\tilde{\Sigma}_3^{-1} - \tilde{\Sigma}_3^{-1}\Sigma_1^{-1}\right)^{1/2}}{\Sigma_1^{-1} + \tilde{\Sigma}_2^{-1} + \tilde{\Sigma}_3^{-1}}$$

12.4

$$\mathscr{I}_{\text{stat}} = M\sum_{\nu}^{\text{occ}}\left\langle\nu\left|y^2 + z^2\right|\nu\right\rangle$$

$$= M\sum_{\nu}^{\text{occ}}\left\langle\nu\left|\frac{\hbar}{2M\omega_2}(a_2^+ + a_2)(a_2^+ + a_2)\right.\right.$$

$$- \frac{\hbar}{2M\omega_3} (a_3^+ - a_3)(a_3^+ - a_3) \Big| v \Big\rangle$$

$$= \sum_v^{\text{occ}} \Big\langle v \Big| \frac{\hbar}{2\omega_2} (a_2^+ a_2 + a_2 a_2^+) + \frac{\hbar}{2\omega_3} (a_3^+ a_3 + a_3 a_3^+) \Big| v \Big\rangle$$

where the summation runs over occupied orbitals. We have used the fact that operators that change the number of quanta give no contribution to the matrix element.

$$\mathscr{I}_{\text{stat}} = \sum_v^{\text{occ}} \Big\langle v \Big| \frac{\hbar}{\omega_2} \Big(a_2^+ a_2 + \frac{1}{2} \Big) + \frac{\hbar}{\omega_3} \Big(a_3^+ a_3 + \frac{1}{2} \Big) \Big| v \Big\rangle$$

Essentially, the same operator is found in Hamiltonian h_{osc}. A transformation to the operators a_α^+ and a_β^+ leads to

$$\mathscr{I}_{\text{stat}} = \sum_v^{\text{occ}} \Big\langle v \Big| \frac{\hbar}{2\omega_2} + \frac{\hbar}{2\omega_3} + \hbar a_\alpha^+ a_\alpha \Big(\frac{1}{\omega_2} \cos^2 \phi + \frac{1}{\omega_3} \sin^2 \phi \Big)$$

$$+ \hbar a_\beta^+ a_\beta \Big(\frac{1}{\omega_2} \sin^2 \phi + \frac{1}{\omega_3} \cos^2 \phi \Big) \Big| v \Big\rangle$$

We have omitted terms of the type $a_\alpha^+ a_\beta$ and $a_\beta^+ a_\alpha$ because $\langle v | a_\alpha^+ a_\beta | v \rangle = 0$, etc. We write $(\hbar/2\omega_2)$ as $(\hbar/2\omega_2)(\cos^2 \phi + \sin^2 \phi)$ and $(\hbar/2\omega_3)$ in an analogous way. The definitions of Σ_α and Σ_β are now used to obtain

$$\mathscr{I}_{\text{stat}} = \hbar \Sigma_\alpha \Big(\frac{1}{\omega_2} \cos^2 \phi + \frac{1}{\omega_3} \sin^2 \phi \Big) + \hbar \Sigma_\beta \Big(\frac{1}{\omega_2} \sin^2 \phi + \frac{1}{\omega_3} \cos^2 \phi \Big)$$

The frequencies are given by $\omega_i = \overset{0}{\omega}_0 (\Sigma_1 \tilde{\Sigma}_2 \tilde{\Sigma}_3)^{1/3} / \tilde{\Sigma}_i$ where $\tilde{\Sigma}_{2,3} = \frac{1}{2}(\Sigma_\alpha + \Sigma_\beta) \mp \frac{1}{2}(I_m^2 - I^2)^{1/2}$. Thus

$$\mathscr{I}_{\text{stat}} = \frac{\hbar}{2 \overset{0}{\omega}_0 \Big[\Sigma_1 \Big(\Sigma_\alpha \Sigma_\beta + \frac{1}{4} I^2 \Big) \Big]^{1/3}}$$

$$\times \Big\{ \Big[(\Sigma_\alpha + \Sigma_\beta)^2 + (\Sigma_\beta - \Sigma_\alpha)(I_m^2 - I^2)^{1/2} \Big] \cos^2 \phi$$

$$+ \Big[(\Sigma_\alpha + \Sigma_\beta)^2 - (\Sigma_\beta - \Sigma_\alpha)(I_m^2 - I^2)^{1/2} \Big] \sin^2 \phi \Big\}$$

$$= \frac{\hbar}{2 \overset{0}{\omega}_0 \Big[\Sigma_1 \Big(\Sigma_\alpha \Sigma_\beta + \frac{1}{4} I^2 \Big) \Big]^{1/3}}$$

$$\times \Big[(\Sigma_\alpha + \Sigma_\beta)^2 - (\Sigma_\beta - \Sigma_\alpha)(I_m^2 - I^2)^{1/2} (\cos^2 \phi - \sin^2 \phi) \Big]$$

We have $\tan 2\phi = p \Rightarrow$

$$\cos^2 \phi - \sin^2 \phi = \cos 2\phi = \frac{1}{(1+p^2)^{1/2}} = \frac{(I_m^2 - I^2)^{1/2}}{I_m}$$

With $I_m = \Sigma_\beta - \Sigma_\alpha$ we finally arrive at

$$\mathscr{I}_{\text{stat}} = \frac{\Sigma_\alpha^2 + \Sigma_\beta^2 - \frac{1}{2}I^2}{\left[\Sigma_1 \left(\Sigma_\alpha \Sigma_\beta + \frac{1}{4}I^2\right)\right]^{1/3}} \left(\begin{matrix} \hbar \\ 0 \\ \omega_0 \end{matrix}\right)$$

12.5   The ground state configuration of ^{24}Mg is determined by some simple reasoning. With the $N = 0$ and $N = 1$ shells completely filled and with eight particles in $N = 2$, one obtains the total number of quanta: $\Sigma = \Sigma_x + \Sigma_y + \Sigma_z = 24 \times 3 \times 1/2 + 12 \times 1 + 8 \times 2 = 64$. The first term comes from '1/2' in the definition of the Σ_i terms. For $I = 0$, the total energy is given by $E = 3\hbar\overset{0}{\omega}_0 \left(\Sigma_x \Sigma_y \Sigma_z\right)^{1/3}$. For a fixed sum, Σ, it is easy to conclude that this expression has a minimum if Σ_z is as large as possible and Σ_x as small as possible (assuming $\Sigma_z \geq \Sigma_y \geq \Sigma_x$). In this case, for the first four particles in the $N = 2$ shell, we can put both quanta in the z-direction while for the next four, we put one quantum in the z-direction and one in the y-direction. Thus, there will be no quanta in the x-direction for $N = 2$ and Σ_x is minimised simultaneously as Σ_z is maximised (the $N = 1$ shell is completely filled so, in this shell, there are as many quanta in all three directions). For the ground state of ^{24}Mg we now find

$$\Sigma_x = 16, \quad \Sigma_y = 20 \quad \text{and} \quad \Sigma_z = 28$$

In more complicated cases, there might be a conflict between maximising Σ_z and minimising Σ_x but it is then straightforward to evaluate E for a few different cases. Furthermore, it might be advantageous to excite particles to higher shells, for example four particles in $N = 2$ and four in $N = 3$ in the present case. It turns out, however, that the configuration given above has the lowest energy for ^{24}Mg.

The ground state configuration of ^{24}Mg is triaxial and three rotational bands are formed from rotation around the x-, y- and z-axes; in practice by putting $\Sigma_1 = \Sigma_x$, $\Sigma_1 = \Sigma_y$ and $\Sigma_1 = \Sigma_z$, respectively. It is then straightforward to insert in the explicit formulae to calculate how ε, γ and E vary with I and which are the maximal spins for the three bands. The solution is given in graphical form in the figures of

Cerkaski and Szymański (1979) and also in fig. 24 of Ragnarsson et al. (1981).

If we instead would confine ourselves to prolate ground states, we would put the four particles outside ^{20}Ne in the $[Nn_z\Lambda] = [211]$ orbital where we would then make an equal distribution of the perpendicular quanta in the x- and y-directions leading to $\Sigma_x = \Sigma_y = 18, \Sigma_z = 28$. The energy of this configuration is only marginally higher than for the triaxial configuration, so in a more realistic potential, ^{24}Mg might very well come out as axially symmetric.

12.6

$$I = \mathcal{J}\omega \Rightarrow \hbar\omega = I\hbar/\mathcal{J} = 5I\hbar/(2MA^{5/3}r_0^2)$$

where the rigid moment of inertia, $\mathcal{J}_{\text{rig}} = (2/5)\,MA^{5/3}r_0^2$, as calculated in problem 11.1 has been inserted. With $(\hbar^2/Mr_0^2) = 28.8$ MeV we have

$$I = 8\hbar, \ A = 20 \Rightarrow \hbar\omega = \frac{40}{2(20)^{5/3}} \cdot \frac{\hbar^2}{Mr_0^2} = 3.9 \text{ MeV}$$

and thus $\omega = 5.9 \times 10^{21}$ s^{-1}. As the moment of inertia scales as $A^{5/3}$, the same rotational frequency leads to

$$I = (160/20)^{5/3} \cdot 8\hbar = 256\hbar$$

for $A = 160$. The radius is proportional to $A^{1/3}$, so the same velocity on the surface leads to

$$I = (160/20)^{4/3}8\hbar = 128\hbar$$

13.1   The kinetic energy is

$$T = \frac{m_1\dot{r}_1^2}{2} + \frac{m_2\dot{r}_2^2}{2}$$

The potential energy is

$$V = V\left(|\mathbf{r}_1 - \mathbf{r}_2|\right)$$

The Lagrangian is

$$\mathcal{L} = T - V$$

The canonical momenta are

$$p_j = \frac{\partial\mathcal{L}}{\partial\dot{q}_j} \Rightarrow \begin{cases} \mathbf{p}_1 = m_1\dot{\mathbf{r}}_1 \\ \mathbf{p}_2 = m_2\dot{\mathbf{r}}_2 \end{cases}$$

Introduce

$$\mathbf{r} = \mathbf{r}_1 - \mathbf{r}_2, \quad \mathbf{R} = \frac{m_1\mathbf{r}_1 + m_2\mathbf{r}_2}{M} \; ; \; M = m_1 + m_2$$

$$p_x = \frac{\partial \mathcal{L}}{\partial \dot{x}} = \frac{\partial \mathcal{L}}{\partial \dot{x}_1}\frac{\partial \dot{x}_1}{\partial \dot{x}} + \frac{\partial \mathcal{L}}{\partial \dot{x}_2}\frac{\partial \dot{x}_2}{\partial \dot{x}}$$

Express $\mathbf{r}$ and $\mathbf{R}$ in $\mathbf{r}_1$ and $\mathbf{r}_2$:

$$\mathbf{r}_1 = \mathbf{R} + \frac{m_2}{M}\mathbf{r} \; ; \; \mathbf{r}_2 = \mathbf{R} - \frac{m_1}{M}\mathbf{r}$$

Then

$$p_x = m_1\dot{x}_1\frac{m_2}{M} + m_2\dot{x}_2\left(-\frac{m_1}{M}\right) = \frac{m_2 p_{1x} - m_1 p_{2x}}{M} \; ; \text{(cycl.)}$$

$$\Rightarrow \mathbf{p} = \frac{m_2\mathbf{p}_1 - m_1\mathbf{p}_2}{M}$$

Similarly:

$$P_x = \frac{\partial \mathcal{L}}{\partial \dot{X}} \Rightarrow \ldots \Rightarrow \mathbf{P} = \mathbf{p}_1 + \mathbf{p}_2$$

The Hamiltonian is

$$H = \sum p_j q_j - \mathcal{L} = T + V = \frac{p_1^2}{2m_1} + \frac{p_2^2}{2m_2} + V$$

We now express $\mathbf{p}_1$ and $\mathbf{p}_2$ in terms of $\mathbf{p}$ and $\mathbf{P}$ and then $T = T(\mathbf{p}, \mathbf{P})$:

$$T = \frac{1}{2m_1}\left(\mathbf{p} + \frac{m_1}{M}\mathbf{P}\right)^2 + \frac{1}{2m_2}\left(\mathbf{p} - \frac{m_2}{M}\mathbf{P}\right)^2 = \frac{1}{2\mu}p^2 + \frac{1}{2M}P^2$$

where μ is the reduced mass, $\mu = m_1 m_2/(m_1 + m_2)$. We now get

$$H = \frac{1}{2\mu}p^2 + \frac{1}{2M}P^2 + V(r)$$

or equivalently

$$H = \frac{1}{2}\mu\dot{r}^2 + \frac{1}{2}M\dot{R}^2 + V(r)$$

If $m_1 = m_2 = m$ then $\mu = m/2$.

13.2 The triplet wave function

$$\chi_m^{S=i} = \sum_{m_1 m_2} C_{m_1 m_2 m}^{\frac{1}{2}\frac{1}{2}1}\phi_{m_1}(1)\phi_{m_2}(2)$$

$$= \sum_{m_1 m_2} \langle 1/2 \; 1/2 \; m_1 m_2 | 1m\rangle \phi_{m_1}(1)\phi_{m_2}(2)$$

where the one-particle spin wave function is denoted by ϕ_m. The $m = 1$ case is trivial ($m = 1 \Rightarrow m_1 = m_2 = 1/2$):

$$\chi_1^1 = \langle 1/2\ 1/2\ 1/2\ 1/2 | 1\ 1 \rangle \phi_{1/2}(1)\phi_{1/2}(2) = \alpha(1)\alpha(2)$$

For $m = 0$, then $m_1 = 1/2$ and $m_2 = -1/2$ or $m_1 = -1/2$ and $m_2 = 1/2$:

$$\chi_0^1 = \langle 1/2\ 1/2\ 1/2\ -1/2 | 1\ 0 \rangle \alpha(1)\beta(2) \\ + \langle 1/2\ 1/2\ -1/2\ 1/2 | 1\ 0 \rangle \beta(1)\alpha(2)$$

We use the tables of appendix 6B and the symmetry relation

$$\langle j_1 j_2 m_1 m_2 | j_3 m_3 \rangle = (-1)^{j_1 + j_2 - j_3} \langle j_2 j_1 m_2 m_1 | j_3 m_3 \rangle$$

to obtain

$$\chi_0^1 = \frac{1}{\sqrt{2}}(\alpha(1)\beta(2) + \beta(1)\alpha(2))$$

The wave functions χ_{-1}^1 and χ_0^0 can be derived in a similar way. We will now instead determine χ_0^1 from

$$\chi = A\alpha(1)\alpha(2) + B\alpha(1)\beta(2) + C\beta(1)\alpha(2) + D\beta(1)\beta(2)$$

and require $S^2\chi = 1(1+1)\chi$; $S_z\chi = 0$; $\langle \chi|\chi \rangle = 1$

$$S_z\chi = 0 \Rightarrow A = D = 0$$

Expand $S^2 = s_1^2 + s_2^2 + 2s_{1z}s_{2z} + s_{1+}s_{2-} + s_{1-}s_{2+}$ and thus e.g.

$$S^2\alpha(1)\beta(2) = \left[\frac{3}{4} + \frac{3}{4} + 2\left(\frac{1}{2}\right)\left(-\frac{1}{2}\right)\right]\alpha(1)\beta(2) + \beta(1)\alpha(2)$$

$$S^2\chi = 2\chi \Rightarrow \begin{cases} B + C = 2B \\ C + B = 2C \end{cases} \Rightarrow C = B$$

The normalisation is $B^2 + C^2 = 1 \Rightarrow C = B = 1/\sqrt{2}$ in agreement with χ_0^1 derived above.

13.3   From the given formula:

$$(\boldsymbol{\sigma}_1 \cdot \boldsymbol{\sigma}_2)(\boldsymbol{\sigma}_1 \cdot \boldsymbol{\sigma}_2) = \sigma_2^2 + i\boldsymbol{\sigma}_1 \cdot (\boldsymbol{\sigma}_2 \times \boldsymbol{\sigma}_2)$$

As $\sigma_2^2 = 3$ and $\boldsymbol{\sigma}_2 \times \boldsymbol{\sigma}_2 = 2i\boldsymbol{\sigma}_2$, we obtain the desired relation

$$(\boldsymbol{\sigma}_1 \cdot \boldsymbol{\sigma}_2)^2 - 3 + 2\boldsymbol{\sigma}_1 \cdot \boldsymbol{\sigma}_2 = 0$$

We now assume

$$\boldsymbol{\sigma}_1 \cdot \boldsymbol{\sigma}_2 \chi^{\sigma} = \lambda \chi^{\sigma}$$

i.e. χ^{σ} is an eigenvector to $\boldsymbol{\sigma}_1 \cdot \boldsymbol{\sigma}_2$ with the eigenvalue λ. The relation above now gives the equation

$$\lambda^2 - 3 + 2\lambda = 0$$

with the solutions $\lambda_1 = 1$ and $\lambda_2 = -3$ (from the main text, we already know that the corresponding eigenvectors are the triplet and singlet states).

13.4

$$P^{\sigma} = \frac{1}{2}(1 + \boldsymbol{\sigma}_1 \cdot \boldsymbol{\sigma}_2) \,; \text{ with } \hbar = 1 \colon \mathbf{s} = \boldsymbol{\sigma}/2$$

$$\boldsymbol{\sigma}_1 \cdot \boldsymbol{\sigma}_2 = 4\mathbf{s}_1 \cdot \mathbf{s}_2 = 2\left(S^2 - s_1^2 - s_2^2\right)$$

$$\Rightarrow \boldsymbol{\sigma}_1 \cdot \boldsymbol{\sigma}_2 \chi_m^1 = 2\left(2 - \frac{3}{4} - \frac{3}{4}\right)\chi_m^1 = \chi_m^1$$

$$\boldsymbol{\sigma}_1 \cdot \boldsymbol{\sigma}_2 \chi_0^0 = 2\left(0 - \frac{3}{4} - \frac{3}{4}\right)\chi_0^0 = -3\chi_0^0 \Rightarrow P^{\sigma}\chi_m^1 = \chi_m^1 \,; \, P^{\sigma}\chi_0^0 = -\chi_0^0$$

The space of the two-spin wave function is spanned by $\alpha(1)\alpha(2)$, $\alpha(1)\beta(2)$, $\beta(1)\alpha(2)$ and $\beta(1)\beta(2)$. Then, for example

$$P^{\sigma}\alpha(1)\beta(2) = \frac{1}{2}(\alpha(1)\beta(2) + \boldsymbol{\sigma}_1 \cdot \boldsymbol{\sigma}_2 \alpha(1)\beta(2))$$

$$= \frac{1}{2}\alpha(1)\beta(2) + (s_{1+}s_{2-} + s_{1-}s_{2+} + 2s_{1z}s_{2z})\alpha(1)\beta(2)$$

$$= \frac{1}{2}\alpha(1)\beta(2) + \beta(1)\alpha(2) + 2\left(\frac{1}{2}\right)\left(-\frac{1}{2}\right)\alpha(1)\beta(2) = \beta(1)\alpha(2)$$

i.e. an exchange of the spin components. The exchange property is shown in a similar way for $\alpha(1)\alpha(2)$ etc.

13.5

$$V(r) = f(r)(1 + a\boldsymbol{\sigma}_1 \cdot \boldsymbol{\sigma}_2)$$

For the singlet state χ_0^0:

$$V(r)\chi_0^0 = f(r)(1 + a\boldsymbol{\sigma}_1 \cdot \boldsymbol{\sigma}_2)\chi_0^0 = f(r)(1 - 3a)\chi_0^0$$

For the triplet state χ_m^1:

$$V(r)\chi_m^1 = f(r)(1 + a\boldsymbol{\sigma}_1 \cdot \boldsymbol{\sigma}_2)\chi_m^1 = f(r)(1 + a)\chi_m^1$$

$$V_{\text{triplet}} = 2V_{\text{singlet}} \Rightarrow 1 + a = 2(1 - 3a) \Rightarrow a = 1/7$$

13.6 (a) From the relations

$$\sigma_1 \cdot \sigma_2 \chi_0^1 = \chi_0^1$$

$$\sigma_1 \cdot \sigma_2 \chi_0^0 = -3\chi_0^0$$

it follows almost immediately that

$$A + B = 1, \quad A - 3B = 0$$

corresponding to

$$A = 3/4 \quad \text{and} \quad B = 1/4$$

Similarly, C and D are obtained as $C = 1/4$ and $D = -1/4$.
The relation $P^2 = P$ follows from the projection properties but
can also be shown directly by help of the formula of problem 13.3.

$$(P(S = 1))^2 = \frac{1}{16} (3 + \sigma_1 \cdot \sigma_2)(3 + \sigma_1 \cdot \sigma_2)$$

$$= \frac{1}{16} \left[9 + 6\sigma_1 \cdot \sigma_2 + (\sigma_1 \cdot \sigma_2)^2 \right] = \frac{1}{4} (3 + \sigma_1 \cdot \sigma_2)$$

(b)

$$|j_1 m_1 j_2 m_2\rangle = \sum_{J(M)} \langle j_1 m_1 j_2 m_2 | JM \rangle |j_1 j_2 JM\rangle$$

$$\Rightarrow \alpha(1)\beta(2) = \langle 1/2\,1/2\,1/2 - 1/2|10\rangle \chi_0^1 + \langle 1/2\,1/2\,1/2 - 1/2|00\rangle \chi_0^0$$

Insert the Clebsch–Gordan coefficients

$$\alpha(1)\beta(2) = \frac{1}{\sqrt{2}} \left(\chi_0^1 + \chi_0^0 \right)$$

Alternatively:

$$P(S = 1)\alpha(1)\beta(2) = \frac{1}{4} (3 + \sigma_1 \cdot \sigma_2)\,\alpha(1)\beta(2)$$

$$= \left(\frac{3}{4} + \frac{1}{4}\sigma_{1z}\sigma_{2z} + \frac{1}{8} (\sigma_{1+}\sigma_{2-} + \sigma_{1-}\sigma_{2+}) \right) \alpha(1)\beta(2)$$

$$= \frac{1}{2} (\alpha(1)\beta(2) + \beta(1)\alpha(2)) = \frac{1}{\sqrt{2}} \chi_0^1 \Rightarrow c = 1/\sqrt{2}$$

(we have used the formulae $\sigma_z \alpha = \alpha$, $\sigma_- \alpha = 2\beta$ etc., which are equivalent to $s_z \alpha = \frac{1}{2}\alpha$, $s_- \alpha = \beta$ etc.).

13.8

$$S_{12} = \frac{3}{r^2} (\boldsymbol{\sigma}_1 \cdot \mathbf{r})(\boldsymbol{\sigma}_2 \cdot \mathbf{r}) - \boldsymbol{\sigma}_1 \cdot \boldsymbol{\sigma}_2$$

We want to show that $\int S_{12} \, d\Omega = 0$

$$\int S_{12} \, d\Omega$$
$$= \int \left(\frac{3}{r^2} (\sigma_{1x}x + \sigma_{1y}y + \sigma_{1z}z)(\sigma_{2x}x + \sigma_{2y}y + \sigma_{2z}z) - \boldsymbol{\sigma}_1 \cdot \boldsymbol{\sigma}_2 \right) d\Omega$$
$$= \int \left(\frac{3}{r^2} (x^2\sigma_{1x}\sigma_{2x} + y^2\sigma_{1y}\sigma_{2y} + z^2\sigma_{1z}\sigma_{2z}) - \boldsymbol{\sigma}_1 \cdot \boldsymbol{\sigma}_2 \right) d\Omega$$
$$= (4\pi - 4\pi)\boldsymbol{\sigma}_1 \cdot \boldsymbol{\sigma}_2 = 0$$

The integrals entering in the expressions above are easily solved from symmetry considerations:

$$\int \frac{xy}{r^2} \, d\Omega = 0 \quad \text{etc.;} \quad \int \frac{x^2}{r^2} \, d\Omega = \int \frac{y^2}{r^2} \, d\Omega = \int \frac{z^2}{r^2} \, d\Omega = \frac{4\pi}{3}$$

where the last equality follows because the sum, $\int d\Omega = 4\pi$.

13.9 Because $(1 - \boldsymbol{\sigma}_1 \cdot \boldsymbol{\sigma}_2)$ projects out the singlet state it is enough to show

$$V_{\text{tensor}} (1 - \boldsymbol{\sigma}_1 \cdot \boldsymbol{\sigma}_2) = 0$$

$$[3 (\boldsymbol{\sigma}_1 \cdot \mathbf{e}_r)(\boldsymbol{\sigma}_2 \cdot \mathbf{e}_r) - \boldsymbol{\sigma}_1 \cdot \boldsymbol{\sigma}_2] (1 - \boldsymbol{\sigma}_1 \cdot \boldsymbol{\sigma}_2) =$$
$$3 (\boldsymbol{\sigma}_1 \cdot \mathbf{e}_r)(\boldsymbol{\sigma}_2 \cdot \mathbf{e}_r) - \boldsymbol{\sigma}_1 \cdot \boldsymbol{\sigma}_2 - 3 (\boldsymbol{\sigma}_1 \cdot \mathbf{e}_r)(\boldsymbol{\sigma}_2 \cdot \mathbf{e}_r)(\boldsymbol{\sigma}_1 \cdot \boldsymbol{\sigma}_2) + (\boldsymbol{\sigma}_1 \cdot \boldsymbol{\sigma}_2)^2$$

With the formula from problem 13.3

$$(\boldsymbol{\sigma}_2 \cdot \mathbf{e}_r)(\boldsymbol{\sigma}_1 \cdot \boldsymbol{\sigma}_2) = (\boldsymbol{\sigma}_2 \cdot \mathbf{e}_r)(\boldsymbol{\sigma}_2 \cdot \boldsymbol{\sigma}_1) = \mathbf{e}_r \cdot \boldsymbol{\sigma}_1 + i\boldsymbol{\sigma}_2 \cdot (\mathbf{e}_r \times \boldsymbol{\sigma}_1)$$

Then (with the same formula)

$$(\boldsymbol{\sigma}_1 \cdot \mathbf{e}_r)(\mathbf{e}_r \cdot \boldsymbol{\sigma}_1) = \mathbf{e}_r \cdot \mathbf{e}_r = 1$$

Furthermore

$$(\boldsymbol{\sigma}_1 \cdot \mathbf{e}_r) i\boldsymbol{\sigma}_2 \cdot (\mathbf{e}_r \times \boldsymbol{\sigma}_1) = i(\boldsymbol{\sigma}_1 \cdot \mathbf{e}_r)(\boldsymbol{\sigma}_2 \times \mathbf{e}_r) \cdot \boldsymbol{\sigma}_1$$

$$= \text{(same formula again)} = i\{\mathbf{e}_r \cdot (\boldsymbol{\sigma}_2 \times \mathbf{e}_r) + i\boldsymbol{\sigma}_1 \cdot [\mathbf{e}_r \times (\boldsymbol{\sigma}_2 \times \mathbf{e}_r)]\}$$

$$= i\boldsymbol{\sigma}_2 \cdot (\mathbf{e}_r \times \mathbf{e}_r) - \boldsymbol{\sigma}_1 \cdot [(\mathbf{e}_r \cdot \mathbf{e}_r)\boldsymbol{\sigma}_2 - (\mathbf{e}_r \cdot \boldsymbol{\sigma}_2)\mathbf{e}_r]$$

$$= -\boldsymbol{\sigma}_1 \cdot \boldsymbol{\sigma}_2 + (\boldsymbol{\sigma}_1 \cdot \mathbf{e}_r)(\boldsymbol{\sigma}_2 \cdot \mathbf{e}_r)$$

$$\Rightarrow V_{\text{tensor}}(1 - \boldsymbol{\sigma}_1 \cdot \boldsymbol{\sigma}_2) = -\boldsymbol{\sigma}_1 \cdot \boldsymbol{\sigma}_2 - 3 + 3\boldsymbol{\sigma}_1 \cdot \boldsymbol{\sigma}_2 + (\boldsymbol{\sigma}_1 \cdot \boldsymbol{\sigma}_2)^2 = 0$$

where for the last equality, the result of problem 13.3 has been used.

13.10

$$\left[S^2, (\boldsymbol{\sigma}_1 \cdot \mathbf{e}_r)(\boldsymbol{\sigma}_2 \cdot \mathbf{e}_r)\right] = \left[s_1^2 + s_2^2 + 2\mathbf{s}_1 \cdot \mathbf{s}_2, (\boldsymbol{\sigma}_1 \cdot \mathbf{e}_r)(\boldsymbol{\sigma}_2 \cdot \mathbf{e}_r)\right]$$

$$= \frac{1}{2}\left[\boldsymbol{\sigma}_1 \cdot \boldsymbol{\sigma}_2, (\boldsymbol{\sigma}_1 \cdot \mathbf{e}_r)(\boldsymbol{\sigma}_2 \cdot \mathbf{e}_r)\right]$$

$$= \frac{1}{2}\left(\boldsymbol{\sigma}_1 \cdot \left[\boldsymbol{\sigma}_2, (\boldsymbol{\sigma}_1 \cdot \mathbf{e}_r)(\boldsymbol{\sigma}_2 \cdot \mathbf{e}_r)\right] + \left[\boldsymbol{\sigma}_1, (\boldsymbol{\sigma}_1 \cdot \mathbf{e}_r)(\boldsymbol{\sigma}_2 \cdot \mathbf{e}_r)\right] \cdot \boldsymbol{\sigma}_2\right)$$

where we have used $[\mathbf{s}, s^2] = 0$; $\mathbf{s} = \frac{1}{2}\boldsymbol{\sigma}$

$$[\boldsymbol{\sigma}_1, (\boldsymbol{\sigma}_1 \cdot \mathbf{e}_r)(\boldsymbol{\sigma}_2 \cdot \mathbf{e}_r)] = \boldsymbol{\sigma}_1 \cdot \mathbf{e}_r [\boldsymbol{\sigma}_1, \boldsymbol{\sigma}_2 \cdot \mathbf{e}_r] + [\boldsymbol{\sigma}_1, \boldsymbol{\sigma}_1 \cdot \mathbf{e}_r]\boldsymbol{\sigma}_2 \cdot \mathbf{e}_r$$

where $[\boldsymbol{\sigma}_1, \boldsymbol{\sigma}_2 \cdot \mathbf{e}_r] = 0$ because all operators of particle '1' commute with those of particle '2'.

We now consider the x-component:

$$[\boldsymbol{\sigma}, \boldsymbol{\sigma} \cdot \mathbf{r}]_x = [\sigma_x, \boldsymbol{\sigma} \cdot \mathbf{r}] = \sigma_x [\sigma_x, x] + [\sigma_x, \sigma_x] x + \sigma_y [\sigma_x, y]$$
$$+ [\sigma_x, \sigma_y] y + \sigma_z [\sigma_x, z] + [\sigma_x, \sigma_z] z = 2i (y\sigma_z - z\sigma_y) = 2i (\mathbf{r} \times \boldsymbol{\sigma})_x$$

$$\Rightarrow [\boldsymbol{\sigma}_1, (\boldsymbol{\sigma}_1 \cdot \mathbf{e}_r)(\boldsymbol{\sigma}_2 \cdot \mathbf{e}_r)] = 2i (\mathbf{e}_r \times \boldsymbol{\sigma}_1)(\boldsymbol{\sigma}_2 \cdot \mathbf{e}_r)$$
$$\Rightarrow [\boldsymbol{\sigma}_1 \cdot \boldsymbol{\sigma}_2, (\boldsymbol{\sigma}_1 \cdot \mathbf{e}_r)(\boldsymbol{\sigma}_2 \cdot \mathbf{e}_r)]$$
$$= 2i [\boldsymbol{\sigma}_1 \cdot (\mathbf{e}_r \times \boldsymbol{\sigma}_2)(\boldsymbol{\sigma}_1 \cdot \mathbf{e}_r) + (\mathbf{e}_r \times \boldsymbol{\sigma}_1)(\boldsymbol{\sigma}_2 \cdot \mathbf{e}_r) \cdot \boldsymbol{\sigma}_2]$$
$$= 2i [\boldsymbol{\sigma}_1 \cdot (\mathbf{e}_r \times \boldsymbol{\sigma}_2)(\boldsymbol{\sigma}_1 \cdot \mathbf{e}_r) + (\boldsymbol{\sigma}_2 \cdot \mathbf{e}_r)\boldsymbol{\sigma}_2 \cdot (\mathbf{e}_r \times \boldsymbol{\sigma}_1)]$$
$$= 2i \{(\mathbf{e}_r \times \boldsymbol{\sigma}_2) \cdot \mathbf{e}_r + i\boldsymbol{\sigma}_1 \cdot [(\mathbf{e}_r \times \boldsymbol{\sigma}_2) \times \mathbf{e}_r]$$
$$+ \mathbf{e}_r \cdot (\mathbf{e}_r \times \boldsymbol{\sigma}_1) + i\boldsymbol{\sigma}_2 \cdot [\mathbf{e}_r \times (\mathbf{e}_r \times \boldsymbol{\sigma}_1)]\}$$
$$= -2 \left\{\boldsymbol{\sigma}_1 \cdot \left[\mathbf{e}_r^2\boldsymbol{\sigma}_2 - (\mathbf{e}_r \cdot \boldsymbol{\sigma}_2)\mathbf{e}_r\right] + \boldsymbol{\sigma}_2 \cdot \left[(\mathbf{e}_r \cdot \boldsymbol{\sigma}_1)\mathbf{e}_r - \mathbf{e}_r^2\boldsymbol{\sigma}_1\right]\right\}$$
$$= 0$$

where we have used the formula of problem 13.3. To prove $[\mathbf{J}, V_{\text{tensor}}] = \mathbf{0}$, we only need to consider

$$[\mathbf{J}, (\boldsymbol{\sigma}_1 \cdot \mathbf{e}_r)(\boldsymbol{\sigma}_2 \cdot \mathbf{e}_r)] = [\mathbf{L} + \mathbf{S}, (\boldsymbol{\sigma}_1 \cdot \mathbf{e}_r)(\boldsymbol{\sigma}_2 \cdot \mathbf{e}_r)]$$

From the calculations above:

$$[\mathbf{S}, (\boldsymbol{\sigma}_1 \cdot \mathbf{e}_r)(\boldsymbol{\sigma}_2 \cdot \mathbf{e}_r)] = i [(\boldsymbol{\sigma}_2 \cdot \mathbf{e}_r)(\mathbf{e}_r \times \boldsymbol{\sigma}_1) + (\boldsymbol{\sigma}_1 \cdot \mathbf{e}_r)(\mathbf{e}_r \times \boldsymbol{\sigma}_2)]$$

$$[\mathbf{L}, (\boldsymbol{\sigma}_1 \cdot \mathbf{e}_r)(\boldsymbol{\sigma}_2 \cdot \mathbf{e}_r)] = (\boldsymbol{\sigma}_1 \cdot \mathbf{e}_r)[\mathbf{L}, \boldsymbol{\sigma}_2 \cdot \mathbf{e}_r] + [\mathbf{L}, \boldsymbol{\sigma}_1 \cdot \mathbf{e}_r]\boldsymbol{\sigma}_2 \cdot \mathbf{e}_r$$

$$[\mathbf{L}, \boldsymbol{\sigma} \cdot \mathbf{e}_r]_x = \ldots = i[\boldsymbol{\sigma} \times \mathbf{e}_r]_x$$

which follows from $[L_x, y] = iz$, etc. This then gives

$$[\mathbf{L}, (\boldsymbol{\sigma}_1 \cdot \mathbf{e}_r)(\boldsymbol{\sigma}_2 \cdot \mathbf{e}_r)] = -[\mathbf{S}, (\boldsymbol{\sigma}_1 \cdot \mathbf{e}_r)(\boldsymbol{\sigma}_2 \cdot \mathbf{e}_r)]$$

and thus $[\mathbf{J}, V_{\text{tensor}}] = \mathbf{0}$.

13.11 The total wave function in a singlet spin state is $\chi_0^0 Y_{LM_L}$. The total wave function must have a good total angular momentum J. Furthermore, in a singlet state, $S = 0$. Because $\mathbf{J} = \mathbf{L} + \mathbf{S}$, this means that also L is well defined, namely $L = J$. The matrix element of $\mathbf{L} \cdot \mathbf{S}$ then becomes

$$\langle |\mathbf{L} \cdot \mathbf{S}| \rangle = \left\langle \left| \frac{1}{2} \left(J^2 - L^2 - S^2 \right) \right| \right\rangle$$

$$= \frac{1}{2} [J(J+1) - L(L+1) - 0] = 0$$

13.12 From appendix 13B:

$$\psi_{M=1}^{J=1} = \frac{u(r)}{r} \frac{1}{(4\pi)^{1/2}} \chi_1^1$$

$$+ \frac{w(r)}{r} \left[\left(\frac{1}{10} \right)^{1/2} Y_{20}\chi_1^1 - \left(\frac{3}{10} \right)^{1/2} Y_{21}\chi_0^1 + \left(\frac{3}{5} \right)^{1/2} Y_{22}\chi_{-1}^1 \right]$$

$$Q = \left\langle \psi_{M=1}^{J=1} \left| \left(\frac{\pi}{5} \right)^{1/2} r^2 Y_{20} \right| \psi_{M=1}^{J=1} \right\rangle$$

$$= 2 \left(\frac{\pi}{5} \right)^{1/2} \int \frac{1}{(4\pi)^{1/2}} u(r) \left(\frac{1}{10} \right)^{1/2} w(r) Y_{20} Y_{20} r^2 \, dr \, d\Omega$$

$$+ \left(\frac{\pi}{5} \right)^{1/2} \int w^2(r)$$

$$\times \left(\frac{1}{10} (Y_{20})^3 + \frac{3}{10} Y_{21}^* Y_{20} Y_{21} + \frac{3}{5} Y_{22}^* Y_{20} Y_{22} \right) r^2 \, dr \, d\Omega$$

where the other terms disappear because of orthogonality. The addition theorem (appendix 6B) is used to calculate

$$\int Y_{2\mu}^* Y_{20} Y_{2\mu} \, d\Omega = \int Y_{2\mu}^* \sum_{LM} \left(\frac{5 \cdot 5}{4\pi(2L+1)} \right)^{1/2} C_{0\mu M}^{22L} C_{000}^{22L} Y_{LM} \, d\Omega$$

$$= \left(\frac{5}{4\pi}\right)^{1/2} C_{0\,\mu\,\mu}^{2\,2\,2} C_{0\,0\,0}^{2\,2\,2}$$

With the Clebsch–Gordan coefficient inserted:

$$Q = \frac{1}{10} \int u \cdot w \cdot \sqrt{2} r^2 \, dr + \frac{1}{2} \int w^2 r^2 \, dr \left[\frac{1}{10} \left(-\frac{2}{\sqrt{14}}\right)^2\right.$$

$$+ \frac{3}{10} \left(-\frac{2}{\sqrt{14}}\right) \left(-\frac{1}{\sqrt{14}}\right) + \frac{3}{5} \left(-\frac{2}{\sqrt{14}}\right) \left(\frac{2}{\sqrt{14}}\right)\bigg]$$

$$\Rightarrow Q = \frac{1}{10} \int \left(u \cdot w \sqrt{2} - \frac{1}{2} w^2\right) r^2 \, dr$$

13.13

$$Q = \frac{1}{10} \int \left(u(r) w(r) \sqrt{2} - \frac{1}{2} w^2(r)\right) r^2 \, dr$$

$$u(r) = \frac{1}{(1+n^2)^{1/2}} (2\kappa)^{1/2} \, e^{-\kappa r} \; ; \; w(r) = \frac{n}{(1+n^2)^{1/2}} (2\kappa)^{1/2} \, e^{-\kappa r}$$

$$Q = \frac{1}{10\,(1+n^2)} \left(\frac{n\sqrt{2}}{2\kappa^2} - \frac{1}{2} \cdot \frac{n^2}{2\kappa^2}\right)$$

$$Q = 0.282 \, \text{fm}^2 \Rightarrow n \simeq 0.25$$

The amplitude of the D-state is

$$P_{\text{D}} = \int w^2 \, dr \simeq \frac{n^2}{1+n^2} \simeq 6\%$$

14.1   The wave function is given as

$$|\ell\ell IM\rangle = \sum_{mm'} C_{m\,m'\,M}^{\ell\,\ell\,I} Y_{\ell m}(\Omega_1) \, Y_{\ell m'}(\Omega_2) \, R_{n\ell}(r_1) \, R_{n\ell}(r_2)$$

$$\langle V_{\text{res}}\rangle = \langle \ell\ell IM \,|\, V_{\text{res}}(1,2) \,|\, \ell\ell IM\rangle$$

$$= \left\langle \sum_{mm'} C_{m\,m'\,M}^{\ell\,\ell\,I} Y_{\ell m}(\Omega_1) \, Y_{\ell m'}(\Omega_2) \, R_{n\ell}(r_1) \, R_{n\ell}(r_2) \,\right| -\kappa \frac{\delta(r_1 - r_2)}{r_1^2}$$

$$\times \, \delta(\Omega_1 - \Omega_2) \left|\sum_{\mu\mu'} C_{\mu\,\mu'\,M}^{\ell\,\ell\,I} Y_{\ell\mu}(\Omega_1) \, Y_{\ell\mu'}(\Omega_2) \, R_{n\ell}(r_1) \, R_{n\ell}(r_2) \right\rangle$$

$$= \int (R_{n\ell}(r))^4 \, r^2 \, dr \sum_{mm'} \sum_{\mu\mu'} C_{m\,m'\,M}^{\ell\,\ell\,I} C_{\mu\,\mu'\,M}^{\ell\,\ell\,I} \int Y_{\ell m}^* Y_{\ell m'}^* Y_{\ell\mu} Y_{\ell\mu'} \, d\Omega$$

$$\propto \sum_{mm'} \sum_{\mu\mu'} \sum_{LM_1} \frac{(2\ell+1)^2}{4\pi(2L+1)} C_{mm'M}^{\ell\ell I} C_{\mu\mu'M}^{\ell\ell I} C_{mm'M_1}^{\ell\ell L} C_{000}^{\ell\ell L} C_{\mu\mu'M_1}^{\ell\ell L} C_{000}^{\ell\ell L}$$

where the addition theorem has been used twice and then the orthogonality of spherical harmonics.

The orthogonality relation for the Clebsch–Gordan coefficients now leads to

$$\langle \ell\ell I M | - \kappa \delta \, (\mathbf{r}_1 - \mathbf{r}_2) \, | \ell\ell I M \rangle \propto -\kappa \left(C_{000}^{I\ell\ell} \right)^2$$

14.3 The basis states are

$$a_m^+ a_{\bar{m}}^+ a_{m'}^+ a_{\bar{m}'}^+ |0\rangle \; ; \; m, m' = 1, 2, 3, m \neq m'$$

The Hamiltonian is:

$$H^G = -G \sum_{k=1}^{3} \sum_{k'=1}^{3} a_k^+ a_{\bar{k}}^+ a_{\bar{k}'} a_{k'}$$

The matrix elements are

$$H_{n'n,m'm}^G = -G \left\langle 0 | a_{\bar{n}'} a_{n'} a_{\bar{n}} a_n \left| \sum_k \sum_{k'} a_k^+ a_{\bar{k}}^+ a_{\bar{k}'} a_{k'} \right| a_m^+ a_{\bar{m}}^+ a_{m'}^+ a_{\bar{m}'}^+ |0 \right\rangle$$

(a) Assume $n = m$ and $n' = m'$. Then, the matrix element is different from zero only if $k = k'$ and if either $k = m$ or $k = m'$.

$$\Rightarrow \langle H_G \rangle_{n'n,n'n} = -G + (-G) = -2G$$

(b) $n = m$ and $n' \neq m'$. Then $\langle H^G \rangle \neq 0$ if $k' = m'$ and $k = n'$:

$$\left\langle H^G \right\rangle_{n'n,m'm} = -G$$

As the order between n and n' is unimportant, these are the only possibilities:

$$\left\langle H^G \right\rangle_{n'n,m'm} = \begin{pmatrix} 2G & -G & -G \\ -G & -2G & -G \\ -G & -G & -2G \end{pmatrix}$$

The rows and columns correspond to values of 1 2, 1 3 and 2 3 for the indices n' n and m' m, respectively. The eigenvalues λ_i are now obtained from the determinant equation:

$$\begin{vmatrix} -2G-\lambda & -G & -G \\ -G & -2G-\lambda & -G \\ -G & -G & -2G-\lambda \end{vmatrix} = 0 \Rightarrow \begin{cases} \lambda_1 = -4G \\ \lambda_{2,3} = -G \end{cases}$$

14.4

$$[A, A^+] = \sum_m \sum_n [a_{\bar{m}}a_m, a_n^+ a_{\bar{n}}^+] = \sum_{m,n} (a_{\bar{m}}a_m a_n^+ a_{\bar{n}}^+ - a_n^+ a_{\bar{n}}^+ a_{\bar{m}}a_m)$$

$$= \sum_{m,n} (a_{\bar{m}}a_{\bar{n}}^+ \delta_{mn} - a_{\bar{m}}a_n^+ a_m a_{\bar{n}}^+ - a_n^+ a_{\bar{n}}^+ a_{\bar{m}}a_m)$$

$$= \sum_{m,n} (\delta_{mn} - a_{\bar{n}}^+ a_{\bar{m}}\delta_{mn} - a_n^+ a_m \delta_{nm} + a_n^+ a_{\bar{n}}^+ a_{\bar{m}}a_m - a_n^+ a_{\bar{n}}^+ a_{\bar{m}}a_m)$$

$$= \sum_m (1 - a_m^+ a_m - a_{\bar{m}}^+ a_{\bar{m}}) = \Omega - \mathcal{N}$$

where Ω is the degeneracy and $\mathcal{N}$ the particle number operator.

$$\left[H^G, A^+\right] = -G\left[A^+A, A^+\right]$$
$$= -G\left(A^+\left[A, A^+\right] + \left[A^+, A^+\right]A\right) = -GA^+(\Omega - \mathcal{N})$$

where we have inserted the commutator derived above and furthermore, of course, $[A^+, A^+] = 0$. We now want to change the order of A^+ and $\mathcal{N}$ and therefore need the commutator $[A^+, \mathcal{N}]$. The general method in deriving such commutators is to put all operators in so-called normal order (as is done above), i.e. with annihilation operators to the right and creation operators to the left. One then finds $[A^+, \mathcal{N}] = -2A^+$ and consequently $[H^G, A^+] = -GA^+(\Omega - \mathcal{N}) = -G(\Omega - \mathcal{N} + 2)A^+$.

14.5 We want to prove that

$$\left[H^G, (A^+)^p\right] = -Gp\left(\Omega - \mathcal{N} + p + 1\right)(A^+)^p$$

We assume that the relation is valid for '$(p-1)$' and use the commutators of problem 14.4 to calculate:

$$\left[H^G, (A^+)^p\right] = A^+\left[H^G, (A^+)^{p-1}\right] + \left[H^G, A^+\right](A^+)^{p-1}$$

$$= -GA^+(p-1)(\Omega - \mathcal{N} + p)(A^+)^{p-1} - G(\Omega - \mathcal{N} + 2)(A^+)^p$$

$$= -G(p-1)(\Omega - \mathcal{N} + 2 + p)\,(A^+)^p - G(\Omega - \mathcal{N} + 2)\,(A^+)^p$$

$$= -Gp(\Omega - \mathcal{N} + p + 1)\,(A^+)^p$$

Thus, the relation holds true for all p.

14.7   The BCS wave function is

$$\Psi_0 = \prod_\nu (U_\nu + V_\nu a_\nu^+ a_{\bar\nu}^+)\,|0\rangle$$

$$\langle \Psi_0 | \Psi_0 \rangle = \langle 0| \prod_\mu (U_\mu + V_\mu a_{\bar\mu} a_\mu) \prod_\nu (U_\nu + V_\nu a_\nu^+ a_{\bar\nu}^+)\,|0\rangle$$

We use $[a_{\bar\mu} a_\mu, a_\nu^+ a_{\bar\nu}^+] = 0$ if $\mu \neq \nu$

$$\langle \Psi_0 | \Psi_0 \rangle = \langle 0| \prod_\mu (U_\mu + V_\mu a_{\bar\mu} a_\mu) \left(U_\mu + V_\mu a_\mu^+ a_{\bar\mu}^+ \right)\,|0\rangle$$

$$= \langle 0| \prod_\mu \left(U_\mu^2 + V_\mu^2 a_{\bar\mu} a_\mu a_\mu^+ a_{\bar\mu}^+ \right)\,|0\rangle = \langle 0| \prod_\mu \left(U_\mu^2 + V_\mu^2 \right)\,|0\rangle$$

$$\Rightarrow \langle \Psi_0 | \Psi_0 \rangle = 1 \quad \text{if} \quad U_\mu^2 + V_\mu^2 = 1.$$

14.8

$$\left\langle \Psi_0 \left| a_\mu^+ a_{\bar\mu}^+ a_{\bar\kappa} a_\kappa \right| \Psi_0 \right\rangle =$$

$$\left\langle 0 \left| \prod_\nu (U_\nu + V_\nu a_{\bar\nu} a_\nu)\, a_\mu^+ a_{\bar\mu}^+ a_{\bar\kappa} a_\kappa \prod_{\nu'} (U_{\nu'} + V_{\nu'} a_{\nu'}^+ a_{\bar\nu'}^+) \right| 0 \right\rangle$$

$$= \left\langle 0 \left| \prod_{\nu \neq \mu} (U_\nu + V_\nu a_{\bar\nu} a_\nu)\, V_\mu V_\kappa \prod_{\nu' \neq \kappa} (U_{\nu'} + V_{\nu'} a_{\nu'}^+ a_{\bar\nu'}^+) \right| 0 \right\rangle$$

$$= \begin{cases} \left\langle 0 \left| (U_\kappa + V_\kappa a_{\bar\kappa} a_\kappa)\, V_\mu V_\kappa \left(U_\mu + V_\mu a_\mu^+ a_{\bar\mu}^+ \right) \right| 0 \right\rangle \\ \qquad\qquad = U_\kappa V_\kappa U_\mu V_\mu & \text{if } \mu \neq \kappa \\ V_\mu^2 & \text{if } \mu = \kappa \end{cases}$$

14.9   The wave function Ψ_2 may have a different number of particles than has Ψ_0. Assume that there is an excess of δN particles in Ψ_2 compared with Ψ_0. We must correct for this. As the 'marginal energy' of the 'marginal particle' is λ, the correction is to first order equal $(-\lambda \delta N)$. The corrected excitation energy of a two-quasiparticle state is thus

$$E_2^{\text{exc}} = \langle \Psi_2 | H | \Psi_2 \rangle - \langle \Psi_0 | H | \Psi_0 \rangle - \lambda\, \delta N$$

With $\delta N = \langle \Psi_2 | \mathcal{N} | \Psi_2 \rangle - \langle \Psi_0 | \mathcal{N} | \Psi_0 \rangle$ we obtain

$$E_2^{\text{exc}} = \langle \Psi_2 | H - \lambda \mathcal{N} | \Psi_2 \rangle - \langle \Psi_0 | H - \lambda \mathcal{N} | \Psi_0 \rangle$$

14.10

$$\Psi_0 = \prod_\nu \left(U_\nu + V_\nu a_{\bar{\nu}}^+ a_\nu^+ \right) |0\rangle$$

$$\Psi_2 = \left(-V_\mu + U_\mu a_\mu^+ a_{\bar{\mu}}^+ \right) \prod_{\nu \neq \mu} \left(U_\nu + V_\nu a_\nu^+ a_{\bar{\nu}}^+ \right) |0\rangle$$

$$H' = \sum_\omega (e_\omega - \lambda)\left(a_\omega^+ a_\omega + a_{\bar{\omega}}^+ a_{\bar{\omega}} \right) - G \sum_{\omega,\omega'} a_\omega^+ a_{\bar{\omega}}^+ a_{\bar{\omega}'} a_{\omega'}$$

$$= H^{\text{sp}} - \lambda \mathcal{N} + H^G$$

The excitation energy is calculated from the expression of problem 14.9. The matrix element of Ψ_0 was given in the main text (cf. problem 14.8)

$$\langle \Psi_0 | H' | \Psi_0 \rangle = \sum_\omega (e_\omega - \lambda)\, 2 V_\omega^2 - G \left(\sum_{\omega,\omega'} U_\omega V_\omega U_{\omega'} V_{\omega'} + \sum_\omega V_\omega^4 \right)$$

This result can be taken over when calculating the matrix element for Ψ_2 if for index μ we make the substitution $U_\mu \to -V_\mu$, $V_\mu \to U_\mu$:

$$\langle \Psi_2 | H^{\text{sp}} - \lambda \mathcal{N} | \Psi_2 \rangle = \sum_{\omega \neq \mu} 2 (e_\omega - \lambda) V_\omega^2 + 2 (e_\mu - \lambda) U_\mu^2$$

$$-\frac{1}{G} \left\langle \Psi_2 \left| H^G \right| \Psi_2 \right\rangle = \sum_{\omega \neq \mu} \sum_{\omega' \neq \mu} U_\omega V_\omega U_{\omega'} V_{\omega'} + 2 (-V_\mu) U_\mu \sum_{\omega \neq \mu} U_\omega V_\omega$$

$$+ U_\mu^2 (-V_\mu)^2 + \sum_{\omega \neq \mu} V_\omega^4 + U_\mu^4$$

$$= \sum_{\omega,\omega'} U_\omega V_\omega U_{\omega'} V_{\omega'} - 4U_\mu V_\mu \sum_\omega U_\omega V_\omega + 4U_\mu^2 V_\mu^2 + \sum_\omega V_\omega^4 - V_\mu^4 + U_\mu^4$$

Terms of order G are now neglected and we obtain

$$E_2^{\text{exc}} = \langle \Psi_2 | H' | \Psi_2 \rangle - \langle \Psi_0 | H' | \Psi_0 \rangle$$
$$= \left(2U_\mu^2 - 2V_\mu^2 \right) (e_\mu - \lambda) + 4U_\mu V_\mu G \sum_\omega U_\omega V_\omega$$

With $\Delta = G \sum_\omega U_\omega V_\omega$, $U_\mu^2 - V_\mu^2 = (e_\mu - \lambda) / E_\mu$ and $2U_\mu V_\mu = \Delta / E_\mu$:

$$E_2^{\text{exc}} = \frac{2 (e_\mu - \lambda)^2}{E_\mu} + \frac{2\Delta^2}{E_\mu} = \frac{2E_\mu^2}{E_\mu} = 2E_\mu$$

14.12   The one-quasiparticle state is $\Psi_1 = a_\mu^+ \prod_{v \neq \mu} (U_v + V_v a_v^+ a_{\bar{v}}^+) |0\rangle$. It is then straightforward to calculate (cf. problem 14.8)

$$\langle \Psi_1 | H - \lambda N | \Psi_1 \rangle = \sum_{\omega \neq \mu} 2 (e_\omega - \lambda) V_\omega^2 + (e_\mu - \lambda)$$

$$\left\langle \Psi_1 \left| H^G \right| \Psi_1 \right\rangle = -G \left(\underbrace{\sum_{\omega \neq \mu} \sum_{\omega' \neq \mu} U_\omega V_\omega U_{\omega'} V_{\omega'}}_{\omega \neq \omega'} + \sum_{\omega \neq \mu} V_\omega^2 \right)$$

The ground state energy is now subtracted, terms of the order G are neglected, the definition $\Delta = G \sum_v U_v V_v$ and the relations $2U_v V_v = \Delta / E_v$, $U_v^2 - V_v^2 = (e_v - \lambda) / E_v$ are used to obtain an excitation energy as

$$E_\mu = \left[(e_\mu - \lambda)^2 + \Delta^2 \right]^{1/2}$$

14.13

$$E(\Delta) - E(0) = \rho \int_{-S}^0 \left(2V^2 - 2 \right) e \, de + \rho \int_0^S 2V^2 e \, de - \frac{\Delta^2}{G}$$

$$2V^2 = 1 - \frac{e}{(e^2 + \Delta^2)^{1/2}} \implies$$

$$E(\Delta) - E(0) = \rho \left[\int_{-S}^0 \left(-\frac{e}{(e^2 + \Delta^2)^{1/2}} - 1 \right) e \, de \right.$$
$$\left. + \int_0^S \left(1 - \frac{e}{(e^2 + \Delta^2)^{1/2}} \right) e \, de \right] - \frac{\Delta^2}{G}$$

The two integrals are identical. Their sum is

$$I = 2\int_0^S \left(1 - \frac{e}{(e^2+\Delta^2)^{1/2}}\right) e\, de$$

$$= S^2 - 2\left[\frac{e(e^2+\Delta^2)^{1/2}}{2} - \frac{\Delta^2}{2}\ln\left[e + \left(e^2+\Delta^2\right)^{1/2}\right]\right]_0^S$$

$$= S^2 - S\left(S^2+\Delta^2\right)^{1/2} + \Delta^2\ln\left(\frac{S+(S^2+\Delta^2)^{1/2}}{\Delta}\right)$$

$$S \gg \Delta \Rightarrow$$

$$I \approx S^2\left[1 - 1 - \frac{1}{2}\left(\frac{\Delta}{S}\right)^2\right] + \Delta^2\cdot\ln\frac{2S}{\Delta} \approx -\frac{1}{2}\Delta^2 + \Delta^2\frac{1}{G\rho}$$

where we have used the relation $(2S/\Delta) = \exp(1/G\rho)$ given in the main text. We now obtain

$$E(\Delta) - E(0) = -\frac{1}{2}\rho\Delta^2 + \Delta^2\cdot\frac{1}{G} - \frac{\Delta^2}{G} = -\frac{1}{2}\rho\Delta^2$$

14.14 We know that the fermion anticommutator relations are fulfilled by the particle operators

$$\{a_\nu, a_\mu\} = \{a_\nu^+, a_\mu^+\} = 0 ; \quad \{a_\nu, a_\mu^+\} = \delta_{\nu\mu}$$

It now follows that

$$\{\alpha_\nu, \alpha_\mu^+\} = \{(U_\nu a_\nu - V_\nu a_{\bar\nu}^+), (U_\mu a_\mu^+ - V_\mu a_{\bar\mu})\}$$

$$= U_\nu U_\mu\{a_\nu, a_\mu^+\} + V_\nu V_\mu\{a_{\bar\nu}^+, a_{\bar\mu}\} = \left(U_\mu^2 + V_\nu^2\right)\delta_{\nu\mu} = \delta_{\nu\mu}$$

if $U_\nu^2 + V_\nu^2 = 1$. The other relations are shown in a similar way.

14.15 Start from the definitions

$$\alpha_\nu^+ = U_\nu a_\nu^+ - V_\nu a_{\bar\nu}$$

$$\alpha_{\bar\nu} = U_\nu a_{\bar\nu} + V_\nu a_\nu^+$$

The first equality is multiplied by U_ν and the second by V_ν. The two equalities are then added:

$$U_\nu\alpha_\nu^+ + V_\nu\alpha_{\bar\nu} = U_\nu^2 a_\nu^+ + V_\nu^2 a_\nu^+ \Rightarrow a_\nu^+ = U_\nu\alpha_\nu^+ + V_\nu\alpha_{\bar\nu}$$

if $U_\nu^2 + V_\nu^2 = 1$. The expressions for a_ν, $a_{\bar\nu}^+$ and $a_{\bar\nu}$ then follow directly.

14.16

$$H'(\Delta) = \sum_v (e_v - \lambda) \left(a_v^+ a_v + a_{\bar{v}}^+ a_{\bar{v}}\right) - \Delta \sum_v \left(a_v^+ a_{\bar{v}}^+ + a_{\bar{v}} a_v\right)$$

$$= \sum_v (e_v - \lambda) \left[\left(U_v \alpha_v^+ + V_v \alpha_{\bar{v}}\right) \left(U_v \alpha_v + V_v \alpha_{\bar{v}}^+\right)\right.$$

$$+ \left.\left(U_v \alpha_{\bar{v}}^+ - V_v \alpha_v\right) \left(U_v \alpha_{\bar{v}} - V_v \alpha_v^+\right)\right]$$

$$- \Delta \sum_v \left[\left(U_v \alpha_v^+ + V_v \alpha_{\bar{v}}\right) \left(U_v \alpha_{\bar{v}}^+ - V_v \alpha_v\right)\right.$$

$$+ \left.\left(U_v \alpha_{\bar{v}} - V_v \alpha_v^+\right) \left(U_v \alpha_v + V_v \alpha_{\bar{v}}^+\right)\right]$$

The terms of type $\alpha^+ \alpha$, etc. are now collected and the commutation relations are used to obtain H_{00}, H_{11} and H_{20} as given in appendix 14A.

14.17

$$|\Delta\rangle \propto \prod_v \alpha_v \alpha_{\bar{v}} |0\rangle = \prod_v \left(U_v a_v - V_v a_{\bar{v}}^+\right) \left(U_v a_{\bar{v}} + V_v a_v^+\right) |0\rangle$$

$$\times \prod_v \left\{U_v^2 a_v a_{\bar{v}} - V_v^2 a_{\bar{v}}^+ a_v^+ + U_v V_v \left(a_v a_v^+ - a_{\bar{v}}^+ a_{\bar{v}}\right)\right\} |0\rangle$$

We now use $a_v a_v^+ = 1 - a_v^+ a_v$ and the fact that $a_{\bar{v}} |0\rangle = a_v |0\rangle = 0$ to obtain

$$|\Delta\rangle \propto \prod_v \left(U_v V_v + V_v^2 a_v^+ a_{\bar{v}}^+\right) |0\rangle = \prod_\mu V_\mu \prod_v \left(U_v + V_v a_v^+ a_{\bar{v}}^+\right) |0\rangle$$

Normalisation of $|\Delta\rangle$ now leads to

$$|\Delta\rangle = \prod_v \left(U_v + V_v a_v^+ a_{\bar{v}}^+\right) |0\rangle$$

14.18 For the linearised pair field, the ground state energy is given by (appendix 14A)

$$E(\Delta) = \sum_{v>0} 2 (e_v - \lambda) V_v^2 - 2\Delta \sum_{v>0} U_v V_v + \frac{\Delta^2}{G}$$

where $\Delta^2/G = G \langle F^+\rangle^2$ is the constant term of H^G. For $\Delta = 0$, this energy becomes

$$E(\Delta = 0) = \sum_{occ} 2 (e_v - \lambda)$$

(note that, for the linearised field, the diagonal pairing energy disappears). The sums are now replaced by integrals and the Fermi energy

λ is put equal zero. The calculations are similar to those in problem 14.13. The result is

$$E(\Delta) - E(0) = \rho \left[S^2 - S(S^2 + \Delta^2)^{1/2} + \Delta^2 \ln \left(\frac{S + (S^2 + \Delta^2)^{1/2}}{\Delta} \right) \right.$$
$$\left. - \Delta^2 \ln \left(\frac{S + (S^2 + \Delta^2)^{1/2}}{(S^2 + \Delta^2)^{1/2} - S} \right) \right] + \frac{\Delta^2}{G}$$
$$\approx \rho \Delta^2 \left[-\ln \left(\frac{2S}{\Delta} \right) - \frac{1}{2} + \frac{1}{\rho G} \right]$$

For $S = 15$ MeV, $\rho^{-1} = 0.35$ MeV, we get as a function of Δ

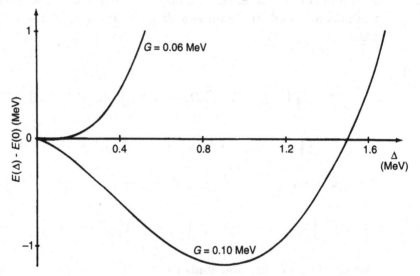

The minima of these curves correspond to the BCS solution. For $G = 0.06$ MeV, the minimum is for $\Delta \simeq 0.06$ MeV and the pairing energy is negligable.

14.19

$$\alpha_\mu^+ |\Delta\rangle = \left(U_\mu a_\mu^+ - V_\mu a_{\bar\mu} \right) \prod_\nu (U_\nu + V_\nu a_\nu^+ a_{\bar\nu}^+) |0\rangle$$
$$= \left(U_\mu a_\mu^+ - V_\mu a_{\bar\mu} \right) \left(U_\mu + V_\mu a_\mu^+ a_{\bar\mu}^+ \right) \prod_{\nu \neq \mu} (U_\nu + V_\nu a_\nu^+ a_{\bar\nu}^+) |0\rangle$$
$$= a_\mu^+ \prod_{\nu \neq \mu} (U_\nu + V_\nu a_\nu^+ a_{\bar\nu}^+) |0\rangle \quad \text{if } U_\mu^2 + V_\mu^2 = 1$$

References

Åberg, S., Flocard, H. and Nazarewicz, W., (1990), *Ann. Rev. Nucl. Part. Sci.* **40**, 439.

Amaldi, E., (1984), *Phys. Rep.* **111**, 1.

Andersson, G., Larsson, S.E., Leander, G., Möller, P., Nilsson, S.G., Ragnarsson, I., Åberg, S., Bengtsson, R., Dudek, J., Nerlo-Pomorska, B., Pomorski, K. and Szymański, Z., (1976), *Nucl. Phys.* **A268**, 205.

Armbruster, P., (1985), *Ann. Rev. Nucl. Part. Sci.* **35**, 135.

Armbruster, P., Agarwal, Y.K., Brüchle, W., Brügger, M., Dufour, J.P., Gäggeler, H., Hessberger, F.P., Hofmann, S., Lemmertz, P., Münzenberg, G., Poppensieker, K., Reisdorf, W., Schädel, M., Schmidt, K.-H., Schneider, J.H.R., Schneider, W.F.W., Sümmerer, K., Vermeulen, D., Wirth, G., Ghiorso, A., Gregorich, K.E., Lee, D., Leino, M., Moody, K.J., Seaborg, G.T., Welch, R.B., Wilmarth, P., Yashita, S., Frink, C., Greulich, N., Herrmann, G., Hickmann, U., Hildebrand, N., Kratz, J.V., Trautmann, N., Fowler, M.M., Hoffman, D.C., Daniels, W.R., von Gunten, H.R. and Dornhöfer, H., (1985), *Phys. Rev. Lett.* **54**, 406.

Atac, A., Piiparinen, M., Herskind, B., Nyberg, J., Sletten, G., de Angelis, G., Forbes, S., Gjørup, N., Hagemann, G., Ingebretsen, F., Jensen, H., Jerrestam, D., Kusakari, H., Lieder, R.M., Marti, G.V., Mullins, S., Santonocito, D., Schnare, H., Strähle, K., Sugawara, M., Tjøm, P.O., Virtanen, A. and Wadsworth, R., (1993), *Phys. Rev. Lett.* **70**, 1069.

Azaiez, F., Kelly, W.H., Korten, W., Stephens, F.S., Deleplanque, M.A., Diamond, R.M., Macchiavelli, A.O., Draper, J.E., Rubel, E.C., Beausang, C.W., Burde, J., Becker, J.A., Henry, E.A., Yates, S.W., Brinkman, M.J., Kuhnert, A. and Wang, T.F., (1991) *Phys. Rev. Lett.* **66**, 1030

Baktash, C., Garrett, J.D., Winchell, D.F. and Smith, A., (1992), *Phys. Rev. Lett.* **69**, 1500.

Baktash, C., Nazarewicz, W. and Wyss, R., (1993), *Nucl. Phys.* **A555**, 375.

Baktash, C., Schutz, Y., Lee, I.Y., McGowan, F.K., Johnson, N.R., Halbert, M.L., Hensley, D.C., Fewell, M.P., Courtney, L., Larabee, A.J., Riedinger, L.L., Sunyar, A.W., der Mateosian, E., Kistner, O.C. and Sarantites, D.G., (1985), *Phys. Rev. Lett.* **54**, 978.

Baran, A., Pomorski, K., Łukasiak, A. and Sobiczewski, A., (1981), *Nucl. Phys.* **A361**, 83.

Bardeen, J., Cooper, L.N. and Schrieffer, J.R., (1957), *Phys. Rev.* **108**, 1175.

Barrett, R.C. and Jackson, D.F., (1977), *Nuclear Sizes and Structure* (Clarendon Press, Oxford).

Belyaev, S.T., (1959), *Mat. Fys. Medd. Dan. Vid. Selsk.* **31**, no 11.

Bengtsson, R., (1975), *Nucl. Phys.* **A245**, 39.

Bengtsson, R., Dudek, J., Nazarewicz, W. and Olanders, P., (1989), *Phys. Scripta* **39**, 196.

Bengtsson, R. and Frauendorf, S., (1979), *Nucl. Phys.* **A314**, 27.

Bengtsson, R. and Garrett, J.D., (1984), *Int. Rev. Nucl. Phys.* **2**, 193.

Bengtsson, T., (1989), *Nucl. Phys.* **A496**, 56.

Bengtsson, T. and Ragnarsson, I., (1983), *Phys. Scripta* **T5**, 165.

Bengtsson, T. and Ragnarsson, I., (1985), *Nucl. Phys.* **A436**, 14.

Bengtsson, T., Ragnarsson, I. and Åberg, S., (1988), *Phys. Lett.* **208B**, 39.

Bengtsson, T., Ragnarsson, I. and Åberg, S., (1991), *Computational Nuclear Physics*, eds. Langanke, K., Maruhn, J.A. and Koonin, S.E. (Springer-Verlag, Berlin), p. 51.

Bentley, M.A., Ball, G.C., Cranmer-Gordon, H.W., Forsyth, P.D., Howe, D., Mokhtar, A.R., Morrison, J.D., Sharpey-Schafer, J.F., Twin, P.J., Fant, B., Kalfas, C.A., Nelson, A.H., Simpson, J. and Sletten, G., (1987), *Phys. Rev. Lett.* **59**, 2141.

Berger, J.F., Girod, M. and Gogny, D., (1989), *Nucl. Phys.* **A502**, 85c.

Bethe, H.A. and Bacher, R.F., (1936), *Rev. Mod. Phys.* **8**, 82.

Bogoliubov, N.N., (1958), *Soviet Phys. JETP* **7**, 41 and (1958) *Nuovo Cimento* **7**, 794.

Bohr, A., (1952), *Mat. Fys. Medd. Dan. Vid. Selsk.* **26**, no 14.

Bohr, A., (1976), *Rev. Mod. Phys.* **48**, 365.

Bohr, A. and Mottelson, B.R., (1953), *Mat. Fys. Medd. Dan. Vid. Selsk.* **27**, no 16.

Bohr, A. and Mottelson, B.R., (1969), *Nuclear Structure*, vol. I (W.A. Benjamin Inc., New York).

Bohr, A. and Mottelson, B.R., (1975), *Nuclear Structure*, vol. II (W.A. Benjamin Inc., New York).

Bohr, A. and Mottelson, B.R., (1977), *Proc. Int. Conf. on Nuclear Structure*, Tokyo, 1977 (*J. Phys. Soc. Jpn* **44**, Suppl.) p. 157.

Bohr, A. and Mottelson, B.R., (1981), *Phys. Scripta* **24**, 71.

Bohr, A., Mottelson, B.R. and Pines, D., (1958), *Phys. Rev.* **110**, 936.

Bohr, N., (1939), *Phys. Rev.* **55**, 418.

Bohr, N. and Wheeler, J.A., (1939), *Phys. Rev.* **56**, 426.

Boleu, R., Nilsson, S.G., Sheline, R.K. and Takahashi, K., (1972), *Phys. Lett.* **40B**, 517.

Brack, M., Damgaard, J., Jensen, A.S., Pauli, H.C., Strutinsky, V.M. and Wong, C.Y., (1972), *Rev. Mod. Phys.* **44**, 320.

Brack, M. and Quentin, P., (1981), *Nucl. Phys.* **A361**, 35.

Brueckner, K.A., Gammel, J.L. and Weitzner, H., (1958), *Phys. Rev.* **110**, 431 and references therein.

Byrski, T., Beck, F.A., Curien, D., Schuck, C., Fallon, P., Alderson, A., Ali, I., Bentley, M.A., Bruce, A.M., Forsyth, P.D., Howe, D., Roberts, J.W., Sharpey-Schafer, J.F., Smith, G. and Twin, P.J., (1990), *Phys. Rev. Lett.* **64**, 1650.

Cerkaski, M. and Szymański, Z., (1979), *Acta Phys. Polon.* **B10**, 163.

Cohen, S., Plasil, F. and Swiatecki, W.J., (1974), *Ann. Phys.* **82**, 557.

Ćwiok, S., Rozmej, P., Sobiszewski, A. and Patyk, Z. , 1989, *Nucl. Phys.* **A491**, 281.

Devons, S. and Duerdoth, I., (1969), *Adv. Nucl. Phys.* **2**, 295.

Dudek, J., Nazarewicz, W., Szymański, Z. and Leander, G.A., (1987), *Phys. Rev. Lett.* **59**, 1405.

Eisenberg, J.M. and Greiner, W., *Nuclear Theory*, vol. 1 (1987), (North-Holland Publ. Comp., Amsterdam).

Elliot, J.P., (1958), *Proc. Roy. Soc. (London)* **A245**, 128 and 562.

Flocard, H., Quentin, P., Vautherin, D., Veneroni, M. and Kerman, A.K., (1974) *Nucl. Phys.* **A231**, 176.

Fock, V., (1930), *Z. Phys.* **61**, 126.

Glas, D., Mosel, U. and Zint, P.G., (1978), *Z. Phys.* **A285**, 83.

Griffin, J.J. and Rich, M., (1960), *Phys. Rev.* **118**, 850.

Grodzins, L. (1962), *Phys. Lett.* **2**, 88.

Gustafsson, C., Lamm, I.L., Nilsson, B. and Nilsson, S.G., (1967), *Ark. Fys.* **36**, 613.

Gustafsson, C., Möller, P. and Nilsson, S.G., (1971), *Phys. Lett.* **34B**, 349.

Haas, B., Janzen, V.P., Ward, D., Andrews, H.R., Radford, D.C., Prévost, D., Kuehner, J.A., Omar, A., Waddington, J.C., Drake, T.E., Galindo-Uribarri, A., Zwartz, G., Flibotte, S., Taras, P. and Ragnarsson, I., (1993), *Nucl. Phys.* **A561**, 251

Hahn, O. and Strassmann, F., (1938), *Naturwissenschaften* **26**, 755.

Hahn, O. and Strassmann, F., (1939), *Naturwissenschaften* **27**, 11 and 89.

Hartree, D.R., (1928), *Proc. Camb. Phil. Soc.* **24**, 89.

Häusser, O., Towner, I.S., Faestermann, T., Andrews, H.R., Beene, J.R., Horn, D., Ward, D. and Broude, C., (1977), *Nucl. Phys.* **A293**, 248.

Haxel, O., Jensen, J.H.D. and Suess, H.E., (1949) *Phys. Rev.* **75**, 1766.

Heisenberg, J. and Blok, H.P., (1983), *Ann. Rev. Nucl. Part. Sci.* **33**, 569.

Herrmann, G., (1980), *Nature* **280**, 543.

Hill, D. and Wheeler, J.A., (1953), *Phys. Rev.* **89**, 1102.

Hofstadter, R., (1963), *Electron Scattering and Nuclear and Nucleon Structure* (W.A. Benjamin, Inc., New York, 1963).

Howard, M. and Nix, J.R., (1974), *Nature* **247**, 17.

Hulet, E.K., Wild, J.F., Dougan, R.J., Lougheed, R.W., Landrum, J.H., Dougan, A.D., Schädel, M., Hahn, R.L., Baisden, P.A., Henderson, C.M., Dupzyk, R.J., Sümmerer, K. and Bethune, G.R., (1986), *Phys. Rev. Lett.* **56**, 313.

Illige, J.D., Hulet, E.K., Nitschke, J.M., Dougan, R.J., Lougheed, R.W., Ghiorso, A. and Landrum, J.H., (1978), *Phys. Lett.* **78B**, 209.

Inglis, D.R., (1954), *Phys. Rev.* **96**, 1059.

Jain, A.K., Sheline, R.K., Sood, P.C. and Jain, K., (1990), *Rev. Mod. Phys.* **62**, 393.

Janssens, R.V.F. and Khoo, T.L., (1991), *Ann. Rev. Nucl. Part. Sci.* **41**, 321.

Jensen, A.S., Hansen, P.G. and Jonson, B., (1984), *Nucl. Phys.* **A431**, 393.

Johansson, S.A.E., (1961), *Nucl. Phys.* **22**, 529.

Kleinheinz, P., Broda, R., Daly, P.J., Lunardi, S., Ogawa, M. and Blomqvist, J., (1979), *Z. Phys.* **A290**, 279.

Kopfermann, H., (1958), *Nuclear Moments* (Academic Press, New York).

Krappe, H.J. and Nix, J.R., (1973), *Proc. Third IAEA Symp. on Physics and Chemistry of Fission*, Rochester, New York (IAEA, Vienna, 1974) vol. 1, p. 159.

Krappe, H.J., Nix, J.R. and Sierk, A.J., (1979), *Phys. Rev.* **C20**, 992.

Larsson, S.E., (1973), *Phys. Scripta* **8**, 17.

Larsson, S.E., Leander, G. and Ragnarsson, I., (1978), *Nucl. Phys.* **A307**, 189.

Larsson, S.E., Leander, G., Ragnarsson, I. and Alenius, N.G., (1976), *Nucl. Phys.* **A261**, 77.

Leander, G.A., Nazarewicz, W., Bertsch, G.F. and Dudek, J., (1986), *Nucl. Phys.* **A453**, 58.

Leander, G.A., Sheline, R.K., Möller, P., Olanders, P., Ragnarsson, I. and Sierk, A.J., (1982), *Nucl. Phys.* **A388**, 452.

Madland, D.G. and Nix, J.R., (1988), *Nucl. Phys.* **A476**, 1.

Mayer, M.G., (1949), *Phys. Rev.* **75**, 1969.

Mayer, M.G., (1950), *Phys. Rev.* **78**, 22.

Meitner, L. and Frisch, O.R., (1939), *Nature (London)* **143**, 239.

Meyer-ter-Vehn, J., (1975), *Nucl. Phys.* **A249**, 111 and 141.

Migdal, A.B., (1959), *Nucl. Phys.* **13**, 655.

Mosel, U. and Schmitt, H.W., (1971), *Phys. Rev.* **C4**, 2185.

Mottelson, B.R. and Nilsson, S.G., (1959a), *Mat. Fys. Skr. Dan. Vid. Selsk.*, **1**, no 8.

Mottelson, B.R. and Nilsson, S.G., (1959b), *Nucl. Phys.* **13**, 281.

Mottelson, B.R. and Valatin, J.G., (1960), *Phys. Rev. Lett.* **5**, 511.

Mustafa, M.G., (1975), *Phys. Rev.* **C11**, 1059.

Myers, W.D., (1977), *Droplet Model of Atomic Nuclei* (IFI/Plenum, New York).

Myers, W.D. and Swiatecki, W.J., (1966), *Nucl. Phys.* **81**, 1.

Myers, W.D. and Swiatecki, W.J., (1967), *Ark. Fys.* **36**, 343.

Möller, P., Leander, G.A. and Nix, J.R., (1986), *Z. Phys.* **A323**, 41.

Möller, P., Nilsson, S.G. and Nix, J.R., (1974), *Nucl. Phys.* **A229**, 292.

Möller, P. and Nix, J.R., (1981), *Nucl. Phys.* **A361**.

Möller, P., Nix, J.R., Myers, W.D. and Swiatecki, J., (1992) *Nucl. Phys.* **A536**, 61, and references therein.

Möller, P., Nix, J.R. and Swiatecki, W.J., (1989), *Nucl. Phys.* **A492**, 349.

Nazarewicz, W., Dudek, J., Bengtsson, R., Bengtsson, T. and Ragnarsson, I., (1985), *Nucl. Phys.* **A435**, 397.

Nazarewicz, W. and Rozmej, P., (1981), *Nucl. Phys.* **A369**, 396.

Nazarewicz, W., Wyss, R. and Johnson, A., (1989), *Nucl. Phys.* **A503**, 285.

Nilsson, B., (1969), *Nucl. Phys.* **A129**, 445.

Nilsson, S.G., (1955), *Mat. Fys. Medd. Dan Vid. Selsk.* **29**, no 16.

Nilsson, S.G., (1978), *Proc. Int. Symp. on Superheavy Elements*, Lubbock, Texas, ed. Lodhi, M.A.K. (Pergamon Press, New York, 1978) p. 237.

Nilsson, S.G. and Prior, O., (1961), *Mat. Fys. Medd. Dan. Vid. Selsk.* **32**, no 16.

Nilsson, S.G., Tsang, C.F., Sobiczewski, A., Szymański, Z., Wycech, S., Gustafsson, C., Lamm, I.-L., Möller, P. and Nilsson, B., (1969), *Nucl. Phys.* **A131**, 1.

Nolan, P.J. and Twin, P.J., (1988), *Ann. Rev. Nucl. Part. Sci.* **38**, 533.

Oganessian, Yu. Ts., Bruchertseifer, H., Buklanov, G.V., Chepigin, V.I., Choi Val Sek, Eichler, B., Gavrilov, K.A., Gaeggeler, H., Korotkin, Yu.S., Orlova, O.A., Reetz, T., Seidel, W., Ter-Akopian, G.M., Tretyakova, S.P. and Zvara, I., (1978), *Nucl. Phys.* **A294**, 213.

Otten, E.W., (1989), *Treatise of Heavy-Ion Science, vol. 8, Nuclei Far From Stability*, ed. Bromley, D.A. (Plenum Press, New York) p. 517.

Pashkevich, V.V., (1988), *Nucl. Phys.* **A477**, 1.

Petrzhak, K.A. and Flerov, G.N., (1940), *Compt. Rend. Akad. Sci. USSR* **25**, 500.

Racah, G., (1950), *Phys. Rev.* **78**, 622.

Ragnarsson, I., (1990), *Nucl. Phys.* **A520**, 67c.

Ragnarsson, I., Åberg, S. and Sheline, R.K., (1981), *Phys. Scripta* **24**, 215.

Ragnarsson, I., Bengtsson, T., Leander, G. and Åberg, S., (1980), *Nucl. Phys.* **A347**, 287.

Ragnarsson, I., Bengtsson, T., Nazarewicz, W., Dudek, J. and Leander, G.A., (1985), *Phys. Rev. Lett.* **54**, 982.

Ragnarsson, I., Nilsson, S.G. and Sheline, R.K., (1978), *Phys. Rep.* **45**, 1.

Ragnarsson, I., Xing, Z., Bengtsson, T. and Riley, M.A., (1986), *Phys. Scripta* **34**, 651.

Rainwater, J., (1976), *Rev. Mod. Phys.* **48**, 385.

Randrup, J., Larsson, S.E., Möller, P., Sobiczewski, A. and Łukasiak, A., (1974), *Proc. 27th Nobel Symp., Super Heavy Elements – Theoretical Predictions and Experimental Generation*, Ronneby, Sweden, eds. Nilsson, S.G. and Nilsson, N.R., *Phys. Scripta* **10A**, 60.

Rasmussen, J.O., (1965), *Alpha-, Beta- and Gamma-ray Spectroscopy*, ed. Siegbahn, K., vol. I (North-Holland, Amsterdam) p. 701.

Rohozinski, S.G. and Sobiczewski, A., (1981), *Acta Phys. Polon.* **B12**, 1001.

Rowe, D.J., (1970), *Nuclear Collective Motion* (Methuen and Co., Ltd, London).

Schmidt, Th., (1937), *Z. Phys.* **106**, 358.

Schuessler, H.A., (1981), *Phys. Today,* February, p. 48.

Sheline, R.K., Ragnarsson, I., Åberg, S. and Watt, A., (1988), *J. Phys. G: Nucl. Phys.* **14**, 1201.

Simpson, J., Riley, M.A., Cresswell, J.R., Forsyth, P.D., Howe, D., Nyakó, B.M., Sharpey-Schafer, J.F., Bacelar, J., Garrett, J.D., Hagemann, G.B., Herskind, B. and Holm, A., (1984), *Phys. Rev. Lett.* **53**, 648.

Sobiczewski, A., (1974), *Proc. 27th Nobel Symp., Super Heavy Elements – Theoretical Predictions and Experimental Generation*, Ronneby, Sweden, eds. Nilsson, S.G. and Nilsson, N.R., *Phys. Scripta* **10A**, 47.

Sobiczewski, A. (1978), *Proc. Int. Symp. on Superheavy Elements*, Lubbock, Texas, ed. Lodhi, M.A.K. (Pergamon Press, New York) p. 274.

Stephens, F.S., (1975), *Rev. Mod. Phys.* **47**, 43.

Stephens, F.S., Deleplanque, M.A., Diamond, R.M., Macchiavelli, A.O. and Draper, J.E., (1985), *Phys. Rev. Lett.* **54**, 2584.

Stephens, F.S., Deleplanque, M.A., Draper, J.E., Diamond, R.M., Macchiavelli, A.O., Beausang, C.W., Korten, W., Kelly, W.H., Azaiez, F., Becker, J.A., Henry, E.A., Yates, S.W., Brinkman, M.J., Kuhnert, A. and Cizewski, J.A., (1990) *Phys. Rev. Lett.* **65**, 301.

Stephens, F.S., Diamond, R.M. and Nilsson, S.G., (1973), *Phys. Lett.* **44B**, 429.

Stephens, F.S. and Simon, R.S., (1972), *Nucl. Phys.* **A183**, 257.

Strutinsky, V.M., (1967), *Nucl. Phys.* **A95**, 420.

Szymański, Z., (1983), *Fast Nuclear Rotation* (Oxford University Press, Oxford).

Szymański, Z., (1990), *Nucl. Phys.* **A520**, 1c.

Taagepera, R. and Nurmia, M., (1961), *Ann. Acad. Sci. Fenn.,* **A78** VI, 1.

Tjøm, P.O., Diamond, R.M., Bacelar, J.C., Beck, E.M., Deleplanque, M.A., Draper, J.E. and Stephens, F.S., (1985), *Phys. Rev. Lett.* **55**, 2405.

Troudet. T. and Arvieu, R., (1979), *Z. Phys.* **A291**, 183.

Twin, P.J., Nyakó, B.M., Nelson, A.H., Simpson, J., Bentley, M.A., Cranmer-Gordon, H.W., Forsyth, P.D., Howe, D., Mokhtar, A.R., Morrison, J.D., Sharpey-Schafer, J.F. and Sletten, G., (1986), *Phys. Rev. Lett.* **57**, 811.

Valatin, J.G., (1956), *Proc. Roy. Soc. (London)* **A238**, 132.

Valatin, J.G., (1958), *Nuovo Cimento* **7**, 843.

Vandenbosch, R. and Huizenga, J.R., (1973), *Nuclear Fission* (Academic Press, New York).
de Voigt, M.J.A., Dudek, J. and Szymański, Z., (1983), *Rev. Mod. Phys.* **55**, 949.
von Weizsäcker, C.F., (1935), *Z. Phys.* **96**, 431.
Wu, C.S. and Wilets, L., (1969), *Ann. Rev. Nucl. Sci.* **19**, 527.
Zelevinskii, V.G., (1975), *Yad. Fiz.* **22**, 1085, *Soviet J. Nucl. Phys.* **22**, 565.

Index